TRENDS IN NEUROCHEMISTRY RESEARCH

TRENDS IN NEUROCHEMISTRY RESEARCH

ROBERT M. COLEMAN
EDITOR

Nova Science Publishers, Inc.
New York

For permission to use material from this book please contact us:
Telephone 631-231-7269; Fax 631-231-8175
Web Site: http://www.novapublishers.com

Library of Congress Cataloging-in-Publication Data
Available upon request

ISBN 1-59454-418-2

Published by Nova Science Publishers, Inc.
New York

Contents

Preface

The brain is an enormously dynamic organ. Even when we sleep connections are made, signals sent and messages delivered. One of the key ways that the brain operates is via chemical stimuli which permits different parts of the brain to communicate between themselves and with the rest of the body. Determining what these chemicals, proteins and molecules are is an important way to not only discover how the brain works, but provide novel targets that may be useful in the treatment of disease, for instance in dealing with memory loss in dementia.. This new book brings together international research in a broad range of topics including molecular and cellular neurochemistry, neuropharmacology and genetic aspects of CNS function, neuroimmunology, metabolism as well as the neurochemistry of neurological and psychiatric disorders of the CNS.

In: Trends in Neurochemistry Research
Editor: Robert M. Coleman, pp. 1-130
ISBN: 1-59454-418-2

Chapter 1

Calcium and the Alzheimer's Ca2+-Tastrophe

András Palotás[1,2,3,*] János Kálmán[1], László G. Puskás[4], Lajos Kemény[5], Zoltán Janka[1], and Botond Penke[2]

[1]Department of Psychiatry, Albert Szent-Györgyi Medical and Pharmaceutical Center, Univeristy of Szeged, Hungary
[2]Department of Medical Chemistry, Albert Szent-Györgyi Medical and Pharmaceutical Center, Faculty of Medicine, University of Szeged, Hungary
[3]Division of Cardiac Surgery, Center for Cardiology, Albert Szent-Györgyi Medical and Pharmaceutical Center, University of Szeged, Hungary
[4]Laboratory of Functional Genomics, Biological Research Center, Hungarian Academy of Sciences, Hungary
[5]Department of Dermatology, Albert Szent-Györgyi Medical and Pharmaceutical Center, University of Szeged, Hungary

To my Beloved Parents, who have Loved, Cared for, and Supported me throughout my life by all means…
and to God, who always gives hope, strength, and shows the right way to everyone…

Key words: Alzheimer's disease, β-amyloid peptide, β-amyloid precursor protein, antipsychotic, calcium, cDNA microarray, gene expression, haloperidol, holistic approach, neural plasticity, neurodegenerative disease, neuroprotection, risperidone, treatment options

* Corresponding author. Address: University of Szeged, H-6720 Szeged, Pécsi u. 4, Hungary; Tel: +(36)(30)255-6225; e-mail: palotas@nepsy.szote.u-szeged.hu

An Introduction to Alzheimer's Disease

Clinically...

Each year millions of people go through a stage of mental deterioration in which their ability to remember gradually fades. At certain stages of the disease a simple task that is taken for granted, such as how to make a phone call, becomes insurmountable. Most of the people who are afflicted with this illness are elderly, but their symptoms are not caused by their aging. After three to ten years from the onset of the disease the patients become unable to recognize close relatives. Eventually fading remembrances slowly destroy their emotional balance, their sense of judgment, and their personal identity. Due to their inability to support themselves they gradually become bedridden. Complications arising from this vegetative state ultimately lead to death.

In the early 1900's the symptoms of this disorder were recognized by a German psychiatrist, Alois Alzheimer, for whom this disease is named [1]. Scientists have recently made major advances in understanding the mechanism which triggers the illness. The ultimate goal of Alzheimer research is to elucidate the pathway underlying the massive neuronal death seen in the brain with this syndrome leading to this specific kind of dementia – a condition denoting a progressive decline in mental function, in memory, and in acquired intellectual skills. About 70% of dementia cases are due to Alzheimer's disease (AD).

AD is a major public health problem secondary to the increased life expectancy of the general population and a better appreciation of the socio-economic consequences of the disease. AD is a disease with several million living victims worldwide [2]. With an annual incidence of about more than half a million cases, and given the aging of the population, improvements in life expectancy, and in treatment of AD, the prevalence of this condition will quadruple in the next 50 years [2].

A definitive diagnosis of this disorder requires pathological examination of the brain, but diagnosis on clinical grounds according to the *Diagnostic and Statistical Manual of Mental Disorders, 4th Edition (DSM-IV)* criteria (Table I) [3] alone is correct in the majority of cases, especially if other causes have been ruled out.

Table I: Diagnostic criteria for dementia of the Alzheimer's type

A. The development of multiple cognitive deficits manifested by both:
   1. Memory impairment (impaired ability to learn new information or to recall previously learned information)
   2. One (or more) of the following cognitive disturbances:
      a. aphasia (language disturbance)
      b. apraxia (impaired ability to carry out motor activities despite intact motor function)
      c. agnosia (failure to recongnize or identify objects despite intact sensory function)
      d. disturbance in executive functioning (*ie.* planning, organizing, sequencing, abstracting)

B. The cognitive deficits in Criteria A1 and A2 each cause significant impairment in social and occupational functioning and represent a significant decline from a previous level of functioning

C. The course is characterized by gradual onset and continuing cognitive decline

D. The cognitive deficits in Criteria A1 and A2 are not due to any of the following:
   1. Other CNS conditions that cause progressive deficits in memory and cognition (*eg.* cerebrovascular disease, Parkinson's disease, Huntington's disease, subdural hematoma, normal-pressure hydrocephalus, brain tumor)
   2. Systemic conditions that are known to cause dementia (*eg.* hypothyroidism, vitamin B12 or folic acid or niacin deficiency, hypercalcemia, neurosyphilis, HIV infection)
   3. Substance-induced conditions

E. The deficits do not occur exclusively during the course of a delirium

F. The disturbance is not better accounted for by another Axis I disorder (*eg.* major depressive disorder, schizophrenia)

Neuropathological examination from autopsy studies suggests a 90% accuracy rate in the clinical detection of AD.

Computerized tomographic (CT) and magnetic resonance imaging (MRI) of the brain of Alzheimer patients show thin cortical gyri of the postero-temporal and postero-parietal lobes (*ie.* brain atrophy), enlarged (lateral) ventricles, and atrophy of the hippocampus. These findings are not specific, however, since other conditions may also produce atrophy of the brain. Furthermore, there is some degree of atrophy in normal aging.

Dementia in AD steadily worsens, with a loss of 3 to 4 points per year on a standard assessment instrument such as the Mini-Mental State Examination (MMSE). Various patterns of deficits are seen, with the most common being an insidious onset, with recent memory loss followed by the development of aphasia, apraxia, and agnosia after several years. Some patients present with irritability and personality changes in the early stages. Later during the course of the disease, patients usually develop gait and motor disturbances, eventually becoming mute and bedfast. On average, AD patients live for 8 to 10 years after they are diagnosed, although the disease can last for up to 20 years [4].

In addition to debilitating the patient, AD impacts on families. This impact is emotional, with caregivers experiencing depression, guilt, anger, and physical stress leading to caregiver frailty, as well as economic consequences, with costs of care in the tens of thousands of dollars per year [5].

Histopathology

The formation of senile plaques in the brain is a main pathological feature of AD. These plaques contain deposits of a 38-43-amino-acid peptide, called the β-amyloid peptide (βAP) [6]. In the presence of senile plaques neurites (*ie.* neuronal processes: the dendrites and axons) degenerate; ultimately brain cells necrotize. The dystrophic neurites and cells surround the senile plaques and together they are referred to as the neuritic plaques [7]. This morphological characteristic has led to the speculation that the amyloid masses, particularly the presence of βAP in the extracellular space cause the death of adjacent neurons.

Apart from forming the senile plaques, β-amyloid fibrils are deposited along the walls of the cerebral and meningeal vascular system, too, causing vascular amyloid angiopathy [8].

Another microscopic hallmark of AD is the profound neurodegeneration in the frontal, temporal and parietal cortices. Depletion of cholinergic neurons in the basal forebrain (particularly in the Meynert's nucleus) and its accompanying cognitive deficits are well described. Granulo-vacuolar degeneration of pyramidal cells along with neural loss and plaques in the hippocampal formation are also characteristic of AD [9].

Cytopathological manifestation of AD is the presence of paired helical filaments (PHF) within the brain cells [10]. These intraneuronal filaments aggregate and form neurofibrillary tangles (NFT) [11]. This amorphous material is assembled from an abnormally phosphorylated and transformed cytoskeletal protein called *tau* (τ) [12]. Normally, *tau* plays an important role in stabilizing microtubules and thus helps to maintain neural vesicular transport. The most conspicuous effect of the phosphorylation of *tau* is the decrease in its binding capacity to the microtubules. This causes a loss of axonal transport, since microtubules provide the tracks for vesicle traffic driven by motor proteins.

As a reflection of the severe neuronal degeneration and decreased activity of *tau*, AD is associated with deficits in several neurotransmitter systems. A 60-90% loss of cholinergic neurons can be observed in AD. In a less conspicuos manner, however, deficiency in noradrenergic, serotonergic, GABAergic, etc. systems is also present in Alzheimer brains [13,14].

Cholinergic Hypothesis

Central Cholinergic Systems

Eight cholinergic cell groups exist that form the origins of projections to other central nervous system (CNS) structures [15]. The medial septal nucleus and the vertical nucleus of the diagonal band constitute the major cholinergic projections to the hippocampal formation, cingulate cortex, olfactory bulb, and hypothalamus. The horizontal limb of the diagonal band nucleus projects to the olfactory bulb, whereas the nucleus basalis of Meynert provides the major innervation of the cerebral cortex and amygdala. The pedunculo-pontine nucleus and the latero-dorsal tegmental nucleus of the brainstem project to the thalamus; the medial habenula innervates the inter-peduncular nucleus; and the para-bigeminal nucleus projects to the superior colliculus.

Cholinergic neurons manufacture the enzyme choline-acetyltransferase (ChAT), which is transported to projection targets, where it catalyzes the synthesis of acetylcholine (Ach) from acetyl-CoA and choline. All of the cholinergic innervation of the human cerebral cortex and thalamus arises from these extrinsic cholinergic sources. All layers of the cerebral cortex receive cholinergic innervation; the density of cholinergic projections is highest in layers 1 and 2 and the upper regions of layer 3. Muscarinic cortical input disinhibits the cortical pyramidal cells that enhance the intralaminar transfer of information between cortical columns; in contrast, nicotinic input augments inhibition [16]. There is regional variability in the cholinergic innervation of the human cortex [17]: limbic areas that include the amygdala and hippocampus have the highest density of cholinergic axons; paralimbic regions have the next highest density of cholinergic fibers; the unimodal and heteromodal association cortices have intermediate densities of cholinergic innervation; and the primary visual cortex has the least abundant cortical cholinergic projections.

Two classes of cholinergic receptors are recognized on the basis of their responses to specific agonists and antagonists [18]: muscarinic and nicotinic. Three types of muscarinic receptors have been identified pharmacologically, and five types have been shown to exist on the basis of molecular cloning experiments. Two major nicotinic receptor types have been identified in the CNS by using α-bungarotoxin and neuronal bungarotoxin. Muscarinic receptors are metabotropic and use G-proteins for signal transduction; nicotinic receptors are ionotropic and use ligand-gated ion channels for signal transduction [18,19].

The M_1 receptor is the most common muscarinic receptor subtype in the cerebral cortex. Its highest concentrations are found in the dentate gyrus, hippocampus, anterior olfactory nucleus, cerebral cortex, olfactory tubercle, and nucleus accumbens. Moderate concentrations are found in the olfactory bulb and amygdala. The M_2 receptor is found in brain areas with abundant cholinergic neurons, including the interpeduncular nucleus and basal forebrain. The M_2 receptor is a presynaptic autoreceptor that governs cholinergic release [20]. M_3 receptors are concentrated primarily in the diencephalic and brainstem regions, and M_4 receptors are found mainly in the striatum and olfactory tubercle [21]. Nicotinic receptors are most abundant in the thalamus, periaqueductal gray matter, and substantia nigra. Intermediate levels are found in the cerebral cortex and striatum. Relatively low levels are found in the hippocampus and amygdala.

When functioning normally, cholinergic neurons release Ach into the synaptic cleft, where it binds to postsynaptic or presynaptic receptors, either muscarinic (mAchR) or nicotinic (nAchR), depending on the specific tract to which the cell belongs. Ach is a neurotransmitter that plays a critical role in learning and memory function. Ach remains active until it is hydrolyzed to choline and acetate by cholinesterases. There are two cholinesterases in the CNS – acetylcholine-esterase (AchE) and butyrylcholine-esterase (BchE) [22]. The former is widely distributed in CNS and peripheral nervous system, while the latter is found also in the liver and plasma.

Cholinergic Abnormalities in AD

AD is a complex neurodegenerative disease with characteristic histological changes. A variety of neurochemical deficits affect the serotonergic, noradrenergic, and cholinergic systems [23]. The cholinergic deficit of AD is well documented; there is a marked loss of neurons in the basal forebrain nuclei that usually exceeds 75% of the total neuronal population at the time of an autopsy [24]. There are abundant NFTs in the remaining neurons of the basal nucleus [25]. The death of cholinergic neurons leads to reductions in ChAT of 80%-90% in the hippocampus and temporal cortex and 40%-75% in the parietal cortex and frontal convexity. Less marked reductions occur in the mammillary bodies, cingulate gyrus, priming motor and sensory cortices, and caudate. Normal levels are present in the pons, midbrain, thalamus, hypothalamus, nucleus accumbens, and cerebellum [26,27]. M_1 receptors are preserved or modestly reduced in AD, presynaptic M_2 autoreceptors are decreased, and M_3 receptor levels are normal or up-regulated [20,28]. nAchRs also are decreased in AD [29]. There is a preferential loss of the membrane-bound form of AchE compared to the soluble form [30]. Patients with the apolipoprotein E4 genotype have less brain ChAT and nAchR binding than patients with the E2 or E3 genotype [31,32].

β-amyloid Peptide

βAP is derived from the highly regulated and sequential cleavage of amyloid precursor protein (APP) as a normal physiological product [33,34]. It is a soluble component at nanomolar concentrations of the plasma and the human cerebrospinal fluid (CSF) [35] as a range of isoforms between 38 and 43 amino acids in length.

The presence of βAP under normal conditions suggests the possibility of a physiological role for βAP. βAP may serve an essential role in synapse and synaptic plasticity that underlie learning and memory [36,37]. A role for βAP in neural and synaptic structure-functional plasticity is additionally supported by an increase of synaptic APP with learning capacity [38], by neuronal activity dependent secretion of natural βAP [39], and up-regulating a synaptic vesicle protein transcript by $\beta AP_{1\text{-}42}$ [40].

Because soluble βAP is apparently a physiological peptide, it was proposed that the key to the cellular toxicity of βAP appears to be proportional to its aggregation state [41]. The aggregation of soluble βAP peptide into fibrillar cross-β pleated-sheet conformation to form oligomers, polymers or protease-resistant fibrils over time is generally considered to be a critical event in the pathology of AD [42].

This is signified by the finding that the predominant carboxyl-terminal isoforms in CSF of healthy individuals are $\beta AP_{1\text{-}40}$ (90%) and the fibrilogenic $\beta AP_{1\text{-}42}$ (10%). In patients with AD, however, the relative proportions of 1-40 and 1-42 change to ~50% each [43]. Moreover, $\beta AP_{1\text{-}42}$ is the major component of amyloid deposits in brain with AD [44].

Mechanism for the gradual cytotoxic effects that apparently follow βAP deposition involves the so-called β-amyloid-associated proteins. In this scenario, βAP, particularly in its more aggregated forms, can serve as a matrix upon which a variety of soluble molecules that are normally made in small amounts by local cells can accumulate. Some of the proteins that appear to be rather tightly associated with βAP have already been identified: α_1-antichymotrypsin, several components of the classical complement cascade, *etc* [45]. Local astrocytes surrounding neuritic plaques may produce not only one or more of these polypeptides, but also cytokines and other acute phase proteins that could mediate the local inflammatory response found in and around mature plaques. The reactive astrocytic and microglial changes that occur within and around the senile plaques are likely to contribute in complex ways to local neuritic alterations.

Interestingly, βAP deposition has been observed in various brain areas without accompanying neurodegeneration [46,47,48], whereas neurodegeneration can occur in areas with no βAP deposition [49]. Therefore, attention is turning away from the deposits of extracellular insoluble aggregated amyloid in plaques and toward soluble, oligomeric and even intracellular βAP [51,52]. Accumulation of intracellular βAP (βAP_i) and preamyloid protrofibrils are being recognized as a contributing factor or initiating event towards neuronal death and synaptic loss in early AD pathogenesis [52].

Toxicity of βAP

The theory that βAP plays a key role in neurodegeneration in AD has been proven both *in vivo* and *in vitro* [53,54,55], leading to the "amyloid cascade" hypothesis [56,57]. The neurotoxicity exerted by aggregated βAP can be mediated by several mechanisms.

Several reports suggest that the neurotoxic effects of βAP are mediated by the disruption of calcium homeostasis, resulting in increased neuronal intracellular calcium levels ($[Ca^{2+}]_i$) [58,59]. βAP is a potent L-type voltage-gated calcium-channel activator [60], and causes reduction in the activity of Na^+/K^+-ATPase [61], Na^+/Ca^{2+} exchanger [62] and K^+-channel [63]. βAP is also proved to directly form non-selective calcium-channels [64]. Neurons exposed to βAP exhibit calcium responses sensitive to excitatory amino acids and membrane depolarization: enhancement of glutamate-mediated excitotoxicity [65] adds to the elevated calcium-levels.

When in an aggregating form, βAP induces membrane lipid peroxidation by generating reactive oxygen species (ROS; hydrogen peroxide and hydroxyl radical) via chemical reactions requiring Cu^+ or Fe^{2+} [66,67,68]. An aldehyde called 4-hydroxy-2,3-nonenal (HNE) is produced during this process [69]. HNE covalently modifies and thereby impairs the function of ion-motive ATPases (Na^+/K^+ and Ca^{2+}-ATPases), glucose and glutamate transporters, resulting in membrane depolarization and reduced energy availability to cells [61,70,71]. Importantly, the adverse effects of βAP on the membrane transporters result in an elevation of basal intracellular Ca^{2+} levels and greatly enhances Ca^{2+} overload upon activation of voltage-dependent calcium channels and ionotropic glutamate receptor channels [61,72]. Loss of membrane integrity resulting from βAP-induced free-radical damage may also lead to disruption of signaling pathways, and activation of nuclear transcription factors, lipid, DNA and protein oxidation [67,73,74]. Antioxidants that inhibit lipid peroxidation, such as vitamin E, estrogen, and uric acid, prevent impairment of the membrane transporters, stabilize cellular Ca^{2+} homeostasis, and prevent death of neurons exposed to βAP [75,76].

The increased ROS and intracellular calcium levels induced by βAP impair mitochondrial function and can trigger a programmed cell death process called apoptosis in which mitochondria undergo a membrane permeability transition (MPT) resulting the disruption of the mitochondrial membrane potential (ΔΨ) and the release of cytochrome-c from the mitochondrial intermembrane space. The released cytochrome-c molecules, together with cytosolic factors, are involved in the activation of caspase family proteases, enzymes that execute DNA fragmentation and other apoptotic (cell-death) events [77,78,79].

βAP has been also shown to induce excessive nitrogen monoxide (nitric oxide, NO) generation [80]. NO is an endogenous water- and lipid-soluble gas. By readily diffusing through the cytoplasm and plasma membranes, it is critical for the normal physiological regulation of the nervous system; it has effects on neuronal transmission, and plays an important role in synaptic plasticity, learning and memory [81,82]. In the vasculature, NO directly and, by enhancing the release of neurotransmitters, indirectly results in smooth muscle relaxation and vasodilation. NO may also be involved in the regulation of protein activity through S-nitrosylation. NO may react with oxygen and water to form nitrates and nitrites (*eg*. peroxynitrite) in the extracellular milieu. These oxidizing free radicals can cause lipid oxidation and DNA fragmentation. In the mitochondria, they act on the respiratory chain (I-IV) complex and manganese superoxide dismutase (MnSOD), to generate superoxide anions and hydrogen peroxide (H_2O_2), respectively. Due to its properties as a free radical, NO can be neurotoxic and it has, therefore, been proposed to have an involvement in normal aging and neurodegenerative processes, including AD [83].

NO is generated by NO-synthase (NOS) from L-arginine. There are three isoforms of NOS: neuronal (nNOS), endothelial (eNOS) and inducible NOS (iNOS or type 2 NOS). In general, nNOS and eNOS are Ca^{2+}-calmodulin-dependent constitutive forms, whereas iNOS is Ca^{2+}-calmodulin independent and is usually only produced in response to pathological stimuli [84]. βAP may have effect on NOS by two distinct mechanisms. Excessive NO is generated by iNOS following exposure to certain cytokines, such as interferon-γ (IFN-γ), that are activated by βAP-induced inflammatory processes in AD brains. On the other hand, nNOS and eNOS are soluble enzymes that, unlike iNOS, require elevated Ca^{2+} levels for activation. βAP-induced increase in the Ca^{2+} concentration may induce these enzymes. Increased production of reactive oxygen and nitrogen species, coinciding with a depletion of antioxidant defenses, might cause oxidative stress in AD brains [85].

Early association of activated microglial cells and reactive astrocytes in neuritic plaques and the appearance of inflammatory markers indicate a state of chronic inflammation in AD. Immune activation and/or inflammatory activity have been shown to be significantly elevated in the brains of AD patients [86]. Microglia and astrocytes would be activated, perceiving βAP oligomers and fibrils as a foreign material, because these kinds of βAP assemblies are apparently never observed during brain development and in the immature nervous system [87]. CD40 expression is increased in AD, which is necessary for βAP-induced microglial activation [88].

Activated microglia participate in triggering of the classic complement pathway. βAP, however, directly activates the complement system [89], and complement proteins are integral components of amyloid plaques and cerebral vascular amyloid in AD brains. Microglia are also involved in the production of a variety of proinflammatory cytokines, chemokines and growth factors, including prointerleukin-1β, interleukin-6, tumor necrosis factor-β, monocyte chemoattractant protein-1, macrophage inflammatory peptide-1β, IL-8, and macrophage colony-stimulating factor [90,91].

The importance of the inflammatory component to the disease mechanism is supported by the consistent finding that anti-inflammatory drugs have a beneficial effect on disease progression.

Extracellular interaction of β-amyloid aggregates with cell membranes could induce a wide variety of alterations through numerous potential target sites that are consistent with the pathological activation of signal transduction. These include the mitogen-activated protein kinase (MAPK), calpain, cyclin-dependent kinase-5 (cdk-5), glycogen synthase kinase-3β (GSK-3β), and tyrosin kinase Src pathways [92,93]; by progressive and sustained activation thereof, βAP induces tyrosine phosphorylation of numerous neuronal proteins, including *tau* and microtubule-associated protein-2c (MAP-2c) [94].

Neuronal death may be the ultimate consequence of the above cellular dysfunctions induced by βAP. The 11-amino acid fragment $\beta AP_{25\text{-}35}$ has been proposed to be the functional domain of βAP [95], of which $\beta AP_{31\text{-}35}$ is crucial for its cytotoxic properties [96,97]. The oxidation state of methionine at residue 35 also plays a critical role in the cytotoxicity of βAP; when oxidized, βAP-Met35^{OX} is less toxic [96,98,99].

Tauopathy

If there has been any real controversy associated with the amyloid cascade hypothesis, this has been with the question of *tau* and NFTs. These neuronal inclusion bodies are a defining feature of AD, and are also found in certain fronto-temporal dementias. NFTs are intraneuronal lesions; the neurons containing them show loss of vital intracellular organization with the loss of normal neuronal cytoskeleton, and there is convincing neuropathological evidence that the presence of NFTs heralds the death of that neuron. The number of NFTs correlates extremely well with dementia severity and show an anatomical localization in those regions where function is lost, occurring first in the transentorhinal region and spreading to hippocampal formations and then to the cortex, but never occurring in the cerebellum. Extracellularily localized plaques, on the other hand, show no such consistent progression, and while they do occur in some quantity in the hippocampus where function is lost, they also occur in cerebellum where no such loss is noted in dementia [100].

Despite the presence of NFTs in AD, no cases of AD secondary to mutations in the *tau* gene on chromosome 17 have been reported. Therefore, the "cytoskeletal hypothesis" eventually confirms the "amyloid cascade" theory. Mutations in APP give rise to both plaques and NFTs, while mutations in *tau* yields tangles only.

Lipid Theory

Cholesterol is accused in atherosclerosis and heart disease, but was largely understudied in relation to brain function and neural structural and functional (*ie.* activity dependent) plasticity. Cholesterol is implicated in basic synaptic function, particularly in trafficking and recycling of synaptic vesicles, in receptor function, activity of accessory synaptic proteins, and in modulation of membrane biophysical properties [101].

Living cells produce their own cholesterol and can receive or donate cholesterol via lipoprotein transit, an attested body lipid transportation system, which operates a number of vehicle classes. These include low density lipoproteins (LDL) that are considered to be "bad", and high density lipoproteins (HDLs) that are referred to as "good" cholesterol transporters. βAP is a normal structural-functional constituent of HDL apolipoproteins (*eg.* ApoE and ApoJ) in plasma and CSF. It modulates cholesterol and phospholipid synthesis, and affects cholesterol esterification [102,103]. In turn, lipoproteins provide the proper thermodynamic environment for [104], and potently inhibit neural toxicity of βAP [105].

Apart from the whole-body-system stands distinct brain lipid transport authority that uses different subtypes of HDLs and normally does not maintain LDLs [104]. This vital service is in charge of cholesterol redistribution inside the brain and cholesterol export out of the CNS to liver for excretion. There is no reported cholesterol import to the brain, an issue that makes brain cholesterol availability entirely dependent on local manufacturing. Brain lipid transportation must have good management and operating capacity, because the quantity of cholesterol in the brain is much higher then anywhere else in the body: having just two percent of the body weight, brain has a quarter of cholesterol present in the whole individual. Conceivable, the break in any element of the harmonized system of brain/neuronal cholesterol transport (caused by genetic defects of one of the enzymes or receptors associated with cholesterol turnover; by pharmacological modulation or environmentally) may result in abnormal homeostasis of cholesterol in the brain and impair fine tuning of synaptic function. Lack of cholesterol supply of neurons via lipoprotein transport impaires two different brain waveforms: paired pulse facilitation (PPF) and long term potentiation (LTP), indicative of neurotransmission and synaptic plasticity, respectively. Moreover, loss of neuronal cholesterol homeostasis by means of impaired lipoprotein transport, change in lipid composition of CNS lipoproteins [106], altered βAP-ApoE or βAP-ApoJ interaction, or increased rate of brain cholesterol synthesis causes AD features, such as failure in neurotransmission, synaptic dysfunction, neurite degeneration, and *tau* protein pathology [101,107]. In this purported cascade the change in *tau* and βAP neurobiochemistry reserves a secondary subordination and aims to compensate impaired cholesterol-dependent synaptic and behavioral plasticity. Therefore, the protective effect of statins, inhibitors of 3-hydroxy-3-methylglutaryl-coenzyme A-reductase, which catalyze the early rate-limiting reaction in cholesterol biosynthesis, should be employed in the treatment of AD.

Increased oxidative stress is implicated in Alzheimer's development, and βAP is considered to be an important pro-oxidant in this process [108]. In order to induce oxidation, βAP must first bind transition metals to its metal-binding site(s) at *His*-residues [109] and then reduce them in its metal-reducing site at Met35 residue [110] in order to produce ROS. However, metals bind to the N-terminal hydrophilic part of βAP, whereas metal reduction occurs at its C-terminal part. Since metals must be placed in the vicinity of the reductant to be reduced, fibrillation is likely to fulfil this task by forming complexes where metal atoms bound to the N-terminal part of one molecule of βAP, at the same time might be available for the reductive Met35 residues belonging to other βAP molecules. The resulting reduced transition metal ions can participate in further redox reactions, generating various free radical species. Due to relatively slow reduction of metals by βAP, the above mechanism can only be operative at high concentrations of the βAP peptide, typically in a micromolar range [111].

In contrast to its pro-oxidative properties, an antioxidative activity of βAP is present at low-nanomolar concentrations (*ie.* those of soluble βAP in CSF and plasma). βAP inhibits metal-catalyzed oxidation of lipoproteins of human CSF and plasma [112,113]. Therefore, soluble βAP protects lipoproteins and other biomolecules from oxidation, particularly toxic action of copper and iron. The effect is observed at the peptide concentration reported for biological fluids (0.1-1.0 nM). At higher concentration of βAP its antioxidant action is abolished. In contrast, all βAP peptides are unable to considerably influence metal-independent lipoprotein oxidation, suggesting that the antioxidative activity of βAP is mainly mediated by chelating transition metal ions. βAP in CSF can also act as an antioxidant, as is suggested by the positive correlation between CSF resistance to oxidation and the CSF level of βAP [114]. Mechanistically, antioxidative activity of βAP can be related to the fact that in lipoproteins, a metal-binding region of βAP expresses greater hydrophilic properties and extends into the outer aqueous phase where it can bind transition metals and inhibit metal-catalysed oxidation [115].

As soon as βAP has antioxidative properties on one hand, and is secreted by cells as a part of lipoprotein complexes on another, it is well possible that βAP is secreted by cells to serve as a natural antioxidant for lipoproteins [104,116]. βAP can bind transition metal ions in inactive form and prevents them from catalyzing oxidation of lipoproteins and other biomolecular complexes. Amphiphilic properties of βAP may allow extracellular chelation of metal ions that escape binding by hydrophilic chelators. Therefore, βAP is a functional player in the activity-dependent cholesterol neurochemical pathways and in synaptic structure-functional plasticity. The change in βAP biochemistry in AD and related disorders might be a functional, but not pathologic compensatory phenomenon aiming to counterbalance impaired cholesterol dynamics and associated neurotransmission and synaptic plasticity. Such cholesterol mediated failure of synaptic function and neural degeneration may represent the cause of the major sporadic form of AD. These hypotheses are consistent with earlier observations that βAP is monomeric, non-toxic and exerts beneficial effect on neuron survival, axonal length and neurite outgrowth at low-nanomolar concentrations [117,118,119]. All these activities may be related to antioxidative properties of the peptide.

The Calcium Theory...

Calcium plays fundamental roles in the development, function, and plasticity of nerve cell circuits. The regulation of Ca^{2+} dynamics in neurons is very complex, involving proteins localized in essentially all major subcellular structures including the plasma membrane, cytosol, mitochondria, and endoplasmic reticulum [121]. The concentration of Ca^{2+} outside the cell (1-2 mM) is 10,000-fold higher than in the cytosol (50-200 nM), and this concentration gradient is maintained largely by removal of Ca^{2+} from the cytoplasm by plasma membrane and endoplasmic reticulum (ER) Ca^{2+}-ATPases.

The plasma membrane contains several types of voltage-dependent Ca^{2+} channels (L, N, T, P, and Q) that differ in their biophysical properties. In addition, glutamate receptors are also known to flux high levels of Ca^{2+}. These receptor-dependent Ca^{2+}-transporters include ionotropic (*ie.* the N-methyl-D-asparate (NMDA) subtype and the α-amino-3-hydroxy-propionate (AMPA) receptor), and metabotropic channels. Activation of metabotropic receptors coupled to GTP-binding proteins results in release of Ca^{2+} from ER stores.

The ER membrane also contains several proteins that control Ca^{2+} movements across its membrane under basal conditions and in response to environmental signals. The Ca^{2+} concentration gradient across the ER membrane (higher inside) is maintained by membrane Ca^{2+}-ATPases. The presence of ER Ca^{2+} pools in synapses suggests important roles for Ca^{2+} release in modulating synaptic plasticity. Calcium release from the ER is regulated by an array of modulatory proteins including calcineurin [122], calmodulin [123], and the cytoskeletal protein ankyrin [124].

Mitochondria are located not only in the cell body of neurons, but also in axons, dendrites, and synaptic terminals, and can respond to Ca^{2+} influx by increasing their Ca^{2+} uptake [125,126]; this may serve as a sink for rapid removal of potentially cytotoxic levels of Ca^{2+} from the cytosol [127]. Therefore, mitochondrial Ca^{2+} cycles may play a pivotal role in maintenance of the cytosolic Ca^{2+} oscillations [128,129]. The proton gradient across the inner mitochondrial membrane is crucial in maintaining the activities of the Ca^{2+} transporters; a decrease in the proton gradient can reverse the direction of Ca^{2+} flow across the membrane. Importantly, alterations in mitochondrial Ca^{2+} homeostasis have been shown to play important roles in neuronal cell death induced by various excitotoxic and metabolic insults [130,131]. Mitochondrial Ca^{2+} accumulation may contribute to the oxidative stress and impaired mitochondrial energy production observed in experimental models of AD.

Pathophysiologically, interfering with the intracellular ionic homeostasis of neurons is thought to be one of the most conspicuous brain cell destroying activities of βAP [58]. Destabilization of Ca^{2+} homeostasis in neurons plays a central role in AD pathogenesis [59]. In the presence of the peptide, the intracellular calcium concentration of cells has been reported to increase through various, but not fully understood mechanisms, including the putative calcium channel activity of βAP [64], or its interaction with membrane structures. Measurements of calcium concentrations in brain tissue from AD patients reveals evidence for increased overall levels (free and protein-bound) of calcium [132]. Levels of activated calcium-dependent enzymes and proteases are also increased [133].

Increase in the calcium concentration due to βAP is balanced by pumping out the excess. This feature expends a large amount of the cells' energy, preventing them from providing for their own physiological and biochemical pathways any longer. Concentration of ROS will then increase which may further damage cells. The effects disable cells to control their fluid and ion balance. Water and charged particles (especially even more Ca^{2+}) that are normally pumped out now stream in. Additionally, alterations in intracellular Ca^{2+}-levels are known to induce cell loss by apoptosis, a degenerative pathway typically associated with gene-directed cellular suicide termed programmed cell death. Neurons that are meant to last a lifetime can be transformed at an early stage to this form of death.

β-amyloid-precursor-protein

β-amyloid peptide derives from the proteolytic processing of its precursor, the β-amyloid-precursor-protein (APP) [134]. Full-length APPs are considered integral membrane glycoproteins containing a large extracytoplasmic domain, a transmembrane region, and a small cytoplasmic sequence. The APP spans the cell membrane between amino acids 700-723. The β-amyloid site is represented by the amino acids 672-713; that is, the C-terminal of βAP takes part in forming the transmembrane region. A heparin-binding domain lies within the NH_2-terminal portion of βAP (VHHQK) [136,137]. A heavily glycosylated region of APP lies NH_2-terminal to the βAP sequence. This sequence is highly conserved.

The human APP cDNA was first cloned in 1987 [138], and other homologous genes were identified soon after. cDNA sequences for APP homologues have been published for monkey, rat, mouse, frog (*Xenopus levis*), two species of puffer fish (*Fugu rubries* and *Tetradon fluviatis*), and electric ray (*Narke japonica*). The amyloid sequence was identified for polar bear, pig, rabbit, guineapig, and sheep [139,140]. Additionally, amyloid-precursor-like proteins (APLPs), lacking the βAP domain, have also been characterised. Two mammalian homologues APLP1 (human and mouse) and APLP2 (human, mouse and rat) are described [141,142]. APP-like cDNAs have also been isolated from non-mammals, *eg.* fruitfly (*Drosophila melanogaster*) and worm (*Caenorhabditis elegans*), known as APPL and APL-1 respectively [143]. Thus APP is part of an extensive gene family and displays a high degree of evolutionary conservation.

In mammals, the APP protein is expressed ubiquitously in all tissues examined including neurons, glias, and non-neuronal cells [144,145], including fibroblasts [146], but there are differences in the expression levels of the splicing variants. Alternative splicing of the APP gene gives rise to at least 10 protein isoforms. Three of them known to contain βAP are APP_{770} (full-length APP), APP_{751} (minus exon 8) and APP_{695} (minus exons 7 and 8). Exon 7 encodes the Kunitz protease inhibitor (KPI) [147]; KPI-containing APP isoforms are more amyloidogenic [148]. Neuronal cells predominantly express APP_{695}, whereas the $APP_{751/770}$ form is mainly expressed in non-neuronal tissues such as heart, lung, liver, kidney, muscle and spleen [149,150]. In the brain, APP is expressed with high levels in the hippocampal formation [151].

Because several controlling elements exist in the flanking sequences of the alternatively spliced exons of APP, its expression is highly regulated. Retinoic acid and thyroid hormone responsive elements are involved in the regulation of alternative splicing [152]. Tri-iodo-thyronine (T_3) represses APP promoter activity by binding of the nuclear T_3-receptor to a specific sequence located in the first exon. The unliganded receptor increases promoter activity, and T_3 reverses that activity to basal levels. T_3 regulates not only APP gene splicing, but also the processing and secretion of the APP peptides [153]. Interestingly and in line with the TSH-dependent expression and secretion of APLPs, AD patients show significant increase in the mean values of anti-thyroglobuline (TgAb) and anti-microsomial (McsAb) auto-antibodies [154].

The strong structural conservation within the superfamily and their ubiquitous expression in the body indicate an important function for APP and APP-like proteins [155]. They are involved in a variety of cellular functions, including cell adhesion, migration, and cell-cell communication. They also play key roles in the regulation of nerve process (neurite) outgrowth, synaptogenesis, intracellular calcium levels, cytokine release, and in promotion of neuronal survival [156,157]. They also participate in wound healing by modulating growth factor activities, and APPs are markers of early axonal injury [158,159]. The mechanism by with APP exerts these effects is thought to be via modifying $[Ca^{2+}]_i$-levels by activating high-conductance ion-channels [160]. As calcium is thought to be a key second messenger controlling most trophic functions, APP may efficiently be involved in tissue growth during both development and repair [161]. Also, isoforms of APP containing a protease inhibitor region of the Kunitz type at their ectodomains may function as inhibitors of certain serine proteases, such as trypsin and chymotrypsin, and factors XIa and Ixa [162].

By being a ligand of a yet unidentified receptor, β-amyloid domain of APP is essential for the axonal transport of various proteins, including AchE and APP itself. Excess of freely floating βAP in AD may saturate the receptor, stalling axonal transport.

Apart from one of its domains implicated as being a ligand, full-length APP might also act as a cell-surface receptor [135] similar to other single membrane spanning polypeptides, such as the transferrin receptor or the insulin-like growth factor II (IGF II) receptor. The ectodomain of APP has specialized regions that are capable of binding heparin, and certain metallic cations, including zinc. However, a specific soluble ligand that binds to surface-inserted APP with receptor-like kinetics has not been identified. Potential support for a receptor-like function of APP is the observation that the precursor can associate with the GTP-binding protein G_o via a motif in the APP cytoplasmic domain [163], confirming its role as a cell-surface G-protein-coupled receptor.

Just as APP is essential for the organism's survival, disturbance of its regulation appears to participate in this astounding human disease by yielding βAP.

Processing of APP

The gene encoding the transmembrane APP, containing 18 exons, is located on chromosome 21 [164]. The pre-$mRNA_{APP}$ undergoes alternative splicing to yield multiple molecular weight forms; at least ten isoforms have been identified. The region encoding βAP comprises part of exons 16 and 17; therefore, βAP cannot be produced solely as an alternatively spliced gene product. Rather, it is a by-product of the proteolytic processing of APP [165].

Several integral membrane proteins are known to be released from the lipid bilayer by proteolysis. These include cell adhesion molecules, receptors, ligands (such as tumor necrosis factor-α, TNFα), and also APP. The enzymes responsible for the cleavage of these proteins are generally referred to as membrane-protein secretases.

α-secretion

APP can undergo at least two post-translational processing pathways. In the non-amyloidogenic or normal pathway, constituvely secreted APP (referred to as sAPPα), lacking the transmembrane and cytoplasmic sequences, is produced after full-length molecules are cleaved by an enzyme called α-secretase between amino acids Lys_{16} and Leu_{17} of the β-amyloid sequence [166] (between the amino acid 687-688 of the full-length precursor protein). Because sAPPα is produced after full-length molecules are cleaved by the α-secretase within the β-amyloid sequence, this cleavage event is incompatible with the production of the neurotoxic β-amyloid. As a by-product, an 83-amino-acid membrane-bound C-terminal fragment (CTF_{83} or CTFα) is also generated (Figures 1, 2).

sAPPα has several neuroprotective properties. sAPPα protects neurons from ischemic injury [167,168], has effects on ion fluxes [169], stimulates neurite outgrowth [170], promotes proliferation [171], cell survival [172], and protects neurons from oxidative, excitotoxic [173] and metabolic, including βAP [174]-induced insults. Regarding signal transudation pathways, sAPPα is implicated to play a role in activation of neuroprotective NF-κB-dependent transcription [175], phospholipase C – protein kinase C (PKC) and inositol trisphosphate (IP_3) [176] pathways.

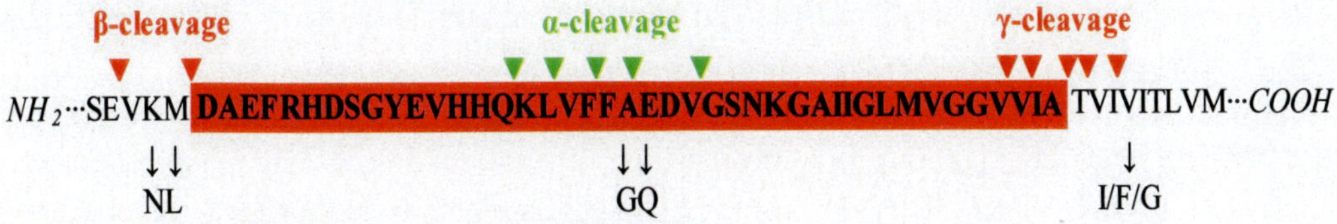

Figure 1: Amino acid sequence of APP in the βAP region
Sites of proteolytic cleavage by α-, β-, and γ-secretases are shown. Mutations in APP which cause familial forms of AD or a related amyloidosis are also presented. βAP region within the precursor protein is highlighted.

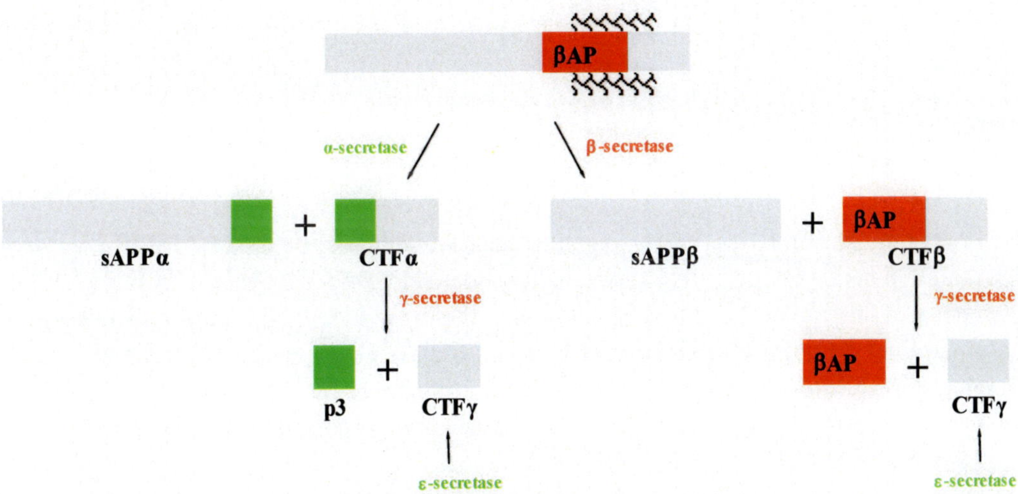

Figure 2: Schematic diagram of APP and its principle metabolic derivatives
The figure depicts the largest of the known APP alternate transcripts, comprising 770 amino acids. Regions of interest are indicated at their correct relative positions. A single membrane-spanning domain at amino acids 700-723 is indicated by the horizontal hatched bars. The βAP fragment (red box) includes 28 residues just outside the membrane plus the first 11-15 residues of the transmembrane domain. The left lower diagram indicates the site of the constitutive proteolytic cleavage that enables secretion of the soluble ectodomain of APP (sAPPα) and retention of the 83-residue C-terminal fragment (~10kDa) in the membrane (CTFα). This non-amyloidogenic processing of APP by α-secretase precludes βAP-formation. The right lower portion of the diagram depicts the alternative proteolytic cleavage by β-secretases that results in the formation of sAPPβ (an ectodomain similar to sAPPα), and a potentially neurotoxic ~12kDa C-terminal fragment (CTFβ), which serves as an intermediate in the generation of βAP. A subsequent γ-cleavage after the α- or β-secretion yields non-toxic p3 and neurotoxic βAP, respectively. As a by-product of the γ-activity, non-toxic CTFγ is formed, which is subject to ε-cleavage.

α-secretase activity has both constitutive and inducible components. The constitutive activity has not yet been identified, but inducible α-secretase activity seems to be under the control of PKC. As part of the inducible pathway, cholinergic stimulation (M_1 and M_3 muscarinic Ach-receptors and AchEIs) enhances sAPPα release, and thus leads to the reduction of toxic βAP generation [177,178]. Stimulation of nicotinic Ach-receptors (nAChR) also protects against βAP- and carboxyl-terminal fragment (CTF)-induced cytotoxicity [179].

α-secretase displays characteristics of certain membrane-tethered metalloproteases. Several proteins have been identified that have α-secretase-like activity. All members display a common domain organisation and possess four potential functions: cell fusion, cell adhesion, intracellular signaling and proteolysis. TNFα-converting enzyme (TACE), also called ADAM17, has been identified as having α-secretase activity. Its primary function is to cleave the TNFα precursor to release its extracellular domain (TNFα) from cells [180] in a manner similar to that of APP. TACE apparently processes a spectrum of type 1 membrane glycoproteins, including TNF-α, L-selectin adhesion molecule, and TGF-α; also, it may be responsible for the majority of regulated α-secretase cleavage of APP [181]. Several other members of the disintegrin and metalloprotease (adamalysin, or ADAM) family, such as ADAM9-10, have also been shown to be capable of cleaving APP at the α-cleavage site [182].

Because it is likely that several proteases contribute to α-secretase activity, it may be difficult to regulate APP processing pharmacologically through this pathway.

β-secretion

βAP is derived from the highly regulated and sequential cleavage of APP by proteases designated β-secretase and γ-secretase. This amyloidogenic β-secretion pathway is an alternative of normal APP processing, involving cleavage of APP by β-secretase adjacent to the N-terminal residues of the βAP sequence. β-secretase generates the N-terminus of βAP, cleaving APP to produce a soluble version of the precursor (sAPPβ) and a 99-amino-acid residue C-terminal fragment that remains membrane bound (CTF_{99}, or CTFβ).

Two β-site APP-cleavage enzymes ($BACE_1$ and $BACE_2$) have been identified as β-secretase [183]. These are membrane-anchored aspartyl proteases, and like many other proteases, they are produced as a pro-enzyme predominantly within the endoplasmic reticulum (ER) and the nuclear envelope. The propeptide does not proteolytically cleave itself; the prodomain is activated by the action of proprotein convertases [184].

β-secretase RNA is highly expressed in the brain and is also found in a variety of human tissues [185,186,187], consistent with the finding that βAP is normally produced by many cell types and in accordance with that expected for β-secretase [188,189,190].

The gene for $BACE_1$ is located on chromosome 11, but no AD-causing mutation in this gene has been identified so far [191]. Its homolog, $BACE_2$ maps to chromosome 21, just as APP itself, raising the possibility that $BACE_2$, along with APP and its derivatives, contribute to Down syndrome (trisomy 21). Down syndrome patients secrete more βAP from birth, which invariably leads to the early emergence of AD neuropathology; therefore, most Down syndrome patients develop AD by age 50 [191]. Although $BACE_2$ cleaves APP in a β-secretase-like manner [192], there is very little of this protease in the brain, suggesting that it may play little if any role in the formation of cerebral plaques seen in AD. Instead, the AD associated with Down syndrome is probably caused by the presence of an extra copy of the APP gene, which is also located on chromosome 21. $BACE_2$ is strongly expressed in heart, kidney, and placenta, suggesting that it may be important in highly vascularized systemic tissues [192]. It will be critical to develop drugs that selectively block $BACE_1$ but not $BACE_2$.

γ-secretion

Cleavage after β-secretion by a subsequent γ-secretase at position 40, 42 or 43 of the βAP sequence within CTFβ results in the release of neurotoxic βAP, a 4kDa derivative from APP [193], and a remaining 57-59 residue C-terminal tail: CTFγ (Figure 2). In addition, cleavage of the by-product of α-secretion, CTF_{83} (CTFα) by γ-secretase yields non-pathogenic derivative referred to as *p3* and CTFγ. In contrast with α-secretion that almost exclusively occurrs in the plasma membrane, APP may be internalized and processed in the endosomes (proteasomes/lysosomes) during sequential β/γ-secretion to produce βAP.

γ-activity is heterogeneous: most of the full-length βAP species produced is a 40-residue peptide ($βAP_{1-40}$), whereas a small proportion is 42-43-residue C-terminal variant ($βAP_{1-42/43}$). The longer and more hydrophobic $βAP_{1-42/43}$'s are much more prone to fibril formation than is $βAP_{1-40}$ [194], and even though $βAP_{1-42}$ is a minor form of βAP, it is the major βAP species found in cerebral plaques [195]. Moreover, γ-secretase has remarkably loose sequence specificity for its substrate because many mutations in APP near the γ-secretase site still allow βAP production.

γ-secretion is a regulated intramembrane proteolysis (RIP) in the middle of the transmembrane domain of APP. The intracellular portion of APP (CTFγ) moves to the nucleus where it may affect the transcription of target genes [196]. Proteins that undergo RIP include APP; Notch, a receptor involved in fate decisions during embryonic development; and ErbB-4, a member of the epidermal growth factor tyrosine kinase receptor family [197]. How hydrolysis takes place in what is otherwise a water-excluded environment is unclear. However, the term RIP may be misleading, because proteolytic cleavage within a membrane has never been demonstrated directly. For example, it is possible that α- or β-secretase cleavage of APP results in movement of the C-terminal protein and exposure of the γ-secretase sites to the aqueous environment [198].

A crucial question is how protein cleavage itself is regulated. Apparently, the triggering event is the removal of most of the extracytoplasmic part of the protein (*ie.* releasing the bulk of APP), which is a prerequisite for the intramembrane cleavage by γ-secretase [199].

Identifying γ-activity revealed that γ-secretase is a multimeric entity: presenilins (PS), nicastrin, Aph1 and Pen2 generate an active γ-secretase complex.

Presenilins (PSs) may be the active center of γ-activity. γ-secretase has been demonstrated to have an acidic pH optimum, consistent with the notion that the presenilins are intramembranous aspartyl proteases [200]. Also, PSs are chaperoning molecules that may bring APP to the proximity of γ-secretase.

The missense mutations in the gene encoding APP, as well as those in the genes encoding PS1 and PS2, share the common feature that they alter the γ-secretase cleavage of APP to increase the production of the amyloidogenic βAP_{1-42} [201].

PS1 and PS2 have been found to be involved in a range of biological processes, such as cell adhesion [202], G-protein mediated signaling [203], and protein trafficking [204]. During development, presenilins appear to also cleave a transmembrane protein called Notch, thereby releasing a fragment that activates the transcription of genes involved in cell-fate determination [205]. The Notch-signaling pathway is critical during cell development, and there is firm evidence that the intracellular portions of Notch modulate gene transcription [206]. Presenilins, therefore, are essential mediators for the proteolytic processing of Notch, which like APP is membrane-associated and undergoes intramembranous proteolysis [207]. Interestingly, TACE and ADAM-10 also process Notch receptor [208].

Nicastrin has been shown to interact strongly with the presenilins; PS1 and nicastrin regulate each other and determine γ-secretase function via complex formation that in turn affects trafficking and processing of APP [210].

Two other genes, *aph-1* and *pen-2*, each encoding multipass transmembrane proteins that interact strongly with presenilins and nicastrin are also required for the activity of the γ-secretase multiprotein complex [211].

ε-secretion

Coordinated with γ-secretion, epsilon-(ε-)cleavage is proposed at residues 49-52 [212]. This protease activity is considered to degrade CTFγ produced from the amyloid intracellular domain (AICD) by γ-cleavage. Alternatively, sequential β- and ε-secretion without γ-cleavage generates a series of stable, potentially neurotoxic CTFs (CT49-99) [213].

Carboxyl-terminal Fragments (CTFs)

Proteolytic processing of APP by α/β/γ/ε-secretases and/or caspases generates a series of carboxyl-terminal fragments (CTFs) of APP. Those CTFs bearing the βAP region (*ie.* all fragments undergoing no α-cleavage) are considered amyloidogenic, and consequently have also been implicated in the pathogenesis of AD [214,215]. Therefore, βAP is not the the sole active fragment in AD, and some other factors are also involved in inducing neuronal loss.

CTF peptides have been found not only in various cultured cells [216,217] but also in paired helical filaments [218], in senile plaques [144], in microvessels [219], in choroid plexus from human brain [220], and in human platelets [221]. CTFs have also been found in media and cytosol of lymphoblastoid cells obtained from patients with early- or late-onset familial AD [222] and Down syndrome [223].

Apart from being an intermediate precursor of βAP, CTFs are directly linked with neurodegeneration [224,225]. CTFs induce a great depression of receptor-mediated synaptic transmission [226] and Ach level [227]; eventually, they cause significant impairments in cued, spatial, and working memory performances [228].

CTFs induce strong nonselective inward currents [229,230]. They block IP_3-sensitive Ca^{2+}-release by altering IP_3-induced receptor-signaling pathway [213]. The channel-inducing and toxic activity of CTF_{105} is much more potent than that of any βAP fragments. CTFs are more effective than βAP in forming channels [229,230]. This nonselective channel-forming effect is reminiscent of the properties of pore-forming proteins, *eg.* perforin, the membrane attack complex of complement. CTF-induced channels are more selective for Ca^{2+} and Na^{+} (permeability sequence $PCa^{2+} > PNa^{+} > PLi^{+} > PCs^{+} > PMg^{2+}$), whereas βAP-induced channels are more selective for Cs^{+} and Li^{+} (with the selective sequence being $PCs^{2+} > PLi^{+} > PCa^{2+} > PK^{+} > PNa^{+}$) [232,233]. Overall, CTF-induced pores have higher selectivity to Ca^{2+} and Na^{+}, the two primary ions involved in various neurodegenerative processes, than βAP-induced pores. These suggest that the neuronal ionic homeostasis in brain with AD can be more easily disturbed by CTF than by βAP.

CTFs may be released from the cell and/or more easily from the damaged neurons into the extracellular space. Extracellular CTFs form *de novo* ion channels, and this may induce neuronal death from outside the cells. Interestingly, cholesterol reduces calcium increase and toxicity induced by CTF, whereas the toxicity of βAP peptide is attenuated by calcium antagonists, such as nifedipine and verapamil [232,234].

CTFs may attack mitochondria, which leads to the increased release of cytochrome c and the activation of caspase-3 and may enter the nucleus to affect genetic regulation of some genes and finally to death.

Extensive neuronal loss with an accompanying increase in reactive astrocytes around the senile plaques is one of the well-established pathological findings in brains with AD [235,236]. CTFs also cause reactive gliosis and neurodegeneration in the neocortex and hippocampus [228]. CTF-induced IL-1β expression upregulates iNOS gene through activating the redox-sensitive nuclear transcription factor (NF-κB) in astrocytes and microglial cells [237]. This leads to increased production of NO, which may induce cell death. In addition, CTFs stimulate the release of chemokines such as MIP-1α, MCP-1 which play roles in accumulation of microglial cells around amyloid plaques [237].

CTFs are also present in the nucleus [238], where they play a role in genomic events contributing to delayed neuronal degeneration. CTFs translocate to the nucleus in a Notch-like manner, in which they bind a protein called Tip60, which is part of a large complex of proteins involved in gene transcription [196,239]. In addition, CTFs could form ternary complex with transcription factors in the nucleus, which induce GSK-3β and *tau* phosphorylation.

The Role of Signal Transduction

APP spans the cell membrane. Receptors, which also pass through the plasma membrane, relay information to a series of intracellular intermediaries, the second messengers, through a process termed signal transduction. The presence of a signal region in the extracellular domain along with a specific intracellular region define the receptor. Evidence suggests that APP can function as a receptor [135] due to the presence of a signal region in the extracytoplasmic domain as well as an intracellular domain made up of the same twenty amino acids that are involved in signal transduction. This cytoplasmic sequence interacts with G_o proteins, a class of guanosine triphosphate (GTP)-binding molecules, further raising the possibility that APP is involved in signal transduction [163]. Since G_o proteins are also involved in the intracellular trafficking of proteins, this interaction may play a role in the regulation of the intracellular routing of the precursor protein through different membrane compartments. Abnormal signal transduction systems have been implicated in the pathophysiology of AD, but their precise role has been difficult to establish.

A major route of APP processing is via the α-secretase pathway. The activity of α-secretase has both constitutive (yielding constitutively secreted sAPPα) and inducible components. The constitutive α-secretase activity is primarily at the cell surface, while the regulated activity is predominantly located within the Golgi apparatus.

A variety of evidence has emerged indicating that inducible activity of α-secretase is regulated by signal transduction pathways. Active second messengers, which can diffuse through the cytoplasm and carry and pass the signal beyond the cell membrane to their final destination (usually in either the cytoplasm or the nucleus), trigger a cascade of molecular reactions that lead to changes in cell behavior. These involve activation of protein kinases which are responsible for intracellular phosphorylations. The role of phosphorylation in APP processing has been the subject of a number of studies. Stimulating protein phosphorylation by activating the phospholipase C (PLC) – protein kinase C (PKC) pathway leads to increased secretory cleavage of APP.

Various first messengers known to activate PLC, including Ach, regulate the metabolism of APP. Activation of PLC leads to hydrolysis of phosphatidyl-inositol-4,5-bisphosphate (PI-P_2), yielding two intracellular signaling molecules. *a)* PLC stimulates PKC through the formation of diacyl-glycerol (DAG). PKC has been advocated to up-regulate the α-cleavage pathway and therefore plays a crucial role in the non-amyloidogenic processing of APP [241]. This mode of regulation could be relevant not only to cells undergoing direct activation of PKC (by *eg.* phorbol esters), but also to the regulation of sAPPα release from neurons in response to a variety of extracellular signals, such as certain first messengers (*eg.* Ach, the activity of which is markedly decreased in AD) or depolarization. *b)* PLC also increases cytoplasmic Ca^{2+} through the action of inositol-1,4,5-trisphosphate (IP_3).

The non-amyloidogenic APP processing can also be stimulated by G-protein-coupled cell surface receptors (also probably APP itself) through the activation of complex signal transduction pathways. G-proteins are associated with various effector molecules, such as guanylate-cyclase (yielding cGMP), adenylate-cyclase (that produce cAMP), tyrosine-kinase, Ca^{2+}-channels, and phospholipase A_2 (that mediate arachidate cascade). These enzymes further activate intracellular (3rd, 4th, *etc.*) messengers, involving ions (just as Ca^{2+} and Mg^{2+}) and phosphorylation cascades that include both PKC-dependent and independent mechanisms [242].

Other reagents can also increase the non-amyloidogenic cleavage of APP including estrogen, testosterone, various neurotransmitters and growth factors.

Disruption of the concentration of the central second messenger, Ca^{2+}, by βAP leads to increased levels of $[Ca^{2+}]_i$ that, in turn, might play a role in the abnormal proteolytic processing of APP resulting in increased production of βAP.

Because calcium is one of the most important second messengers, long-term increase in its concentration triggered by the cell's receipt of external or internal messages (such an effect is the presence of βAP and/or $βAP_i$) can interfere with various intracellular mechanisms and signal transduction pathways including several classes of protein kinases. Sustained activation of kinases induces hyperphosphorylation of intracellular proteins, including *tau*, leading to the formation of PHFs and later to NFTs seen in AD.

Disturbed Ca^{2+} homeostasis and βAP itself also induce apoptosis [243,244]. The central component of the apoptotic machinery is a proteolytic system involving a family of 14 cysteine-dependent aspartate direct-proteases called caspases [245]. Caspases that either bind and activate other caspases or recruited by death adaptor molecules to signal a cascade of caspase activation are termed initiator caspases. These are thought to act upstream of the executioner caspases that contain death effector domain. Among them, caspase-3 has been involved as a key executor protease in neuronal death [246], which has been shown to play a role in mediating βAP neurotoxicity [247].

Caspases cut APP yielding βAP [248]; this cleavage of APP releases a second apoptosis-promoting peptide called CTF_{31}. If brain caspases attack APP to release the toxic products βAP and CTF_{31}, then there may be a vicious cycle in which βAP, by triggering caspase activation, fosters its own production, and thus further caspase activation and cell death. Overexpression of CTF_{31} triggers selective increase of $βAP_{1\text{-}42}$ but not $βAP_{1\text{-}40}$ production and elicits a concomitant cell toxicity that is caspase-independent [249].

Actin, another prominent protein of the cytoskeleton, undergoes caspase cleavage in brain with AD. Caspase-cleaved actin is localized to the degenerating nerve terminals [250].

Caspases, therefore, would be potential targets for drugs aimed at stopping or at least slowing Alzheimer's development. However, developing such drugs will not be easy, given the fact that caspases play key roles in cell regulation throughout the body. For example, by helping eliminate cells with damaged DNA, the caspases protect against cancer.

Elimination of βAP

βAP is continuously produced in low concentrations, and is rapidly turned over in the normal brain [251]. The steady-state level of βAP represents a balance between its biosynthesis from APP through the action of the β- and γ-secretases and its degrading enzymes. Reduced catabolism may account for AD development [252].

It is necessary to distinguish between proteases that can degrade βAP only in its monomer state and those that can degrade oligomeric and/or highly aggregated fibrillar forms of βAP. βAP becomes resistant to proteases as a result of structural changes associated with its polymerization into fibrils. Pharmacologically up-regulating certain βAP-degrading proteases or interfering with the production or processing of their natural inhibitors could have great therapeutic potential.

Neprilysin (or encephalinase) has been described as part of an efficient clearance mechanism to prevent βAP accumulation and subsequent aggregation [253]. Neprilysin is a membrane-anchored proteinase with wide tissue distribution and substrate specificity that inactivates a broad range of circulating peptides including glucagon, enkephalins, endothelin, substance P, neurotensin, oxytocin, bradykinin, oxidized β-chain of insulin, atrial natriuretic peptide and βAP [254]. Neprilysin mRNA levels are lowest in the hippocampus and temporal gyrus of brains with AD, which are vulnerable to senile plaque development, but levels are highest in the caudate and peripheral organs, which are resistant to senile plaque development [255]. Work is underway to find evidence of linkage between AD and markers on chromosome 3, where the neprilysin gene is found.

Insulin-degrading enzyme (IDE, or insulysin) is an extracellular metalloprotease capable of degrading a number of peptides including insulin, glucagon, atrial naturetic peptide, TGF-α, and βAP [256]. The cleavage products of βAP by IDE are not neurotoxic and are not prone to depositing amyloid plaques. Therefore, IDE efficiently reduces βAP toxicity [257]. The protein not only occurs in soluble form in the cytoplasm, but has also been observed abundantly as secreted and membrane-associated isoforms in neurons and microglia [258]; therefore, it can participate in the degradation of both soluble and membrane-associated forms of βAP. In general, IDE recognizes a substrate with β-sheet conformation [259]. IDE may also be the protease responsible for the clearance of the CTFs following liberation of βAP [260]. In contrast with neprilysin which has little role in degrading soluble βAP but can cleave insoluble peptides, IDE mediates degradation of soluble, monomeric βAP but has less ability to eliminate βAP once it becomes insoluble or oligomeric.

The plasmin proteolytic cascade, known to be crucial for many physiological processes such as fibrinolysis, cell migration, tissue remodelling, has recently been implicated in βAP clearance, as well as in pathophysiology of a large number of human disorders [261]. βAP is a potent stimulator of tissue-type plasminogen activator (tPA) [262], which triggers plasmin mediated proteolytic cleavage of fibrin and other substrates, including monomeric and fibrillar βAP specifically [263], inferring that plasmin-mediated proteolytic activity is involved in amyloid plaque degradation. Also, plasmin can activate metalloproteases [264], and all candidate α-secretases ADAM9-10 and TACE [265]; therefore it can increase sAPPα levels and in turn decrease βAP production.

The integral membrane zinc metalloprotease endothelin-converting enzyme can cleave the endothelin precursors, bradykinin, substance P, and the oxidized insulin β-chain. The enzyme also directly proteolyses βAP [266]. Other proteases that have been reported to digest synthetic βAP include matrix metalloproteinase-9 and cathepsin-D and E, gelatinase A and B, aminopeptidase, and trypsin- or chymotrypsin-like endopeptidase [252]. This latter is signified by α_1-antichymotrypsin (or serine proteinase inhibitor A3, ACT), the serine protease inhibitor, which can increase βAP deposition [267,268]. ACT is an acute-phase protein which is mainly synthesized by hepatocytes, but it also produced by activated astrocytes [269]. ACT is found biochemically associated with the senile plaques in AD brain and is over-expressed in the astrocytes surrounding them [270]. ACT forms complexes with βAP, comparable in specificity and stability to a proteinase-inhibitor interaction, therefore inhibiting its degradation. It can also accelerate polymerization of βAP [271]. ACT-βAP complex can have multiple cellular effects on lipid metabolism and on expression of the inflammatory transcription factors NF-κB [272].

The LDL-receptor related protein (LRP) has the capacity to influence both the production and the clearance of βAP. LRP is highly expressed not only in the liver, but also in virtually all neurons in brain. The diversity in biological functions of LRP is mediated by the interactions of LRP with a wide range ligands and a major function of LRP lies in the removal of proteinase and proteinase inhibitor complexes [273]. LRP is a member of the LDL receptor gene family, and is identical to the α_2-macroglobulin (α_2M) receptor (α_2M-R or

CD91), and the ApoE receptor (ApoE-R). ApoE, α_2M, secreted APP and βAP are all ligands of LRP. In addition, βAP also binds avidly to LRP ligands, such as α_2M and ApoE, and LRP-mediated clearance of these ligands complexed to βAP contributes to a reduction in βAP level [274]. Hence transport of βAP out of the brain may be influenced by the LRP ligands α_2M and ApoE [275].

Human α_2M is an abundant serum pan-protease inhibitor; it entraps and isolates virtually all proteases before binding to their receptors and subsequently being internalised [276]. α_2M has been found associated with neuritic plaques in AD [277], and βAP can bind to α_2M subunits tightly regardless of the activation state of the molecule [278]. α_2M has the ability to mediate the clearance and degradation of βAP either directly in an LRP-independent pathway when complexed with an unidentified serine protease [279], or indirectly by means of clearance through LRP mediated endcoytosis [280].

ApoE also has the ability to interact with βAP and influence its conformation and clearance plays a major role [281]. In contrast with α_2M, however, ApoE4 may enhance the steady-state levels of βAP by decreasing its clearance from the brain tissue [282]. The mechanism by which ApoE4 protein leads to increased βAP deposition is not clear.

Environmental Factors…

A variety of environmental factors have been proposed to influence risk of AD. Fisrt of all, this study would like to point out that studies demonstrate an inverse relationship between educational attainment (*eg*. having been granted a PhD degree…) and risk for AD [283]. General risk factors for AD also include age, gender (females are at greater risk), previous head injury, and cardiovascular disease [284].

Mild cognitive impairment (MCI) is a sublclinical form of cognitive dysfunction, in which no concurrent difficulties with activities of daily living are experienced and there is normal intellectual functioning. MCI may be consiered a prodromal AD, since patients with MCI have a four-fold greater risk of progressing to dementia (particularly of the Alzheimer-type).

Epidemiological data supporting a link between calorie intake and neurodegenerative disorders were obtained in studies showing that those with the lowest calorie intakes were at the lowest risk of AD [285]. This is supported by the well-established observation that dietary restriction increases lifespan [286] and suppresses age-related molecular alterations in the brain [287], and can counteract the adverse effects of aging on learning and memory [288]. Dietary restriction can induce the expression of several neurotrophic factors in the brain, prominent among which is brain-derived neurotrophic factor (BDNF) [289,290]. BDNF can suppress oxyradical production, stabilize mitochondrial function, and protect neurons against various insults [72,291]. The increase in BDNF levels induced by dietary restriction may enhance the plasticity of neurons.

Dietary restriction also induces a mild cellular stress in neurons, which manifests increased expression of heat-shock proteins (HSPs) in neurons [292]. These are stress-responsive protein chaperones that have been shown to protect neurons against excitotoxic, metabolic, and oxidative insults relevant to AD [293]. Moreover, HSP70 has been shown to stabilize intracellular calcium levels in neurons [294]. Therefore, and based on the findings that neurons in dietary restricted rats are resistant to damage caused by kainic acid, a toxin whose action is mediated by excessive calcium influx, it is likely that the neuroprotective mechanism of dietary restriction involves stabilization of neuronal calcium homeostasis [295]. By inducing a mild stress response, dietary restriction can protect synapses and neurons against dysfunction and death induced by βAP and other insults relevant to AD. These findings suggest that dietary restriction may decrease risk of AD in humans.

The risk of several different age-related diseases, including coronary artery disease, stroke, and several types of cancer, is increased by a deficiency of dietary folic acid [296,297]. The mechanism whereby low level of folic acid endangers neurons may involve an increase in levels of homocysteine because folic acid normally plays a key role in homocysteine metabolism [298], and because homocysteine can promote neuronal apoptosis [299]. Exposure of neurons to homocysteine can increase their vulnerability to calcium-mediated excitotoxicity [299], and βAP-induced death [300]. Homocysteine endangers neurons by a mechanism involving increased DNA damage, p53, and caspases [299].

Studies of animals maintained in enriched environments support the hypothesis that intellectual activity is neuroprotective. Rats raised in enriched environments exhibit increased complexity of dendritic arbors and synapses [301,302], and do not exhibit age-related decreases in synaptic density [303] compared to control animals maintained in the typical "boring" cages of most animal facilities. Environmental enrichment can also increase neurogenesis, improve learning and memory ability, increase resistance of neurons to injury, and promote functional recovery [304,305]. Environmental enrichment upregulates the expression of BDNF and other neurotrophic factors and may thereby enhance neuronal calcium homeostasis.

Exercise may protect the brain against age-related dysfunction and disease, just as it benefits the cardiovascular and musculoskeletal systems [306]. Exercise is associated with enhanced structural plasticity in the motor cortex [307]. As with dietary restriction and environmental enrichment, it has been shown that exercise can increase production of neurotrophic factors in the brain [308,309].

Post-menopausal women who take estrogens are at reduced risk of AD [310]. Estrogens have been shown to enhance dendritic growth and synaptogenesis [311], and can protect neurons against excitotoxic and oxidative insults relevant to AD, including βAP toxicity [312]. Estrogens can enhance mitochondrial function and suppress oxidative stress in synapses [313]. The neuroprotective effects of estrogen are associated with enhanced cellular calcium homeostasis [312]. Estrogens may also promote non-amyloidogenic processing of APP [314].

Genetic Background...

Currently there is no effective treatment for preventing or delaying the progression of AD. A major difficulty in developing rational therapies for the disorder is the lack of knowledge of its primary cause. While the identification of specific environmental agents causing human neurodegenerative diseases can be a daunting task, in at least some cases of AD the primary cause lies not in the environment but in the genome of the affected individual. Therefore, AD is a disease with multiple etiologies; environmental or hereditary factors may predispose individuals to AD.

Mutations in three different genes (PS1, PS2, and APP) all serve to transmit AD via autosomal-dominant inheritance [315]. This form of AD is referred to as familial Alzheimer's disease (FAD) and is characterized by early-onset of symptoms; in contrast, sporadic forms of AD are considered to be of later onset.

APP Mutations

The most compelling evidence for the role of APP and βAP in the pathogenesis of AD comes from genetic studies on familial forms of the disease. Simple overproduction of APP may cause AD pathology. The extra copy of the APP gene, which occurs in Down's syndrome (trisomy 21), may lead to an increase in $mRNA_{APP}$; as a consequence, βAP levels are also elevated. This explains why middle-aged Down-patients develop brain pathology similar to that seen in AD [316]. By the fourth decade of life, individuals with Down's syndrome display many of the same neuropathological features as do individuals with AD, and many of these individuals develop dementia early in life [317,318].

Mutations can happen in or near to the β-amyloid region of APP. If these alterations exist just upstream or just beyond of βAP sequence of the precursor, affinity of the α-secretase to APP decreases. Meanwhile, other proteases, such as the β- and γ-secretases metabolize the amyloid precursor, thus βAP is released.

Several point mutations have been identified in the APP gene, clustering around the region encoding the βAP sequence. Mutations at codon 717 results in substitutions of a valine residue for isoleucine, phenylalanine or glycine [319] (Figure 1). The Swedish double mutation at codons 670 and 671 involves the substitution of two amino acid residues, an asparagine and a leucine for methionine and aspartate at the N-terminus of the βAP sequence [320]. Mutations in the APP gene have been found to cause an increase in the production of βAP_{1-42}, a more amyloidogenic agent then the more commonly produced βAP_{1-40}. Mutations at codon 717 increase the ratio of $\beta AP_{1-42}/\beta AP_{1-40}$, while the Swedish mutation increases the production of both forms [321].

Presenilins

The APP gene defects account for only a small proportion (5-7%) of all FAD cases. AD is also associated with gene defects linked to specific loci within chromosomes 1 and 14 encoding homologous proteins of approximately 50kDa known as presenilins (PS) [322]. They contain 6-9 transmembrane domains, and have an important role in cellular trafficking [323]. Mutations in both PS1 (located on chromosome 14) and PS2 (of chromosome 1) have been found to increase the proportion of βAP_{1-42} production [324].

PSs are ubiquitously expressed membrane proteins [325]. Their structure comprises eight membrane-spanning regions with cytoplasmic orientation for both the N- and C-termini [326]. The unstable holoproteins of PSs undergo constitutive endoproteolysis by presenilinase and quickly converted into stable and functional high-molecular-mass heterodimers [327]. The endoproteolysis occurs within the cytoplasmic loop between the sixth and seventh transmembrane domains in the endoplasmic reticulum vesicles with subsequent stabilisation of the fragments in the Golgi apparatus [328]. The transmembrane structure of PSs could call for structural, receptor, or channel functions.

Presenilins directly or indirectly control the physiopathological maturation of APP. PSs are functional components of γ-secretases, and may act upstream as a chaperoning protein bringing APP to a cell compartment containing γ-secretase activities. The fact that PSs could physically interact with APP, β-catenin, G_o-proteins [329], and *rab11*, a GTPase activity involved in vesicular transport [330] argues in favor of a PS-dependent chaperoning of APP. This would likely occur early in the routing of APP through the constitutive secretory pathway, as suggested by the fact that PSs display favored interaction with immature APP [331]. Also, PS1 brings the GSK-3β and cdk-5 to the proximity of *tau* proteins that behave as one of the kinase substrates [332]. Interestingly, GSK-3β displays a higher affinity for FAD-linked PS1 [333]. Therefore, this higher affinity leads to a higher recruitment (chaperoning activity) and therefore phosphorylation of endogenous *tau*. This is very important when considering that hyperphosphorylation of *tau* proteins likely underlies the development of NFTs, a major histopathological hallmark in AD [334].

Alterations in APP processing that result from PS mutations may result from altered Ca^{2+}-mediated regulation of the enzyme activities that mediate cleavage of the protein. PSs bind calsenilin, a calcium-binding protein [335] and directly interact with Ca^{2+}-channels. Also, the pathogenic action of PS mutations cause an overfilling of ER Ca^{2+} stores resulting in a greater Ca^{2+} response to various stimuli that promote Ca^{2+} release [336,337]. A primary defect in ER Ca^{2+} signaling caused by PS mutations may lead to global abnormalities in the regulation of cellular Ca^{2+} homeostasis.

Synaptic dysfunction and degeneration of synapses are thought to be relatively early and pivotal events that underlie cognitive function and initiate the cell death process in AD and other neurodegenerative disorders [338]. APP and PSs are each present at high levels in synaptic terminals, and it is therefore likely that both their physiological functions and pathogenic actions are exerted in synapses. Ca^{2+} plays important roles in regulating synaptic functions including learning and memory, and βAP [339], sAPPα [340], and PS1 [341,342] have each been shown to affect synaptic plasticity. Therefore, APP and presenilin mutations can promote cell death cascades by perturbing cellular Ca^{2+} homeostasis, which may be initiated in synapses [338].

PSs also interact with ubiquilin and ubiquitin, which are subunits of the proteasome [343]. Proteasome controls the degradation of various proteins, including wild-type and mutated PS1 and PS2 and modulates the production of both βAP and APP [344,345]. PSs are also involved in the Erb-4 tyrosine kinase-mediated intranuclear signaling [197,346]. PSs also interact with cytoskeletal-associated proteins and actin filament network, epithelial (E-)cadherin [347], telencephalin [347], and Notch [349], which are in line with their putative role in both adhesion processes and embryogenesis. Their additional role in programmed cell death (apoptosis) participates in cell fate and tissue organization at early stages of development. Disturbed functioning of PSs caused by mutations, therefore, may lead to cytoskeletal changes similar to that observed in AD.

Apart from processing by presenilases, PSs undergo additional alternative cleavages by caspases and caspase-like proteases [350]. C-terminal PS2-fragments derived from alternative transcription or caspase-3 cleavage (referred to as Ccas-PS2) act as a proapoptotic factor by triggering a caspase-3-like activation leading to an increase of the tumor suppressor apoptotic effector p53; the effect of caspase-3 is also accompanied by a drastic reduction of PS1 [351]. PS1, on the other hand, increases antiapoptotic oncogene bcl-2 levels [352]. Therefore, PS1 and PS2 could have distinct and even opposite functions in the control of cell death and/or other functions, and may compensate each other during cell division [353]. These distinct functions likely contribute to an explanation of various phenotypic forms of AD having distinct onset and evolution.

ApoE – The ε4 Allele

There is now considerable evidence that the ε4 allele of the apolipoprotein E (ApoE4) gene on chromosome 19 is a major susceptibility factor for AD. ApoE is a glycoprotein which is principally involved in lipid transport and related functions. Genetic heterogeneity leads to 3 common isoforms designated ApoE2, E3 and E4, encoded by 3 alleles named ε2, ε3 and ε4. The ε4 allele frequency is increased in AD individuals, and those patients with one or two copies of the ε4 allele are at higher risk of developing AD [354].

Cerebrovascular effects of ApoE4 might contribute to its AD-promoting action, but more direct effects of ApoE on neurons might account for a link to AD. ApoE possesses isoform-specific antioxidant and neuroprotective activities [355,356]. ApoE isoforms differ in two amino acids such that E2 contains two cysteine residues, E3 contains one cysteine residue, and E4 contains none. Interestingly, the isoforms differ in the amount of HNE that they can bind (E2>E3>E4), which is directly related to their ability to protect neurons against βAP-toxicity [356]. In this way, ApoE2 and ApoE3 may stabilize cellular Ca^{2+} homeostasis during oxidative stress; ApoE4, on the other hand, induces Ca^{2+} influx and promotes death [357]. Moreover, patienst with ε4 allele have higher amounts of βAP immunoreactivity in their brains [358]. Reason behind this finding might be the fact that ApoE4 promotes βAP aggregation and amyloid fibril formation, and is less capable of clearing βAP from the neuropil.

There are other genes that are considered susceptibility or risk factors for AD. These include α_2-macroglobulin [359], a gene for a component of α-ketoglutarate dehydrogenase [360], the K-variant of butyryl-cholinesterase [361], serotonin-transporter, MAO-A, COMT, NOS-3, and several mitochondrial genes [284].

Although age and the inheritance of predisposing genetic factors appear to play a major role, more recent evidence suggests that the development and progression of AD is subject to a wide variety of both environmental and genetic modifiers.

Co-morbidity and Other Complications

α-synucleinopathy

Besides the presence of βAP and degenerated neurites, the major component of senile plaques is referred to as non-amyloid component (NAC) which is a main fragment composed of α-synuclein and its proteolytic derivatives [362].

α-synuclein is a 140-amino acid peptide of 18 to 20 kDa that is encoded by a single gene consisting of seven exons borne by chromosome 4 [363,364]. Several members of the synucleins family have been described, including β- and γ-synucleins encoded by chromosomes 5 [364] and 10 [365], respectively. α-synuclein is a protein that, just as $\beta AP_{1-42/43}$, is extremely prone to aggregation; it accumulates as full-length and truncated aggregated species into Lewy bodies. These intracytoplasmic inclusions occur in several neurodegenerative diseases, including Parkionson's disease (PD) and dementia with Lewy bodies (DLB) [366]. A deleterious cross-talk between βAP and α-synuclein could lead to a mixed pathology corresponding to AD with Lewy bodies. Moreover, several human cases with clinical signs of AD and PD have been reported [367,368].

Human cases with clinical and neuropathological features of both AD and PD raise the possibility that these diseases involve overlapping pathways. Approximately 25% of patients with AD develop frank PD [369], and α-synuclein-immunoreactive Lewy-body-like inclusions develop in most cases of sporadic AD and FAD, as well as in Down syndrome [371,371]. Therefore, AD pathology can be classified into two groups according to the existence or absence of α-synuclein aggregation [372]. Synucleins are present in high concentrations in the neocortex, hippocampus, and substantia nigra, three brain zones reminiscent of those affected early in AD and PD [373].

α-synucleins undergo few post-translational events, and are located in the presynaptic nerve endings [374]. At endogenous levels and physiological conditions, α-synuclein exerts an antiapoptotic control. This function would be abolished by all mechanisms triggering exacerbated aggregation and leading to Lewy bodies. This hypothesis is reinforced by the observation that aggregated α-synucleins induce apoptotic cell death [375]. The aggregation promoting events could be caused by increased phosphorylation or altered structural features associated with PD-related α-synuclein mutations [376]. These aggregates could be ubiquitinated and subsequently degraded by the proteasomal machinery. When aggregated α-synuclein occurs in excess, the clearing process could be insufficient to avoid accumulation.

The interaction of α-synuclein with proteins involved in the MAP kinase pathway [377], *Erk*-cascade and PKC suggests a role in cell death. On the other hand, its ability to lowering p53-dependent caspase activation, and its direct interaction/sequestration of proapoptotic oncogenes including *bad* and *bax* [378] implicate an antiapoptitic activity of α-synuclein [379]. α-synuclein also displays chaperoning properties [380]: it binds to *tau*; this α-synuclein/*tau* association seems functional because it leads to enhanced protein kinase A-mediated *tau* phosphorylation [381].

α-synuclein also interacts with proteasomal activity [382]. A small proportion of "abnormal" α-synuclein variants undergo ubiquitination by parkin, an E3 ubiquitin ligase protein [383]. Parkin mutations responsible for recessive juvenile Parkinsonism are characterized by an impairment of the intrinsic E3-ligase activity argues in favor of a deleterious influence of an altered degradation process. Proteasome inhibitors lead to the accumulation of α-synuclein without any ubiquitination process [384].

Both dopamine and L-dopa inhibit α-synuclein fibril formation and stabilize the potentially toxic intermediate protofibril α-synuclein species [385]. This could underlie the molecular explanation to the dopaminergic region-specific α-synuclein associated dysfunction occurring in PD. β-synuclein could also reverse some of the α-synuclein-associated phenotypes. It seems to block α-synuclein to aggregate, and may act as an endogenous regulator of α-synuclein functions [386].

Depression

Although cognitive and functional decline is the clinical hallmark of AD, psychiatric or non-cognitive mental disturbances affect as many as 90% of patients with AD over the course of their illness. They take on different forms, including agitation, delusions, hallucinations, depression, sleep disturbance, and problem behaviors, and are a major focus of treatment. Depression is one of the most frequent psychiatric complications of AD, affecting as many as 50% of patients [387,388].

Several risk factors for depression in AD have been described [389]. Most consistent is the presence of family history of mood disorder in first-degree relatives. In addition, prior personal history of depression, female gender, and younger age of onset of AD may be risk factors [390,391]. The natural history of depression in AD is poorly known [392]. It might be a psychological reaction to the knowledge of having AD or to the associated losses and disability. Pathologic consequences of the neuropathologic features and brain damage brought about by AD might also be important to its development. Depression has been associated with selective loss of noradrenergic cells in the locus ceruleus of the brain [393,394]. Depression is also associated with loss of dorsal raphe serotonergic nuclei in AD [393]. These findings have implications for the treatment of depression, specifically, about the role of manipulation of the relevant neurotransmitter systems in treating AD-associated depression.

Depression is often one of the first symptoms of AD, and mood symptoms fluctuate over time [392]. Epidemiologic data indicate that depressive symptoms are common in mild to moderate dementia and less prevalent in severe dementia [395]. One of the major causes of additional disability for Alzheimer's patients and their caregivers, depression in patients with AD has been associated with greater impairment in quality of life, greater disability in activities of daily living [396], greater probability of physical aggression, greater likelihood of being discharged from an assisted living facility, earlier entry into a nursing home, and greater caregiver depression and burden [397,398]. In addition, depression has been associated with a slight increase in mortality in the presence of cognitive impairment or dementia and may also be associated with an increased risk of suicide. Depression in AD should be treated aggressively, with careful monitoring of congnitive function.

Behavioral and Psychological symptoms of Dementia

AD is characterized clinically by decline in memory and cognitive functions, however manifestations of behavioral and psychological symptoms of dementia (BPSD), including psychosis, delusions, hallucinations, agitation, *etc.* during the course of the disease are significant events. Their frequency ranges from 50 to 90% and they can occur at any point in the disease process, often being the presenting symptom. The number and severity of behavioral problems increase with worsening of cognitive impairment [399]. BPSD can be mainly grouped into behavioral and psychological symptoms of dementia disorders. The former are usually identified on the basis of patient observations, while the latter are mainly assessed on the basis of interviews with patients and relatives (Table II.A) [400].

Table II: Behavioral and psychological symptoms of dementia (BPSD)

	Group 1 ***Most common, most distressing***	**Group 2** ***Moderately common/distressing***	**Group 3** ***Less common, manageable***
Psychological	Delusions, hallucinations, depressed mood, sleeplessness	Misidentifications	
Behavioral	Physical aggression, restlessness	Agitation, culturally inappropriate behavior, screaming, wandering	Crying, cursing, lack of drive, repetitive questioning, anxiety, shadowing behavior and disinhibition

A: Groups 1-3 of BPSD

Signs and symptoms	**Medication**
Hallucinations, hostility, aggressiveness, violent behavior, sleep disturbances	Atypical antipsychotics (risperidone, clozapine, olanzapine, quetiapine)
Depression, obsessive-compulsive disorder	SSRIs (sertraline, fluoxetine, paroxetine, fluvoxamine, citalopram)
Anxiety, agitation, sleep disturbances	Benzodiazepines
Recurrent agitation, sleep disturbances, disruptive behavior	Anticonvulsants (carbamazepine, valproic acid, gabapentin)

B: First-line agents in the treatment of BPSD

Among the other neurotransmitters involved in AD, the relative increase in monoaminergic systems (dopamine, noradrenaline, serotonin), secondary to significant decrease in cholinergic activity, have been correlated with cognitive functions and BPSD. The imbalance between the glutamate and dopaminergic systems may lead to dysfunction in cortical-striatal-thalamic circuit, resulting in psychotic symptoms. Reduced levels of noradrenaline are associated with higher rates of depressive symptoms in AD patients, whereas higher levels of the neurotransmitter have been found in the substantia nigra of patients with AD with concomitant psychotic symptoms instead than in patients without psychosis [394]. Abnormalities of serotonergic system have been specifically implicated in some BPSD, namely depressed mood, anxiety, agitation, restlessness and aggressiveness. Reduced serotonin concentrations in the presubiculum have been found in AD patients with psychotic symptoms [394].

Compared with control subjects, AD patients with BPSD have less neurons in parahippocampal gyrus, CA_1-region of the hippocampus, dorsal raphe and locus ceruleus [402]. Also, the process leading to the formation of the NFTs has been related to the production of psychotic symptoms [403]. A significant increase in mean density of NFTs in the middle frontal cortex was reported in the group of AD patients with psychotic symptoms. This may explain the association between increased rate of cognitive decline and psychotic symptoms in AD patients [394,404]. Functional neuro-imaging studies suggest that BPSD are associated with dysfunction in brain cortical metabolism [405]. AD patients with behavioral disturbances have higher rates of cortical hypoperfusion with respect to AD patients without BPSD in the same stage of disease, with highest cortical hypoperfusion rates being evident in frontal areas.

A great part of BPSD is not drug-responsive. With respect to treatment, non-pharmacological interventions are first-line in dealing with milder BPSD, while for moderate to severe BPSD, medication is clearly indicated, often in conjunction with non-pharmacological interventions (Table II.B).

Interestingly, AD pathology is infrequent in patients with schizophrenia (SCH) [406,407]. It has been proposed that medications used to treat SCH, such as haloperidol (HAL), might have a protective effect against developing AD neuropathology, which would account for a low frequency of these changes in elderly SCH. The mechanism by which antipsychotic drugs might exert such an effect is not known.

A Systemic Approach...

β-amyloid is reported to be ubiquitously present in the body [408], and has every now and then been demonstrated to form depositions in the skin of Alzheimer patients [409], presumably interfering with intracellular biochemical pathways of peripheral tissues. In deed, several alterations in the blood of AD patiens have been reported. In lymphocytes, changes in ionic and signal transduction pathways [410], elevated number of apoptotic peripheral mononuclear cells [411], increased vulnerability to programmed cell death [412] and raised levels of oxidative DNA damage [413] was reported. With respect to platelets, disrupted membrane fluidity [414,415], hypofunction of mitochondrial enzymes [416], and modified activation of cells [417] have been found. Dysfunctions of red blood cells including decreased erythrocyte AchE activity are also observed [418]. In case of fibroblasts, disturbances of Ca^{2+}-uptake [419], ionic homeostasis [420], glucose metabolism [421], proteolysis [422], signal transduction [423] and DNA repair [424] are some of the biochemical changes reported in AD.

Apart from brain-specific changes, abnormalities attributed to AD seen in many peripheral cell types such as erythrocytes, lymphocytes, platelets and fibroblasts has lead to the speculation that AD might be a systemic disorder, with the most prominent pathology in the cognitive functions of the CNS.

A substantial body of data exists indicating that biochemical changes of brain cells in various neurobiological diseases with systemic manifestation are reflected in peripheral tissues, such as fibroblasts and lymphocytes [425,426]. In the course of surveying a wide variety of these mirror changes, previous reports have proved that fibroblasts and lymphocytes can serve as models and diagnostic markers of several neurodegenerative disorders, including metachromatic leucodystrophy, Tangier's disease, and generalized glycogenosis [425,427].

Lymphocytes and cultured fibroblasts offer many advantages for the study of various pathobiochemical alterations in AD. These pathways are more robust in these living cells than in postmortem material, such as brain, and the results are not complicated by agonal or postmortem events [428].

Objective of the Study

Involvement of the Periphery in AD

The identification of the mirror shifts in AD described in the *introduction* has been of my interest. My aim was to study whether readily available tissues such as fibroblasts and lymphocytes may serve as simple supportive systems to monitoring intracellular ionic imbalance, particularly that of the basal intracellular free calcium levels ($[Ca^{2+}]_i$) in AD. Broadly in line with this hypothesis, the present communication summarizes techniques which have been used to study the biochemistry and physiology of skin fibroblasts and lymphocytes isolated from patients with certain neurometabolic disorders including AD. Consistent with this, alterations in calcium homeostasis of fibroblasts and lymphocytes associated with AD are discussed in the present study.

Pharmacological Studies

Several investigators have found a low frequency of AD in patients with SCH, and it has been proposed that antipsychotic medications may be responsible [406,407].

The prototype classical antipsychotic HAL is dopamine-2 (D_2) and sigma-1 (σ_1) receptor antagonist with apoptotic activity [429,430,431]. HAL is involved in neurotoxicity, causing clinically troublesome adverse events such as extrapyramidal and cardiac side-effects [432]. The "gold-standard" neuroleptic HAL has been reported to be a calcium and calmodulin antagonist [433], and was also demonstrated to efficiently inhibit βAP-formation from APP [434]. The mechanism by which it might exert such an effect is not known. However, HAL and other traditional antipsychotics have several adverse effects. Therefore, the use of atypical antipsychotics is taking over medical indications of conventional neuroleptics with very much success without notable side effects [435,436]. The widely used high-potential atypical antipsychotic RISP is a potent D_2-, $5HT_2$- and NA-antagonist with less adverse effects. RISP has no known effect on AD pathology, including APP metabolism.

In light of evidence that HAL, one of the most frequently prescribed antipsychotic medications used in the treatment of SCH [437], interferes with Ca^{2+}-homeostasis [433] and APP processing [434], we evaluated it for its effects on βAP-induced Ca^{2+}-imbalance *in vitro* and its role on APP metabolism. To gain more insight on its cellular effects at the molecular level, we have also studied its role on gene expression *in vivo*. In the course of surveying a wide variety of agents directed against βAP-toxicity, we also examined RISP with respect to $[Ca^{2+}]_i$, APP and gene expression profile.

Experimental Procedures

Patients

Demographics

Primary cultures of human fibroblasts and suspension of lymphocytes were obtained and prepared from forearm skin-biopsies and blood samples after informed consent from participants. Sporadic AD patients of late-onset and age-matched controls volunteered as donors (Table III).

Table III: Demographic data of participants

Sporadic late-onset AD patients and age-matched controls volunteered as donors. Selection of individuals was made after their informed consent. Fibroblasts were harvested from forearm excisional biopsies; lymphocytes were separated from venous blood samples drawn in the usual fashion.

	Studies on fibroblasts		Studies on lymphocytes	
	Age of donors (mean±SD)	*Number of volunteers*	*Age of donors (mean±SD)*	*Number of volunteers*
Control	71±8.8	8	66.8±10.2	12
AD	73±10.11	8	71±8.5	10

At the time of obtaining samples, every single individual was outpatient. The diagnosis of AD was made according to DSM-IV criteria [3]. All AD donors met the DSM-IV definition of dementia of the Alzheimer's type (Table I), which requires evidence of cognitive deficits (criterion A) and a process of decline from previous levels (criterion C); none of them had history of familial AD. Control individuals were tested negative for any forms of dementia.

Medications

None of the participants in our study received any medication known to interfere with calcium metabolism, such as antihypertensive drugs, calcium-antagonists, or antidepressants. None of the patients was on AchEIs before and during the measurements.

Culture Methods

Fibroblast Culturing

For outgrowth of fibroblasts, small pieces of skin material were placed in cell culture flasks (25cm^2) and grown for 3-5 weeks in Dulbecco's modified Eagle's medium (DMEM) containing 5% of heat inactivated fetal calf serum (FCS) as well as penicillin (100 U/mL), streptomycin (100 µg/mL) and 2mM glutamate in a humidified atmosphere of 95% air and 5% of CO_2. Cells were detached for serial passaging using 0.01% of trypsin and 0.02% of

EDTA. Stocks of each cell line (passage number: 5) were frozen to -80°C and stored at -130°C until use. After thawing and/or trypsinization, cells were used to seed on glass coverslips, 10mms in diameter, pretreated with collagene, in a 24-multiwell plate in the DMEM medium containing 5% of FCS. Cultures maintained in this way were generally confluent by day 7 and were used directly for measurements of $[Ca^{2+}]_i$.

To monitor the long-term effect of βAP, a single sample was used for chronic culturing. Here, the seeding-and-passaging cycle was repeated as needed. In short, after 7 days *in vitro*, cells developed a monolayer, and subcultures were provided by trypsinization.

All fibroblasts had a characteristic spindle-shaped appearance and were attached firmly to the coverslips.

Separation of Lymphocytes

Blood samples were gently layered on Ficoll solution, and were centrifuged at 1800 rpm for 20 mins. Lymphocyte-containing bands were rinsed with phosphate-buffered saline (PBS) three times. Cell concentration was estimated using Bürker's chamber. Lymphocytes in PBS were used for measurements of $[Ca^{2+}]_i$.

Viability

Viability of cells was estimated by the method of intravital staining with Trypan Blue after exposition of cells to the preparation (0.2% Trypan Blue in Hank's balanced saline solution, HBSS). After 5 minutes of incubation in the solution, less than 2-3% of fibroblasts and lymphocytes absorbed the dye. In line with this finding we concluded that more than 97-98% of cells present in the cuvettes were viable.

Treatment

Cells were cultured in the presence of 10^{-7}M, 10ng/mL [438,439] and/or 50μg/L [440,441] final concentrations of βAP, HAL and RISP, respectively. HAL and RISP – designated for intravenous use – were purchased commercially (Haloperidol®; *Gedeon Richter Ltd* and Risperdal®; *Ris-Int-63, Ref.No.: 336341; Janssen-Cilag, division of Johnson&Johnson Ltd., Hungary*, respectively) and were used directly on the cells without further purification or sterilisation. Final concentrations of these antipsychotics were calculated in accordance with their therapeutic blood levels.

Fibroblasts and lymphocytes used for control study were free of any treatment.

Measurements of $[Ca^{2+}]_i$

Fluorescence Imaging

$[Ca^{2+}]_i$'s were quantified by fluorescence ratio imaging of the calcium indicator dye Fura-2AM. Cells were incubated for 20-30mins at 37°C in the presence of 1μM of the acetoxy-methylester form of Fura-2AM. Cultures loaded with the dye were then washed twice with PBS solution and were transferred to cuvettes containing PBS at 25°C for measurements. Fluorescence imaging in standard 1×1cm cross section quartz cuvettes were performed with Hitachi F-2000 spectrofluorimeter. The ratio of the fluorescence emission of the cultures using two different excitation wavelengths of 340nm and 380nm, at emission wavelength of 495nm was applied to determine $[Ca^{2+}]_i$. Values of the 340/380 fluorescence excitation ratio at 495nm emission wavelength ($FL_{340/380}$), which represents the $[Ca^{2+}]_i$, were computed after each fluorescence measurement. In line with several papers examining calcium levels [58,425,442], the authors found the rate of the fluorescence ratios to be more informative of the change of the calcium-level than the absolute values of $[Ca^{2+}]_i$.

Validity of Fura-2AM Labeling

To check that loading of the cells with the dye and intracellular cleavage of the ester occurred, we sampled the fluorescence excitation ratios before and after loading the cultures with the dye. In case of fibroblasts, ratios of unloaded cells observed at 495nm were 10.67±3.26, around 3-times lower when compared to the ratio of 35.43±8.03 after incubation with the dye. As for lymphocytes, sampling of the fluorescence excitation ratios after incubation with the dye revealed 79.89±10.11, which represents a roughly 10-fold increase from the unlabeled ratio of 7.68±2.2. This is characteristic of Fura-2AM fluorescence.

No significant alteration was detectable among labeled cultures in Fura-2AM fluorescence using the calcium-insensitive excitation wavelength of 367nm, indicating that the observed response reflects a real change in the calcium-level of the cells.

Protein Content of Cultures

Because gross alterations in the cell number/cell mass actually present in the cuvettes may interfere with the fuorescence intensities, the total protein content of the samples was determined according to Lowry et al [443]. Coverslips with control fibroblasts were found to contain 146.9±8.4μg protein/coverslip, whereas Alzheimer cells contained 139.1±5.1μg protein/coverslip. Protein content of control lymphocytes revealed 252.1±11.6μg protein/cuvette, and cultures of AD leucocytes contained 239.9±14.0μg protein/cuvette. This difference is not significant in statistical terms, suggesting that the total protein content of Alzheimer and control cells did not differ considerably.

Rationale of the Effect of βAP, HAL and RISP

During the experiments, the treated and control cell populations were cultured in media of identical composition except for the presence or absence of βAP, HAL and/or RISP. Therefore, it is reasonable to conclude that the observed difference in the 340/380 fluorescence excitation ratio of Fura-2AM-loaded cells is only associated with the presence or absence of the agents used, and hence can be regarded as an effect induced by βAP, HAL and/or RISP itself.

Synthesis and Purification of βAP

Amino Acids

All amino acid derivatives, solvents, chemicals and the resins were obtained commercially and were used without further purification. Protected amino acids used in the syntheses were of the L-configuration. The α-amino function group was protected with tert-butoxy-carbonyl (Boc) group, and the reactive side chain functional groups were protected as follows: para-toluene-sulfonyl for Arg, cyclohexyl for Asp and Glu, benzyl-oxy-carbonyl for His, 2-chlorobenzyl-oxycarbonyl for Lys, benzyl for Ser, and 2-bromobenzyl-oxycarbonyl for Tyr. The side chains of Asn and Gln were unprotected.

Solid Phase Peptide Synthesis

βAP was prepared by solid phase methodology [444] using ABI-430A automated peptide synthesizer. For peptides with an amidated carboxy-terminus, para-methyl-benzhydryl-amine (MBHA) resin (0.57mmole/g) and for peptides with free carboxy-terminus Merrified resin (1mmole/g) was used. The coupling reactions were achieved with 3-fold excess of Boc-amino acids using DCC or DCC/HOBt as activating agent in dichloro-methane, dimethyl-formamide or a combination of thereof. Boc-Asn and Boc-Gln were coupled with preformed 1-hydroxy-benzotriazole ester to avoid side reactions. After a coupling time of 2 hours, the completeness of acylation was monitored at each stage by the standard ninhydrin test [445]. In cases where incomplete coupling was found, the coupling procedure was repeated, or the acetylation was carried out before the removal of the N-α-amino protecting group prior to the coupling of the next amino acid. Acetylation was performed with 30% of acetic anhydride in dichloro-methane for 10 and 20 minutes. Intermediate deblocking was achieved with 50% of trifluoro-acetic acid in dichloro-methane containing 0.5% of dithio-threitol followed by neutralization with 10% of triethyl-amine in dichloro-methane. Final deprotection as well as the cleavage of the peptides from the resin were performed with anhydrous hydrogen fluoride in the presence of 8% of anisol, 2% of dimethyl-sulfide, 2% of p-cresol, and 2% of thio-cresol at 0°C for 60 mins. After the removal of the hydrogen fluoride under a steam of nitrogen and *in vacuo*, the free peptides were precipitated with diethyl ether, filtered, washed with diethyl ether and extracted with 95% of TFA, diluted with water, and lyophilized.

Purification

Crude peptides were purified by using a Shimadzu LC-8A preparative HPLC system equipped with Backbond WP-C4 column (300×47 mm, 300Ĺ pore size, 15-20 μm particle size). The column was eluted with a solvent system consisting of (1) 0.1% of aqueous trifluoro-acetic acid and (2) 0.1% of trifluoro-acetic acid in 80% of aqueous aceto-nitrile in a linear gradient mode for 30 minutes at 80mL/min flow. The eluent was monitored at 220nm. The fractions were checked by analytical HPLC, and those with purity higher than 95% were pooled and lyophilized. The HPLC analyses of the peptides were carried out with a Hewlett-Packard Model 1050 liquid chromatograph using a Nucleasil 300 C4 (280×4 mm) reversed-phase column and a gradient elution with two solvent systems described above at a flow rate of 1mL/min. The peaks were monitored at 220nm. Amino acid analyses of the purified peptides were carried out on a Hewlett-Packard Amino-Quant amino acid analyzer after hydrolysis of the samples in 6 M of hydrochloric acid at 110°C for 48hrs in tubes sealed under vacuum. Amino acid analysis of the hydrolysates of peptides showed the expected amino acid composition (data not shown). Data of electrospray mass spectrometry (ES-MS, FinniganMat TSQ 7000 mass spectrometer) were in accordance with the calculated average molecular masses.

"Aging" of the Peptides

Freshly prepared aqueous βAP solutions were "aged" for 1 hour before use.

Treatment of Animals

Male, Sprague-Dawley rats weighing 200-250g were maintained in plastic cages (three per cage) containing wood shavings and were given rodent chow and filter-purified tap water *ad libitum*. Murines were housed in rooms of our animal facility maintaining temperature of 25±2°C and a 12-h light/dark cycle.

All rodents were in good condition, and showed no signs of stress. Each was allowed to acclimate to the housing conditions for 1 week prior to beginning the experiment. With respect to the traditional antipsychotic haloperidol (Haloperidol®; *Gedeon Richter Ltd, Hungary*), groups of six rats were injected intra-peritoneally (i.p.) on a daily basis with therapeutic (0.05mg/kg) or toxic (0.5 mg/kg) doses dissolved in saline [446,447]. In case of the atypical drug risperidone (Risperdal®; *Ris-Int-63, Ref.No.: 336341; Janssen-Cilag, division of Johnson&Johnson Ltd., Hungary*), therapeutic (0.1mg/kg) or toxic (1.0 mg/kg) doses, dissolved in water, were administered [448,449]. Control rats received the vehicle alone. No other animals were housed in the room or allowed contact with the study animals.

Experiments were performed in accordance with a protocol approved by the university ethics committee on laboratory animals.

Preparation of Rat Cortical Samples

After acute (6, 12, 24 and 96 hours) and chronic (1, 2, 3 and 4 weeks) administration of the antipsychotics in both therapeutic and toxic doses, rats were decapitated under ether anesthesia. The cerebellum was removed and the fronto-temporo-parietal cortices were dissected. In order to eliminate post-mortem decay, preparation was performed on dry ice. The samples were homogenized in 50mM of Tris buffer (pH 7.5) containing 0.15M NaCl, 2mM phenyl-methyl-sulfonyl-fluoride, 2mM EDTA, 2 μg/mL leupeptin, 1 μg/mL pepstatin, 1% Nonidet–P-40 and 0,1% sodium deoxycholate by using a glass-teflon potter (1500 rpm, 1 min). The homogenates were centrifuged at 10,000 × *g* for 30mins at 4°C. The supernatants were used for the detection of total APP. Protein concentrations of samples were determined by the method of Folin as modified by Hess et al. [450].

Western Immunoblot

Proteins (30μg/lane) were separated on a 9% of sodium-dodecyl-sulfate (SDS)-polyacrylamide gel and electroblotted onto nitrocellulose membranes by using the BioRad Mini-Protean II system. After protein quenching with 5% non-fat dry milk in 50mM of Tris-buffered saline (TBS, pH 7.5) containing 0.2% of Tween-20, the blotted samples were incubated overnight at room temperature in monoclonal 22C11 antibody (5 μg/mL, against residues 68-81 of APP). After being washed (5×), the membranes were incubated with horse-radish-peroxidase (HRP)-conjugated anti-mouse-IgG (1:500, Sigma-Aldrich, USA). For the detection of blots, the Renaissance Western Blot Chemiluminescence Reagent (Pierce, USA) was employed, followed by exposure to an autoradiographic film. Optical densities of immunoreactive bands were detected and quantified by means of the NIH-Image Analyser Program (NIH, USA). The levels of APPs in the control group were taken as 100%, and changes were calculated with respect to this value (data are given in relative units).

Microarray Protocols

Samples and RNA Preparation

Total RNAs were extracted and purified from rat fronto-temporo-parietal cortices (40mg/animal) using NucleoSpin RNA purification kit (Macherey-Nagel, Dürren, Germany) according to the manufacturer's instructions. The quantity and the quality of all RNA preparations were assessed by gel electrophoresis and spectrophotometry (optical density: OD_{260}/OD_{280} ratios) (NanoDrop, Rockland, DE, USA). RNA samples were aliquoted and stored at -80°C. Total RNA was used for microarray analysis as well as for Real-time quantitative reverse transcription-PCR (QRT-PCR).

Construction of Microarrays

8,000 cDNA inserts (amino-modified oligo-nucleotides) from rat mixed-tissue cDNA libraries (peripherial ganglion, brain, heart, *etc.*), containing named genes, expressed sequence tags (ESTs), housekeeping genes and genomic control spots (Sigma-Operon) were resuspended in 50% dimethyl-sulfoxide/water at a final concentration of 300fmol/µL. Oligonucleotides were arrayed from 384-well plates onto PXM oligo-nucleotide slides (Full Moon Biosystems, Sunnyvale, CA, USA) by using a MicroGrid Total Array System (BioRobotics, Cambridge, UK) spotter with 16 pins in a 4×4 format. DNA elements were deposited in duplicate in distinct area of the array. The diameter of each spot was approximately 200 µm. After printing, slides were incubated in a humid chamber for 14h at 42°C. Prior to hybridization, the slides were processed as described previously [451,452].

Microarray Probe Preparation and Hybridization

For probe preparation, 2µg of total RNA was reverse transcribed using poly-dT primed Genisphere Expression Array 900 Detection system (Genisphrere, Hatfield, PA, USA) in 20µL total volume using 20 units of RNAsin (Fermentas, Vilnius, Lithuania), 1× first strand buffer and 200 units of RNAse H(-) point mutant murine myeloid leukemia virus (M-MLV) reverse transcriptase (Fermentas, Vilnius, Lithuania). All other probe preparation steps were done according the manufacturer's instructions (Genisphere, Hatfield, PA, USA). Both the first step cDNA hybridization and the second step capture reagent hybridization were carried out in a Ventana hybridization station (Ventana Discovery, Tucson, AZ, USA) by using the "antibody" protocol. First hybridization was performed at 40°C for 6h in "FGL2" hybridization buffer (10× Denhart solution, 0.25M sodium phosphate buffer pH: 7.0, 1mM EDTA, 1× SSC, 0.5% SDS), then 2.5µL of each Cy5 and Cy3 capture reagents were added to the slides in 200μL "Chiphyb" hybridization buffer (Ventana Discovery, Tucson, AZ, USA) and incubated at 42°C for 2h. After hybridization, the slides were washed in 0.2×SSC twice at RT for 10min, then dried and scanned.

Scanning

Each array was scanned under a green laser (543nm for Cy3 labeling) or a red laser (633nm for Cy5 labeling) using a ScanArray Lite (GSI Lumonics, Billerica, MA) scanning confocal fluorescent scanner with 10μm resolution (Laser power: 85% for Cy5 and 90% for Cy3, Gain: 80% for Cy5 and 75% for Cy3) [453]. Scanned output files were analyzed using GenePix Pro5.0 software (Axon Instruments Inc., Foster City, CA, USA). Each spot was defined by automatic positioning of a grid of circles over the image. For each channel the median values of feature and local background pixel intensities were determined [453,454]. The background-corrected expression data were filtered for flagged spots and weak signal. Technical replicates on the same array were averaged. Data were excluded in cases where technical replicates were significantly different. Normalization was performed using the print-tip Lowess method [455,456], followed by Student's *t*-test in order to determine the genes to be regarded as regulated in response treatment. Logarithm was taken from each expression ratio to fulfill the *t*-test's requirement for a normal distribution. Genes for which the mean of log-ratios across the biological replicates was equal to zero at a significance level $\alpha=0.05$ are considered to have an unchanged expression. On the other hand, genes having a P value smaller than α and the average-fold change (increase or decrease) of the four data points were at least 2.0-fold were considered as regulated genes.

Expression Profile Verification

QRT-PCR was performed on a RotorGene 2000 instrument (Corbett Research, Sydney, Australia) with gene-specific primers and SybrGreen protocol to confirm the gene expression changes observed by microarrays. 2μg of total RNA from each sample was reverse transcribed in the presence of poly(dT) sequences in a total volume of 20μL. After dilution with20 μL of water, 1μL of the diluted reaction mix was used as template in QRT-PCR. The 20μL reaction volume contained 0.2mM of dNTP, 1× PCR reaction buffer (ABGene, Epsom, UK), 6mM of each primer, 4mM of $MgCl_2$, 1× SYBR Green I (Molecular Probes, Eugene, Oregon) at final concentration, and 0.5 units of thermostart Taq DNA polymerase (ABGene, Epsom, UK). The amplification was carried out with the following cycling parameters: 600 sec heat start at 95°C, 45 cycles of denaturation at 95°C for 25sec, annealing at 60°C for 25sec and fluorescence detection at 72°C for 15sec. Relative expression ratios were normalized to β-actin. Non-template control sample was used for each PCR run to check the genomic DNA contaminations of cDNA template. Analysis of results was performed by Pfaffl method [457]. Using this calculation method differences between the amplification efficiencies of reactions could be corrected.

Table IV: Primers used in QRT-PCR analysis

Gene product	Forward primer	Reverse primer
β-actin	GGAAATCGTGCGTGACATTAAA	TGCGGCAGTGGCCATC
Midkine	ATGGGAGCACTGGCACCA	GGGCATTGTACCGAGCCTT
Translin	TGCTTCAGCGCCTGGTCT	TCGGGTCACCAGAGTCTCTGT
Proteasomal ATPase	AAATTGATGCCATTGGAGGG	CCTCTGCACCTCATTGTCACC
Integrin-β	GCACTGGTGGTGGCAGG	ACTGTGTGGAATGCTCGCTG
ATPase H^+-pump	TGCCTATGGCACAGCCAA	TCTGGCCTCATGACCGACA

Data Analysis

Results presented are "arithmetic means ± standard deviation" of experiments performed in triplicate. Data were analyzed using Excel 98 (Microsoft, Redmond, WA) utilizing Student's *t*-test to determine the significance of observed changes. All differences stated in the text are statistically significant ($P<0.05$).

Results and Discussion of Findings

Involvement of the Periphery in AD

$[Ca^{2+}]_i$ of Lymphocytes

Lymphocytes were successfully harvested and purified on Ficoll gradient from both Alzheimer and control donors. Cells were 97-98% viable as confirmed by intravital staining with Trypan Blue, and were competent to be labeled with Fura-2AM at 37°C by passive diffusion during the experiments. The ratio of fluorescence at 340 and 380nm ($FL_{340/380}$) representing $[Ca^{2+}]_i$ of control cells was 1.001±0.117, whereas Alzheimer lymphocytes exhibited a ratio of 1.131±0.100 (Table V). These findings indicate that AD cells demonstrate increases in free cell calcium when compared to leukocytes of age-matched controls.

Table V: $FL_{340/380}$ of lymphocytes

Samples	$FL_{340/380}$ (mean±SD)	Number of cuvettes
Control	1.001±0.117	23
Alzheimer	1.131±0.100*	18
Control+βAP	1.507±0.368*	23
Alzheimer+βAP	1.130±0.075	18

Comparative fluorimetric measurements on lymphocytes of human Alzheimer and control donors on $FL_{340/380}$ using Fura-2AM. Resting $[Ca^{2+}]_i$ appeared to be higher in AD cells when compared to that seen with control lymphocytes. After incubating cells in the presence of βAP, the $[Ca^{2+}]_i$ of the control cells elevated, while that of AD lymphocytes did not differ considerably. (*$P<0.05$)

Lymphocytes of our sporadic AD patients were distinguished from normal ones in our experiments, which clearly shows that detectable biochemical alterations are present in Alzheimer lymphocytes. Considering βAP central to AD pathomechanism, elevated $[Ca^{2+}]_i$ of AD cells is broadly in line with the notion that βAP has the potential to increase Ca^{2+}-levels [58,59,64,132,133]. However, serum βAP-level in AD patients does not significantly differ from that seen in healthy individuals to convincingly explain the findings. Moreover, βAP is considered a physiological peptide [33-35]; it exerts pathologic effects either in its aggregated form [41], or its cytotoxicity has been attributed to attending intracellularily [50-52]. Nevertheless, even though blood concentrations of βAP is not significantly higher, its level might be slightly elevated enough to disrupt Ca^{2+}-homeostasis in lymphocytes, and AD-specific changes in βAP aggregation or distribution inside the cells might be present to further the calcium-imbalance.

To justify the Ca^{2+}-rising effect of βAP, measurements of $[Ca^{2+}]_i$ with the calcium indicator dye Fura-2AM in control lymphocytes maintained in the presence of 10^{-7}M βAP for 2 hours revealed an increase in the 340/380 fluorescence excitation ratio (FL_{LYMPH}: 1.507±0.368), the mark of an increase in free calcium level during the 2-hour exposure period. By directly monitoring $[Ca^{2+}]_i$, we found that Alzheimer cells, maintained in the same conditions, showed no significant change in the free cell calcium (FL_{LYMPH}: 1.130±0.075). AD lymphocytes appeared to be resistant to βAP.

These findings might be explained by our hypothesis which assumes that slightly, but chronically elevated serum βAP levels disrupt Ca^{2+} in a tonic manner, causing continuous Ca^{2+}-overload into the neurons. This may indirectly confirm why resting $[Ca^{2+}]_i$ of untreated AD lymphocytes is higher than that of control ones. The incessant amyloid-stimulus will ultimately make lymphocytes resistant to βAP. Furthermore, βAP can cause sustained alterations in cell membrane fluidity [482]. Aggregated and intracellular βAPs interact with membrane structures (both protein-peptide and lipid-peptide interactions), which can cause permanent change in the structure of the cell membrane and in $[Ca^{2+}]_i$. These alterations might promote the resistance of lymphocytes to βAP with respect to $[Ca^{2+}]_i$.

Because Ca^{2+} is a central secondary messenger, tonic elevation of its concentration has various effects. In blood, Ca^{2+} plays role in activation of lymphocytes. Two sequentially coupled mechanisms contribute to the generation of the Ca^{2+} signal in T-cells. The first event is the binding of peptide-loaded MHC to the T-cell receptor – CD_3 complex (TCR/CD_3), which results in the recruitment and activation of protein tyrosine kinases, and the concomitant activation of PLCγ, which cleaves PI-P_2 into DAG and IP_3, and in turn, initiates two signaling pathways in lymphocyte activation. DAG activates the PKC pathway, which leads to the phosphorylation of several intracellular substrates and transcription factors. The second pathway, initiated by the generation of IP_3, governs the sustained elevation of the cytosolic $[Ca^{2+}]_i$ required for efficient signal transduction. IP_3 results in the release of Ca^{2+} stored in the sarco/endoplasmic reticulum (SR/ER) through two types of receptor-operated calcium channels: the IP_3-receptors (IP_3Rs) and the ryanodine receptors (RyRs) [458,459], causing a rise in the $[Ca^{2+}]_i$ from a resting level of ~100 nM to a peak concentration of ~500 nM [460]. The calcium signal activates the Ca^{2+}-calmodulin-dependent phospahatase calcineurin, which then dephosphorylates several transcription factors, including the nuclear factor of activated T-lymphocytes (NF-AT) thereby enabling it to accumulate in the nucleus and bind to the promoter element of the central cytokine, the interleukin-2 (IL-2) gene. Activation of IL-2 gene and IL-2 expression is a critical commitment point beyond which further T-cell activation becomes antigen independent. A sustained Ca^{2+} signal is required to keep NF-AT in the nucleus in the transcriptionally active state.

The second phase of the biphasic Ca^{2+} signal generated in T-lymphocytes, following the activation of the IP_3-pathway, is governed by Ca^{2+} influx from the extracellular space across the plasma membrane. In response to stimulation of the TCR/CD_3 a long-lasting Ca^{2+} entry can be observed. This Ca^{2+} entry is necessary for clonal expansion of T-cells and therefore essential for the function of the immune system. Ca^{2+} influx is triggered by the depletion of intracellular Ca^{2+} stores, rather than by the direct action of the increased IP_3 concentration generated during signal transduction. Ca^{2+} influx, therefore, is based on the capacitative Ca^{2+} entry, whereby store-depletion opens store-operated (*ie.* Ca^{2+}-release-activated) Ca^{2+}-channels (CRAC) in the plasma membrane in order to facilitate store refilling. Other than CRAC, the incessant Ca^{2+} signal relies on the operation of several other ion channels in T-cells, especially the voltage-gated and the Ca^{2+}-activated K^+-channels.

Elevation of intracellular free Ca^{2+} is one of the key triggering signals for T-cell activation by antigen. A remarkable variety of Ca^{2+} signals in T-cells, ranging from infrequent spikes to sustained oscillations and plateaus, derives from the interactions of multiple Ca^{2+} sources and sinks in the cell. The amplitude and dynamics of the Ca^{2+} signal are shaped by several mechanisms, including voltage, receptor and store-operated channels, membrane potential, slow modulation of the plasma membrane Ca^{2+}-ATPase, and mitochondria that buffer Ca^{2+} and prevent the inactivation of CRAC channels. Ca^{2+} signals have a number of downstream targets occurring on multiple time scales. At short times, Ca^{2+} signals help to stabilize contacts between T-cells and antigen-presenting cells (APCs) through changes in motility and cytoskeletal reorganization. Over periods of minutes to hours, the amplitude, duration, and kinetic signature of Ca^{2+} signals increase the efficiency and specificity of gene activation events. The complexity of Ca^{2+} signals contains a wealth of information that may help to instruct lymphocytes to choose between alternate fates in response to antigenic stimulation.

βAP is known to interfere with various (including Ca^{2+}- and K^+-) channels, and is able to form non-selective cation pores in the cell surface, both known to yield elevated intracellular free Ca^{2+}-levels. Tonic βAP-trigger may lead to incessant antigen-independent activation and clonal expansion of lymphocytes, which is required to generate an efficient generalized immune response to a specific antigen (*eg.* βAP). However, the immune system is complex enough that activation of CD_4^+ T-cells can lead to several different functional outcomes: activation and cytokine secretion, death, or functional non-responsiveness (anergy).

The induction of activated CD_4^+ T-cells occurs following the recognition of an antigenic peptide complexed to gene products of major histocompatibility complex class II (MHC II) on the surface of APCs, and other co-stimulatory ligands (*eg.* CD28-B7). Typically, peptides are derived from exogenous protein, which are internalized by the APC and cleaved in endosomal/lysosomal compartments by intracellular proteases. Subsequently, the peptides are loaded onto MHC II in a specialized MHC compartment and subsequently transported to the cell surface. Following antigen recognition, CD_4^+ T-cells differentiate into helper T-cells (T_H cells) that are capable of producing large amounts of cytokines. The production of interleukins and IFNγ by type 1 T_H (T_{H1}) cells is critical for the activation of cellular immunity through cytotoxic CD_8^+ cells and macrophages. T_{H2} responses and the production of IL-4 and IL-5 promote the development of the humoral arm of the immune response, directing the differentiation of B-cells into antibody-producing plasma cells secreting IgE and IgA. Consequently, a T_{H2} response predominates for the control of the production of antibodies that neutralize toxins – such a toxin can be βAP aggregate.

Like CD_4^+ T-cells, CD_8^+ T-cells also recognize peptide antigen in the context of MHC molecules (MHC Ia) on the surface of APCs. However, these are generally of cytoplasmic origin, either self-peptides or those generated by the proteasome from pathogens that reside in the cytoplasm of cells such as viruses or intracellular bacteria. Regardless of the antigen-presenting molecule used, following antigen recognition and the appropriate co-stimulatory events, including cytokines produced by CD_4^+ T_H-cells, CD_8^+ T-cells acquire two major effector functions. First, CD_4^+ T-cells contribute to macrophage activation via the production of IFNγ and TNF; and second, CD_8^+ T-cells can directly lyse infected cells by one of two pathways – the perforin/granzyme pathway and a second pathway that induces apoptosis via the interaction of Fas. Such a cytoplasmic peptide that activates this pathway might be intracellular βAP.

A special type of antigen-specific CD_4^+ T-cell (suppressor T-cell, T_S) that secretes IL-10, but not other T_{H2} cell-derived cytokines (*eg*. IL-4), can suppress antigen-specific proliferation of other T-cells and, quite importantly, may prevent autoimmunity.

Activation of lymphocytes in AD by Ca^{2+} might be of critical importance. When Alzheimer first described the clinical and pathologic features of the dementing disorder in early 1900s, he noticed signs of an inflammatory reaction; however, only in the last decade has the notion that the brain is "immunologically privileged" begun to be dismissed as an oversimplification, supported by mounting evidence that senile plaques and NFTs in brains of patients with AD are associated with a unique endogenous inflammatory reaction that is orchestrated by microglial cells. Activated microglia, the residing macrophage of the brain, in conjunction with astrocytes, which sustain the functional integrity of neurons, contribute extracellular matrix proteins, and can be considered to be cells of the innate immune system, are central to the inflammatory hypotheses for the neurodegenerative processes that are associated with AD. Microglial activation may be induced by soluble βAP fragments through causing Ca^{2+}-imbalance, plaque deposition, degenerating neurons, or all three. Activated microglia may release neurotoxins, toxic inflammatory mediators, and free radicals, but may also be involved in scavenging βAP. Inflammatory stress may increase generation of βAP through upregulation of APP. Apart from directly activating microglia by increasing $[Ca^{2+}]_i$, βAP might also circuitously stimulate microglia. This activatory scenario might not uniquely happen in the CNS.

It is well established that cross-talk takes place between the brain and the peripheral immune system. CD_4^+ helper cells are triggered by peptide antigens bound to class II MHC molecules. A class II allele that can bind to a given self-antigen (*eg*. amyloid-aggregate) may facilitate an autoimmune response. T-cell antigen receptor (TCR) is comprised of a ligand-binding subunit, the α and β chains, and a signaling subunit, namely the CD_3 units. The physiologic ligand for the TCR is foreign peptide bound to the MHC expressed on APCs, including dendritic cells, macrophages, and B cells. βAP may bind to either units, and therefore fibrillar βAP deposits and βAP fibrils may serve as activator of the immune system both in and outside of the CNS. Activation of autoreactive (amyloid-reactive) T-cells in periphery leads to release of cytokines/adhesion molecules and passage of cells through altered blood-brain-barrier into CNS.

Activated lymphocytes do penetrate into the normal brain in small numbers and, during CNS inflammatory reactions, such as in AD, their migration is dramatically upregulated [462]. Leptomeningeal infiltrates primarily around amyloid-laden blood vessels with T-lymphocytes (mostly CD_4^+ type) suggests that lymphocytes freely enter the brain in AD [463]. In order for leukocytes to perform their immune function in tissue surveillance they must be capable of leaving the vasculature and entering tissues. The way in which specific leukocyte subsets may achieve this in different vascular beds is in many cases unresolved. In the CNS, extravasating leukocytes must penetrate the highly impermeable tight junctions, which link the vascular endothelial cells, or use an alternative unresolved parajunctional mechanism. However it is apparent that whichever mechanism is used for leukocyte diapedesis, the cells of the blood-brain barrier are not passive in this respect and are actively involved in facilitating lymphocyte migration into the CNS. Ca^{2+}-mediated intracellular events are essential for lymphocyte migration through the blood-brain barrier [464], as triggered by βAP.

T-cells cross the intact blood-brain-barrier, home to neuritic plaques, and cause local inflammation either directly, or by activating microglia. However, relatively few T- and B-lymphocytes are detected in the vicinity of plaques. Nonetheless, an increase in the levels of IFNγ, the principal T_{H1} cytokine, is correlated with the severity of AD [465] by synergistically enhancing the βAP-dependent microglial activation, resulting in neuronal cell injury [466]. This hypothesis is confirmed by the finding that the activity of certain T_{H2} cytokines, such as IL-10, which seem to be potent in reducing the activity and neurotoxicity of microglia, is reduced in patients with AD, and a decrease in the numbers and activity of suppressor T-cells has been reported [465]. Therefore, local inflammatory up-regulation seen in the brains of patients with AD may arise from an inadequately regulated systemic immune response [467], presumably triggered by βAP.

Autoimmunity is unregulated hyperimmunity to organ-specific proteins, inducing rapid turnover of antigen-specific T-cells of the acquired immune system with ultimate exhaustion and loss of acquired immunity, IL-2 and IFNγ production and proliferative decline, conforming to the limited capacity of clonal division. In AD, the primary degenerative process of βAP precedes a cascade of events that ultimately leads to a local brain inflammatory response. Unregulated systemic immune processes are secondary but important as a driving-force role in AD pathogenesis [467]. Brain-specific autoantibodies, including βAP-specific antibody, have been described in the blood and CSF of patients with AD. Antibodies against βAP in serum and CSF may be regarded as a sign of ongoing protective autoreactivity that is self-initiated to defend against disease and to repair damage by enhancing the catabolism of βAP aggregates [468]. The presence of immunoglobulins and complement in plaques in the brains of patients with AD, however, has lead to the suggestion that autoimmune mechanisms may be involved in the pathology of AD [469]. Nevertheless, the immune activation by βAP seems to be impaired in patients with AD, who express IL-2 receptor but not T-cell proliferation in response to stimulation with βAP [470]. Patients with

AD may have been exposed to higher concentrations or different epitopes of βAP due to altered APP mechanisms, resulting in T-cell anergy rather than activation [468]. Even though acquired antigen-specific immunity against βAP may have a protective role against the development of AD, patients with this disease may suffer because of their reduced ability to recognize βAP as an autoantigen and to eliminate it as a consequence.

Although the actions of T-lymphocytes have commonly been associated with autoimmune processes that induce CNS damage such as demyelination in experimental autoimmune encephalomyelitis, there is increasing evidence that processes viewed as inflammatory or autoimmune are neuroprotective in specific contexts. T-lymphocytes trafficking into the brain and spinal cord can enhance neuronal recovery in response to some forms of injury. Circulating immune cells such as T-lymphocytes have the ability to make classic nerve growth factors (*eg.* NGF, BDNF, NT-3) [471], and under certain conditions may curb inflammatory and other events that otherwise would lead to further (secondary) neurodegeneration [472].

A complete understanding of these signaling mechanisms may lead to the identification of target pathways for the therapeutic treatment of brain inflammatory disorders, including AD. In addition, by understanding the mechanisms responsible for triggering lymphocyte migration it may be possible to manipulate peripheral blood lymphocytes to deliver therapeutic agents, that are normally excluded by the blood-brain barrier, to the CNS particularly in inflammatory diseases where these cells would be self targeting. During an influx of leukocytes, the initial interaction of the infiltrating cells is with the endothelial lining of the vessels. After migration through the endothelium, however, the principal interaction of leukocytes is with fibroblasts. This interaction is mediated through the increased expression of intercellular adhesion molecule-1 (ICAM-1). An immediate response to the cross-linking of ICAM-1 on fibroblasts or endothelial cells is the triggering of a rise in $[Ca^{2+}]_I$ [473], and activation of NF-κB, which is a transcription factor ubiquitously associated with the induction of a range of pro-inflammatory cytokines and adhesion molecules [474]. Therefore, the Ca^{2+} homeostasis of Alzheimer fibroblasts has also been examined in this study.

$[Ca^{2+}]_i$ of Fibroblasts

Seven days after seeding, cultured fibroblasts from both Alzheimer and control donors were 80% confluent. More than 97-98% of cells on coverslips were viable as confirmed by intravital staining using Trypan Blue. Cells were competent to be labeled with Fura-2AM at 37°C by passive diffusion during the study. With respect to control cultures, $FL_{340/380}$ revealed 2.48±0.162, whereas Alzheimer fibroblasts exhibited a ratio of 2.052±0.207 (Table VI). Just as with lymphocytes, fibroblasts of our sporadic AD patients were also distinguished from normal ones in our experiments, demonstrating detectable biochemical alterations in Alzheimer fibroblasts.

Table VI: $FL_{340/380}$ of fibroblasts – short-term study

Cultures	$FL_{340/380}$ (mean±SD)	Number of coverslips
Control	2.48±0.162	16
Alzheimer	2.052±0.207*	42
Control+βAP	2.666±0.08*	16
Alzheimer+βAP	2.055±0.125	42

Comparative fluorimetric studies on human control and Alzheimer fibroblasts on $FL_{340/380}$ using the Ca^{2+}-indicator dye Fura-2AM. Cells were maintained with or without βAP for 16 hours at 37°C. The fluorescence excitation ratios were calculated from fluorescence intensities observed at 495nm using dual wavelength spectrofluorimetry (excitation wavelengths: 340 and 380nms). Alzheimer cells exhibited lower $[Ca^{2+}]_i$ when compared to the control cultures. Exposure of fibroblasts to βAP resulted in increased $[Ca^{2+}]_i$ of the control cells, but not of AD cultures. (*P<0.05)

These findings indicate that cultured cells from donors of AD demonstrate significant decreases in resting free cell calcium when compared to that of age-matched controls. However, when taken together with leukocytes, the resting free calcium of AD lymphocytes is apparently opposing that of AD fibroblasts, in that basal $[Ca^{2+}]_i$ of fibroblasts from AD patients is significantly lower, whereas $[Ca^{2+}]_i$ of lymphocytes is higher than that of the matched controls. To reconcile these opposite changes in AD cells, βAP-induced biochemical alterations were also studied in fibroblasts.

Exposure of cells to βAP for 16 hours caused a rise in free cell calcium only in control cultures ($FL_{340/380}$: 2.666±0.08). βAP apparently disrupts calcium-regulating processes in control cells resulting in elevations of free cell calcium. This finding is in line with previous reports [58,420,425,442], demonstrating elevated calcium level as an underlying pathomechanism of βAP-induced cellular degeneration. However, AD cells do not show this response. Alzheimer cultures maintained in the same conditions showed little or no change in $[Ca^{2+}]_i$, with the fluorescence ratio being 2.055±0.125 (Table VI). Alzheimer fibroblasts, like lymphocytes, are therefore refractive to βAP. These results are in line with findings demonstrating a decreased βAP-sensitivity in Alzheimer cells [472,476].

Rise in $[Ca^{2+}]_i$ results in triggering of various pathways of a particular cell. As discussed with the immune system, B-cell receptor (BCR)- and TCR-signaling is initiated through the actions of non-receptor tyrosine kinases in conjunction with B- and T-cell-specific adaptor proteins, transduce signals required for the activation of the downstream signaling pathways, including Ca^{2+} [477]. Just as with lymphocytes, fibroblasts are also activated by Ca^{2+}-signals. In this scenario, fibroblasts undergo activation and secrete cytokines. Upregulation of mRNA encoding for the inflammatory cytokines IL-1β and IL-6 as well as for chemokines IL-8 and MCP-1 might therefore represent potent activators of local inflammatory cytokine and chemokine production [478] to activate lymphocytes. Signaling through ICAM-1 provides a co-stimulatory signal with that of CD_3 leading to T-cell proliferation, migration, and the induction of cytokines [479]. Migration is primarily dependent on the stage and mode of lymphocyte activation, and activated T-cells, irrespective of their antigen specificity, are able to penetrate the blood-brain barrier [480]. ICAM-1, overexpressed by activated fibroblasts and lymphocytes, is central to the migration of T-cells into the CNS [481]. Production of cytokines, moreover, may indirectly lead to microglial activation in the brain.

The most parsimonious explanation for the role of fibroblasts in AD is, therefore, activating members of the immune system, including T-cells and microglia. Both processes may be initiated by βAP. In order to suppress the βAP-induced activation of both fibroblasts and lymphocytes in AD, several pharmacological interventions should be considered. The use of Ca^{2+} channel blockers (L-type calcium blockers, such as verapamil, diltiazem and nifedipine; non-selective calcium blocker amlodipine, *etc.*), activating plasma membrane Ca^{2+}-ATPase, overexpression of SR/ER Ca^{2+}-ATPase, mitochondrial targeting, and NF-AT signaling pathway targeting may be promising.

βAP-induced rise in the $[Ca^{2+}]_i$ leads to activation of fibroblasts and therefore participates in lymphocyte migration. Interestingly, however, AD fibroblasts demonstrate decreases in the resting free Ca^{2+} when compared to control cells.

Long-term Effect of β-amyloid

Short-term (8-16 hours') exposure of cultures to βAP resulted in elevation of the fluorescence ratio (FL: 2.691±0.121), which is an indicative of the increase in the $[Ca^{2+}]_i$ (Table VII). This finding is in line with previous reports, demonstrating that a sustained rise in the basal calcium level is one of the underlying pathomechanisms of neural degeneration. The βAP-induced increase in the $[Ca^{2+}]_i$ peaked in fibroblasts on the 4th day of exposure to the peptide (FL: 2.798±0.101), and turned into a plateau for around 16 weeks (Figure 3). After the approximately 4 months' time the calcium-level of fibroblasts gradually started to decrease. Within 11 weeks, the $[Ca^{2+}]_i$ became significantly lower in the cells (FL: 2.164±0.199) when compared to that of the resting levels seen with cultures before exposure to βAP (FL: 2.479±0.221). The decreased calcium-level appeared to be constant for 10 weeks. In contrast with fibroblasts exposed chronically to βAP, $[Ca^{2+}]_i$ of control cultures did not change significantly over time (Table VII, Figure 3).

Treatment of control fibroblasts with βAP has resulted in a transient rise in the calcium-concentration. Elevation in the $[Ca^{2+}]_i$ appeared to be constant for 16 weeks, after which period the calcium level of cells gradually started to fall. Chronic exposition of fibroblast cultures to βAP, therefore, causes a decline in the calcium-level. This decrease was apparently sustained in our experiments. Thus, cells in our study have responded to βAP treatment in a time-dependant manner. Consistent with previous findings [58,420,425,442], $[Ca^{2+}]_i$ is increased as a result of βAP; this, however, seems to be only a short-term effect of the peptide. Piling evidence suggests that βAP interacts with cell membrane structures and also causes changes in membrane fluidity [482]. Exposure to βAP elicits sustained activation of Ca^{2+}-permeable receptor channels, which results in a pathological enhancement of inward Ca^{2+}-currents and a subsequent increase in the $[Ca^{2+}]_I$ [442,483,484]. Elevation of $[Ca^{2+}]_i$ may further progress through the activation of second messengers (*eg.* IP_3) that mobilize intracellular calcium ions resulting in Ca^{2+}-release from its intracellular stores [485,486]. We hypothesize that the acute rise in $[Ca^{2+}]_i$ is attributable to these interactions of βAP.

Table VII: $FL_{340/380}$ of fibroblasts – long-term study

Duration of treatment	βAP	Control
resting level	2.479±0.221	2.479±0.221
8 hrs	2.691±0.121*	2.483±0.107
16 hrs	2.666±0.080*	2.480±0.162
~half-week	2.798±0.101*	2.475±0.200
~16 weeks	2.802±0.157*	2.489±0.113
~27 weeks	2.164±0.199*	2.469±0.132
~37 weeks	2.165±0.176*	2.489±0.206

Comparative fluorimetric studies on human fibroblasts on $FL_{340/380}$ utilizing the Ca^{2+}-indicator dye Fura-2AM. The cells were cultured with βAP for 37 weeks at 37°C. The table marks a rapid increase in the $FL_{340/380}$, representing elevation in $[Ca^{2+}]_i$. This elevation continues to be significant for 16 weeks, after which a decline in the calcium-level develops. The decrease in the fluorescence-ratio appears to be permanent. $[Ca^{2+}]_i$ of untreated cultures did not change significantly during 37 weeks. (*$P<0.05$)

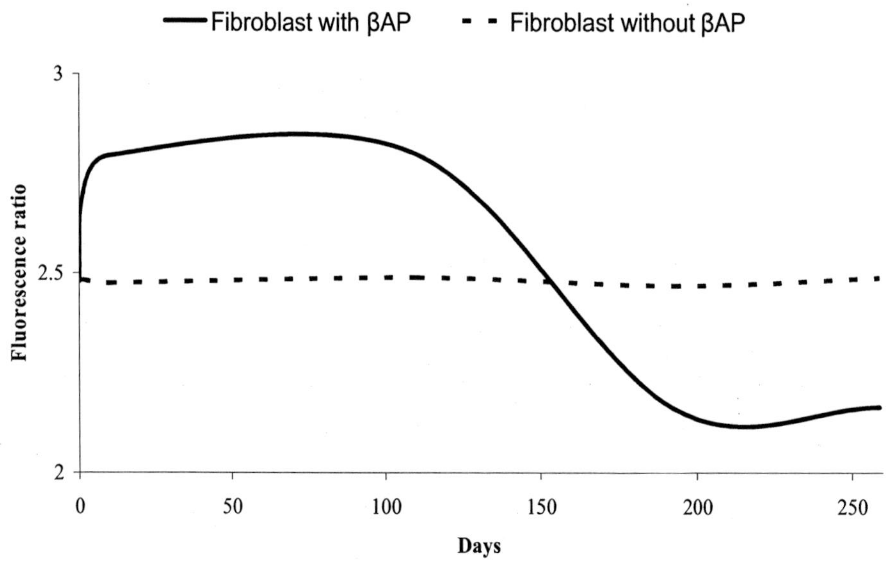

Figure 3: Time-course of βAP on $[Ca^{2+}]_i$ of human fibroblasts.Short-term exposition of fibroblasts to βAP results in an increased $[Ca^{2+}]_i$. The rise in the calcium-level peaks after an approximately half-a-week treatment with the peptide. The flow-chart demonstrates that the reversible elevation of $[Ca^{2+}]_i$ turns into a fall at around the 16th week of exposition to βAP, and yields a sustained decrease in the $[Ca^{2+}]_i$. Calcium-level of control cultures with no βAP treatment appeared to be unchanged during the experiment.

Chronic exposition of cells to βAP apparently decreases $[Ca^{2+}]_i$. This observation is a real *novum*, and the exact pathomechanism behind this finding is yet unknown. Falling of the calcium-level may indicate that βAP causes a total ionic imbalance. Initially, βAP peptides can increase calcium influx through voltage-gated calcium channels (N- and L-type channels); form a cation-selective ion channel after βAP incorporation into the cellular membrane; reduce magnesium blockade of NMDA receptors to allow increased Ca^{2+} influx; and inhibit K^+ channel and Na^+/Ca^{2+} exchanger. Opening elevation of $[Ca^{2+}]_i$ triggers Ca^{2+}-

mediated Ca^{2+}-release from intracellular stores [487]. Also, based on the channel hypothesis [64], βAP may further the calcium imbalance by directly forming Ca^{2+}-channels. As a result, $[Ca^{2+}]_i$ may reach pathological levels which exhausts the buffering capacity of intracellular Ca^{2+}-pools, particularly that of the mitochondria and endoplasmic reticulum. Due to the altered concentration gradient, Ca^{2+} is expelled from the cell through outward channels to compensate the Ca^{2+}-overload. As a result of their pathological interaction with βAP, the function of Ca^{2+}-exchangers and pumps that translocate Ca^{2+} to the interstitium may become hyperactive and uncontrolled. These pumps transport Ca^{2+} from the cytoplasm to the extracellular fluid against large electrochemical potential gradients using energy from ATP, which causes an incessant outward calcium flow leading to a Ca^{2+} loss inside of the cell. This can not be balanced by the cells' internal buffer systems on the long term, yielding a sustained decrease in the $[Ca^{2+}]_I$ [488] and loss of energy. At later stages, damage to the mitochondria results in the enhanced production of free radicals, and activation of factors involved in apoptotic cell death. Accordingly, chronic fall in Ca^{2+} predisposes cells to self-degeneration [489]. As a result, derange of the intracellular calcium homeostasis ensues, and cells ultimately succumb to the chronic ionic imbalance.

The biphasic Ca^{2+}-response to βAP may play an important role in the pathomechanism of AD. Initially fibroblasts, activated by Ca^{2+}, assist lymphocytic migration. A decresed calcium-level on the long-run, however, might deactivate fibroblasts. Dormant fibrocytes with no cytokine and ICAM-1 expression chronically arrest lymphocytic activation and migration to the brain through blood-brain-barrier. This may explain why relatively few T- and B-lymphocytes are detected in the vicinity of neuritic plaques.

Despite the major pathological findings in the central nervous system of Alzheimer patients, the present study, as well as previous ones, indicates that metabolic deficits occur in non-neuronal tissues, including fibroblasts and lymphocytes [408-420]. Interestingly, however, basal $[Ca^{2+}]_i$ of fibroblasts is opposing that of lymphocytes. The difference between an increased $[Ca^{2+}]_i$ in lymphocytes and decreased $[Ca^{2+}]_i$ in fibroblasts may be attributable to the specific ion-regulation of the respective cell-type, or to an altered affinity of βAP to different cells of peripheral tissues. To reconcile these opposite changes in AD cells, the fact that the turnover rate of cells in the blood is way higher than that of the fibroblasts may give a more convincing explanation. βAP initially increases the $[Ca^{2+}]_i$ of cells, which has been demonstrated herewith and earlier [58,420,425,442], this correlates with an increase in the $FL_{340/380}$ of the lymphocytes in this study. βAP, however, completely disrupts calcium homeostasis on the long run, resulting in a decreased resting calcium level, as seen with fibroblasts. On theoretical grounds as explained above, the $[Ca^{2+}]_i$ would also be decreased in white blood cells, had they lived longer (*ie.* had their turnover rate been slower), like fibroblasts.

Importantly, both AD fibroblasts and lymphocytes are resistant to βAP. Alzheimer cells – in contrast to control ones – could be rendered resistant to βAP because of the chronic, long-term expositions, lasting several weeks (with respect to lymphocytes) or years (in case of fibroblasts), of cells to the peptide [409]. A tonic βAP-load seen in AD ultimately decreases $[Ca^{2+}]_i$ and makes cells insensitive to β-amyloid on the long run.

βAP apparently has a time-dependent biphasic effect: acutely, it increases the calcium-concentration of cells; in contrast, on the long-run, βAP acts as a calcium-antagonist.

Therefore, the idea that βAP leads to neural degeneration solely by increasing cells' calcium concentration must be replaced with a more complex view of its dual function in intracellular ionic homeostasis. The neurotoxicity of βAP is therefore due to its dual effect, but mainly to its long-term calcium-antagonistic activity. The observed changes can be considered as long-term cell physiological action and can be used as a sensitive marker for testing for the long-term activities of βAP.

These data signify that Alzheimer-specific biochemical alterations are present in the peripheral tissues. It is yet unknown whether these changes reflect the very dysfunctions in the brain, or these changes are specifically present only in the cell-types studied. Because fibroblasts express metabolic alterations that reflect dysfunction in other organs, such as the brain, it is reasonable to suggest that these changes might also be present in the CNS.

These findings therefore may shed new light on understanding the pathomechanism of AD. This new observation on Ca^{2+} homeostasis might be used as a springboard for studying AD pathology, and for developing novel treatment options in AD. Moreover, fluorimetric differences in AD fibroblasts and lymphocytes might serve as useful models for establishing diagnostic tools in Alzheimer dementia, monitoring progression of the disease and responsiveness to treatments by screening biochemical alterations in the periphery.

Role of Antipsychotics in the Periphery

Effect of Haloperidol on $[Ca^{2+}]_i$ of Fibroblasts

$[Ca^{2+}]_i$ of neither control, nor AD fibroblasts have changed significantly as a result of HAL treatment ($FL_{340/380}$: 2.41±0.119, and 2.049±0.202, respectively). Preincubation of cultures with HAL together with βAP has also resulted in no elevation in $[Ca^{2+}]_i$ (Table VIII).

In the brain, HAL has been previously reported to interfere with calcium metabolism through various mechanisms, including interaction with D_2 and σ_1-receptors [422,490,491]. Moreover, HAL has been shown to induce apoptosis of brain cells, which involves activation of σ-receptors, and also translocation of calcium into the cytoplasm [491,492,493]. Because differences certain to be encountered with respect to tissue-specific macromolecules, lack of various neuronal receptors on fibroblasts may give an explanation. Since fibroblasts do not express D_2 and σ_1-receptors, HAL is not potent in interfering with Ca^{2+}-homeostasis in cutaneous cells. However, other than receptor-mediated pathways may also lead to HAL-induced changes in Ca^{2+}-levels. Therefore, evaluating HAL for its effects on βAP-induced Ca^{2+}-imbalance on fibroblasts in this study was inevitable.

Co-administration of both βAP and HAL yielded unchanged Ca^{2+}-levels of the cultures (Table VIII). This finding is important in light that βAP has proved to change $[Ca^{2+}]_i$, and might be explained on the basis that HAL has been shown to block Ca^{2+}-channels [490,494]. Evidence from a number of studies has demonstrated that βAP can form ion channels in lipid bilayers, liposomes, neurons and other cells [64,495]. βAP-channels are heterogeneous in size, selectivity, blockade, and gating. They exhibit multiple cation selectivity, admitting Ca^{2+}, Na^+, K^+, Li^+, *etc.*, leading to unregulated Ca^{2+}-influx via βAP-channels. Based on the Ca^{2+}-channel hypothesis of βAP, HAL may efficiently block the channel activity of βAP.

Table VIII: $FL_{340/380}$ of fibroblasts: treatment with haloperidol

Cultures	$FL_{340/380}$ (mean±SD)	Number of coverslips
Control	2.48±0.162	16
Alzheimer	2.052±0.207	42
Control+βAP	2.666±0.08*	16
Alzheimer+βAP	2.055±0.125	42
Control+HAL	2.41±0.119	16
Alzheimer+HAL	2.049±0.202	16
Control+βAP+HAL	2.46±0.123	16
Alzheimer+βAP+HAL	2.054±0.198	16

Comparative fluorimetric studies on human control and Alzheimer fibroblasts on $FL_{340/380}$ utilizing the Ca^{2+}-indicator dye Fura-2AM. Cells were cultured with or without βAP and/or HAL for 16 hours at 37°C. Elevation in $FL_{340/380}$ after βAP-treatment was attenuated by co-administration of HAL. (*$P<0.05$, when samples compared to their untreated counterparts only)

Also, because HAL was demonstrated to efficiently inhibit βAP-formation [434], the protective effect of HAL might be achieved through two connected mechanisms. HAL decreases the concentration of βAP; accordingly, less Ca^{2+}-channels, formed of βAP, will be present. However, even these few channels are blocked by HAL. Moreover, HAL may act as a β-sheet breaker. This hypothesis strongly supports the theory that both regulating βAP production and blocking βAP-channels/βAP aggregation may underlie the molecular mechanism of HAL (neuro)protection. Therefore, HAL is a multimodal antagonist of βAP.

Chronic treatment with HAL may result in decreased βAP-induced neurotoxicity in the brain over time through various, but not fully understood mechanisms as discussed above. This may explain why several investigators have found a low frequency of AD neuropathology in SCH, as many patients with SCH are likely to have been chronically treated with HAL, a commonly prescribed drug for this disorder.

Because AD is frequently associated with BPSD, which is treated with antipsychotic medications such as haloperidol [496,497], future randomized controlled studies should target AD patients with BPSD with a history of chronic HAL treatment for neuropathologic examination. The prediction is that these patients should show a lower progression rate than untreated subjects or subjects treated with structurally divergent antipsychotic drugs.

Effect of Risperidone on $[Ca^{2+}]_i$ of Fibroblasts

Exposure to the atypical antipsychotic RISP alone did not significantly alter the $[Ca^{2+}]_i$ of both control and AD fibroblasts, leaving the $FL_{340/380}$ technically unchanged. This suggests that RISP itself does not interfere with Ca^{2+}-homeostasis of these cutaneous cells. No significant alteration was observed when cells were co-treated with RISP and βAP (Table IX).

Table IX: FL340/380 of fibroblasts: treatment with risperidone

Cultures	$FL_{340/380}$ (mean±SD)	Number of coverslips
Control	2.48±0.162	16
Alzheimer	2.052±0.207	42
Control+βAP	2.666±0.08	16
Alzheimer+βAP	2.055±0.125	42
Control+RISP	2.479±0.200	16
Alzheimer+RISP	2.054±0.172	16
Control+βAP+RISP	2.482±0.114	16
Alzheimer+βAP+RISP	2.053±0.099	16

Comparative fluorimetric measurements on fibroblasts of Alzheimer and control donors on $FL_{340/380}$ using Fura-2AM. Cells were maintained with or without βAP and/or RISP for 16 hours at 37°C. βAP-induced increase in $FL_{340/380}$, representing a rise in $[Ca^{2+}]_i$, was efficiently antagonized by RISP. (*$P<0.05$, when samples compared to their untreated counterparts only)

Contrary to HAL, there is no information on the effect of RISP on Ca^{2+}-homeostasis, nor on its ability to interfere with βAP production. However, activation of dopamin-receptors leads to transient increase in $[Ca^{2+}]_i$. Atypical antipsychotics, such as RISP, have been shown to stabilize $[Ca^{2+}]_i$ through antagonizing dopamine-receptors. Furthermore, atypical but not classical neuroleptic drugs reduce astroglial dopamine-sensitivity [498]. Moreover, dopamine-receptor antagonists such as RISP, just as antioxidants, have been shown to protect neurons against the damaging effects of βAP and cathecolamines [499].

RISP did not change $[Ca^{2+}]_i$, and alleviated βAP-induced Ca^{2+}-imbalance. These very same phenomena were found in this study with respect to typical antipsychotic HAL, and therefore they may share common targets. When taken together with the hypothesis that fibroblasts may serve as models of several neurometabolic disorders including AD [425-427], administration of RISP and HAL in the treatment of BPSD appears to be safe with respect to $[Ca^{2+}]_i$. Moreover, both RISP and HAL have proved to efficiently antagonize Ca^{2+}-disrupting cytotoxic effects of βAP. In spite of their inability to directly interfere with Ca^{2+} in the periphery, the observation that RISP and HAL efficiently attenuate βAP-induced Ca^{2+}-imbalance further confirms the possibility that RISP and HAL directly interact with βAP. However, further studies are needed to evaluate the putative neuroprotective effect of both agents. These might include selective pre- and postincubation of cultures with RISP or HAL and βAP; comparing the findings on fibroblasts with results on cultures from central nervous system that do (*eg.* neurons) and those do not (*eg.* glias) express dopamine and sigma receptors; and biophysical studies on the direct interaction between these antipsychotics and βAP. In light of these evidences, RISP and HAL may be useful leads in the development of effective AD therapeutic agents.

Apart from directly modulating calcium levels and blocking Ca^{2+}-channels, RISP and HAL may also alter the metabolism of APP and expression of genes involved in Ca^{2+}-regulation.

Antipsychotics in the Brain –The ALZHEIMER Approach

Effect of Haloperidol and Risperidone on APP Metabolism

Using monoclonal 22C11 antibodies against residues 68-81 of APP, we were able to detect wide, APP containing bands at 110-120 kDa (Figures 4, 5). Traditional antipsychotic HAL efficiently elevated cortical APP-levels in a concentration-dependent manner (Table X, Figures 4, 6). Significant rise in APP was evident as a result of acute administration of HAL in therapeutic and toxic doses within 24 and 12 hours, respectively. Significant changes in APP concentrations were between 2-6% when compared to that seen with control levels. On the other hand, chronic HAL treatment slightly increased APP concentrations, however, did not change APP levels considerably.

Table X: Semi-quantitative evaluation of rat cortical APP-Western-blot after or without acute and chronic intra-peritoneal HAL administration

Duration of treatment	Control	Therapeutic dose	Toxic dose
6 hrs	100	101	100.5
12 hrs	101	103	105.8*
24 hrs	99.4	107.9*	104.7*
96 hrs	99.8	103.9*	103.1
1 week	100	102.8	101.7
2 weeks	100.5	101.6	102.8
3 weeks	101	100.4	102.1
4 weeks	100.3	101.2	101.3

Rat cortical APP-levels were approximated utilizing Western-immuno-blotting, and chemi-luminescent visualization of bands were quantified by densitometry. Acute HAL-treatment revealed significant increase both in therapeutic and toxic doses within 24 and 12 hours, respectively, with regards to APP when compared to that seen with control animals. Chronic (1-4-week) administration of HAL resulted in incessant APP-levels in SPRD rat cortices *in vivo*, as detected by chemiluminescent Western-blot techniques. Optical densities are depicted in relative units, with the controls taken as 100%. (*$P<0.05$)

Table XI: Time-course of acute and chronic RISP treatment on rat cortical APP concentrations

Duration of treatment	Control	Therapeutic dose	Toxic dose
6 hrs	100	100.8	99.6
12 hrs	100.4	99.8	101
24 hrs	100.6	101.1	101.2
96 hrs	101.2	102.1	102.3
1 week	100	100.8	99.6
2 weeks	100.4	99.8	101
3 weeks	100.6	101.1	101.2
4 weeks	101.2	102.5	102.9

Semi-quantitative evaluation of cortical APP-Western-blot after or without acute (6, 12, 24, and 96-hour) and chronic (1-4-week) RISP treatment, administered through the intra-peritoneal route, delineates no significant changes both in therapeutic and toxic doses with respect to APP when compared to that seen with untreated SPRD rats. APP-containing bands were assessed by chemiluminescent Western-immunoblot techniques, and optical densities are displayed in relative units (controls being 100%).

The atypical antipsychotic risperidone slightly, but not significantly increased cortical APP levels in both therapeutic and toxic doses during all treatment periods (Table XI, Figures 5, 7). In case of this atypical antipsychotic drug there was neither time-, nor concentration-dependence with respect to APP levels.

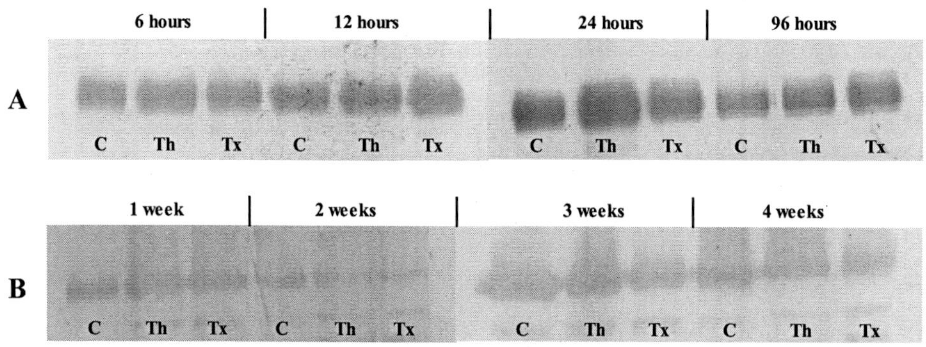

Figure 4: APP immunoblots after haloperidol administration *in vivo*
Quantification of immunoblot analysis by densitometry delineates A: a marked increase after acute haloperidol treatment. B: Chronic haloperidol reveals no significant changes either in therapeutic (Th) or toxic (Tx) doses when compared to that seen with controls (C).

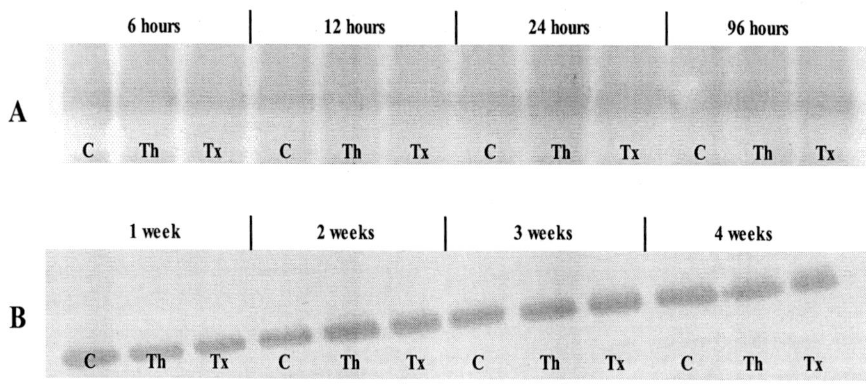

Figure 5: APP immunoblots following risperidone administration *in vivo*
Rat cortical APP immunoblots show technically unchanged densities after acute (A) or chronic (B) risperidone treatments, respectively.

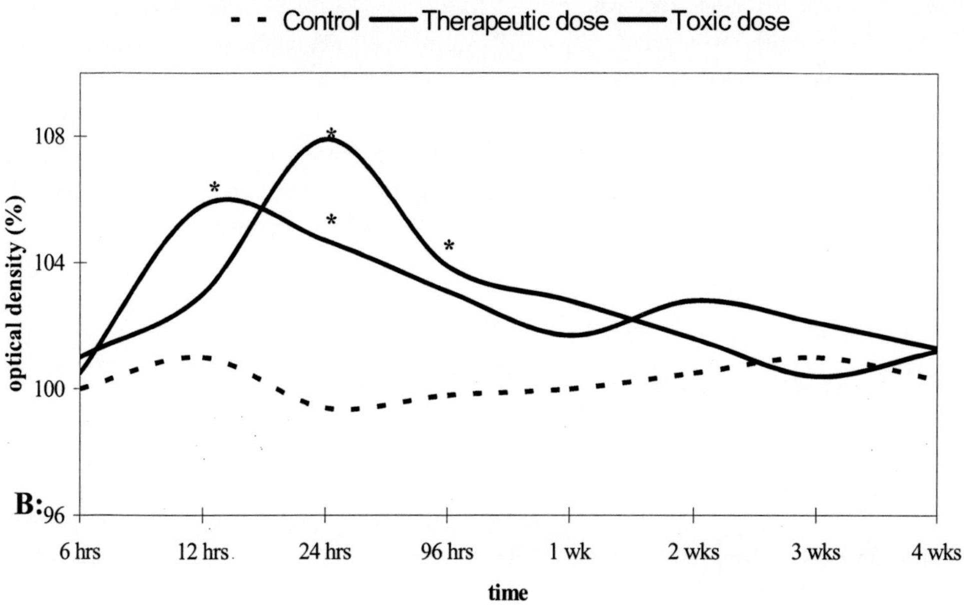

Figure 6: Time-course of HAL on rat cortical APP concentrations
Intra-peritoneal administration of HAL resulted in significant increase in rat cortical APP-level in therapeutic dose within 24 hours. This elevation was seen even in 12 hours in toxic dose of the drug. Both therapeutic and toxic HAL leaved APP concentrations unchanged on the long-run. ($P<0.05$)

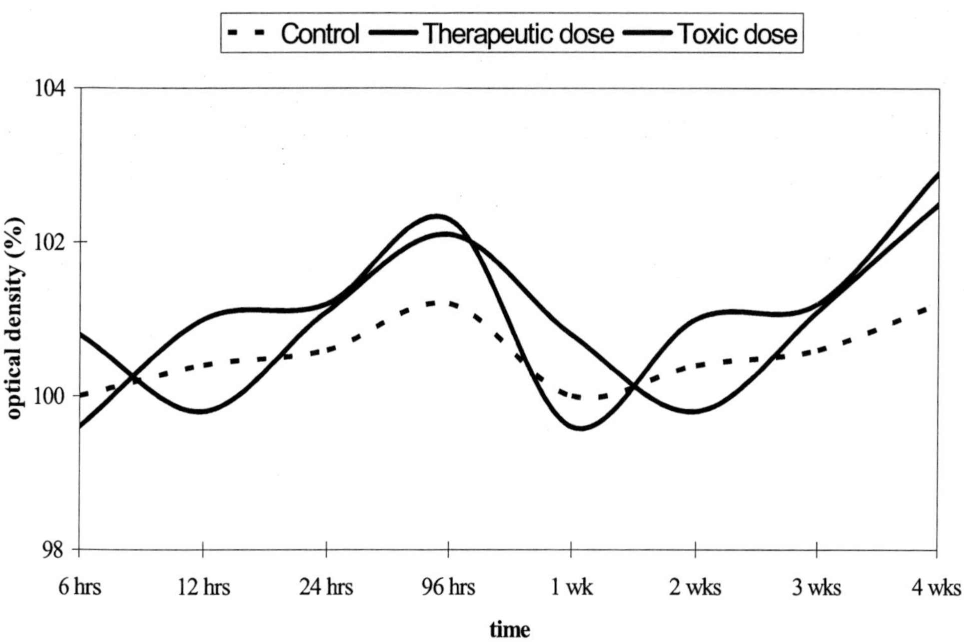

Figure 7: Time-course of RISP on rat cortical APP concentrations
The flow-chart delineates the effect of the typical antipsychotic drug on the APP metabolism in an acute or chronic time-frame. RISP did not significantly interfere with the APP-levels either in therapeutic or toxic doses.

Effect of HAL on APP levels

Acute administration of HAL apparently yields APP elevation in both therapeutic and toxic doses. Because HAL acts as an inhibitor of select proteinases [500], including secretases that cleave APP [434], acute HAL leads to increase in the non-metabolized form of APP. This is supported by the finding that APP metabolism is increased by environmental effects that activate Ca^{2+}-signaling pathways [59], by being a calcium antagonist [433], HAL efficiently inhibits APP processing, leading to elevated full-length APP levels.

Neurotoxicity and Neuroprotection

HAL has been proved to be neurotoxic, and the involvement of oxidative free radicals and apoptotic factors is strongly implicated [431,432].

σ_1-ligands, such as neurosteroids, play a role in neuroprotection [501,502]. σ_1-agonists modulate ionotropic, metabotropic and voltage-gated Ca^{2+}-signals, and enhance intracellular Ca^{2+}-mobilization [503]. Ca^{2+} increases APP metabolism in general [59], and also stimulates sAPPα secretion [504]. sAPPα can reduce neuronal cytosolic Ca^{2+}-levels and have demonstrated neuroprotective efficacy against a variety of insults [173]. Moreover, σ_1-receptor agonists play an important role in learning and memory. They have been demonstrated to possess neuromodulatory effects; they potentiate cholinergic neurotransmission and the glutamatergic system [505,506]. Also, neuroprotection against βAP-induced neurotoxicty was reported [507]. Because the traditional neuroleptic HAL acts as sigma σ_1-antagonist [430], it inhibits this complex neuroprotective pathway to yield sAPPα, and therefore increases full-length APP levels.

By being a D_2-receptor antagonist [429], HAL decreases PKC activity and level [508]. Activating PKC would lead to secretion of neuroprotective sAPPα [240,509], and to modulation of Ca^{2+}-currents by regulating Ca^{2+}-channels through second messengers [510,511]. Ca^{2+}, one of the most important second messengers in the brain, decreases APP mRNA levels [512]. Taken together, decreased PKC by HAL increases APP mRNA.

Even though increased APP level is a precondition to excessive neurotoxic βAP formation, one of the inducible neuroactive (trophic) factors at site of neural injury is the APP that may participate in cell-to-cell recognition, re-establishment of synaptic contacts, alleviation of glutamate-induced membrane depolarization and over-excitation, and stabilization of the intracellular Ca^{2+} homeostasis [484,501,513,514]. Therefore, it is assumed that APP is an important protein for tissue repair and that its expression is upregulated following nerve damage [515], as might be expected to occur following amyloid fibril formation or administration of toxic agents, such as HAL. In addition, persistent induction of APP expression without subsequent accumulation of βAP in response to acute brain injury substantiates the neuroprotective potential of APP [513,516]. This may account for the elevated APP level seen in these experiments after HAL treatment. Because HAL is neurotoxic, acute increase in APP after HAL-treatment might be considered as such a compensatory mechanism. This is supported by the observation that the presumably non-toxic RISP does not exert such effects. Further hypothesizing the advantageous roles of HAL, elevated APP levels might be a preconditioning effect of HAL to other nocuous agents, such as βAP.

Moreover, HAL is known to increase the activity of NF-κB [517]. Activation of NF-κB by either sAPPα [518] or HAL stabilizes cellular Ca^{2+} homeostasis by modulating the expression of genes that encode Ca^{2+}-binding proteins, ionotropic glutamate receptor subunits, and anti-apoptotic *bcl-2* family members [519]. This is in line with HAL-induced cellular protection with respect to neurotoxic APP metabolites, *ie.* βAP, in peripheral tissues (Table VIII).

HAL and Ca^{2+} in AD

Abnormal signal transduction systems have been implicated in the pathophysiology of AD and in APP metabolism, but their precise role has been difficult to establish. It is therefore reasonable to hypothesize that the action of HAL on APP processing is a fine-tuning mechanism involving various second messengers, such as Ca^{2+}.

Calcium, as discussed above, plays central role in HAL-induced neurotoxicity and neuroprotection both in general and in AD. However, HAL modulates extracellular Ca^{2+}-influx and intracellular Ca^{2+}-mobilization through σ_1-receptors, therefore stabilizes Ca^{2+}. This fine-tuning mechanism leads to constitutive sAPPα secretion, which precludes βAP formation. This is in line with the finding on peripheral cells that HAL attenuates βAP-induced Ca^{2+}-imbalance and stabilizes $[Ca^{2+}]_i$ (Table VIII). Moreover, because it has been shown to non-selectively block Ca^{2+}-channels [433], HAL may also directly protect in AD against pores formed in the membrane by βAP that act as cation (*eg.* Ca^{2+}) channels [64].

Yielding an elevated APP concentration in the aforementioned Ca^{2+}-dependent manner by HAL is two-fold: suppressing its cleavage or elevating $mRNA_{APP}$-level. HAL has been reported to block the amyloidogenic processing of APP, which, in turn, precludes βAP production [434], and therefore may stimulate neuroprotective sAPPα secretion which may attenuate both self-induced (*ie.* direct toxicity of HAL) and βAP-derived neurotoxicity.

Nevertheless, chronic treatment with HAL results in stable APP levels, which leads to decreased deposition of βAP in the brain over time by inhibiting βAP formation [434] and alleviating βAP-induced toxicity as presented above. When previous reports are taken together with our findings, the impact of HAL in AD is 3-fold: it stabilizes Ca^{2+}, stimulates sAPPα secretion and inhibits βAP formation; these effects are inter-connected. This may further confirm why a low frequency of AD is present in SCH patients, as many of them are likely to have been treated chronically with antipsychotics, including HAL.

APP-receptor

Various ligand-operated ion channels that control intracellular Ca^{2+}-levels may also regulate APP processing [59]. Both sAPPα and βAP have been demonstrated to modulate cytoplasmic Ca^{2+} concentration [58,173], raising the possibility that these compounds may regulate their own formation. This fine-tuning mechanism may be characteristic of AD because of the involvement of various APP-related substances. However, by being a putative cell-surface receptor, APP itself might also be directly involved in this scenario [135] APP might be a single metabotropic receptor coupled to GTP-binding proteins [163], or it may be a part of a larger receptor complex potentiating D_2- and/or σ_1-receptors. Elevated APP levels after acute HAL treatment argue for this hypothesis, in that inhibitory ligands (*eg.* HAL) result in compensatory up-regulation of their receptors (*eg.* APP itself). Therefore, HAL-APP receptor-kinetics should be investigated in the future.

Effect of RISP on APP Levels

Atypical RISP has common action with HAL on D_2-receptors. However, it also acts as serotonergic-antagonist, of which 5-HT_2 and 5HT_4-receptors have been shown to stimulate sAPPα secretion [520]. Moreover, lesions of cholinergic, serotonergic, and adrenergic neurotransmitter systems all result in the induction of APP levels [521]. By antagonizing both 5-HT and NA (*ie.* adrenergic) receptors, slight elevation in APP concentration elicited by RISP treatment is apparently evident. This suggests that RISP may induce neuroprotection against various internal or external nocuous stimuli by slightly stimulating APP expression [484,513,516]. Furthermore, considering that increase in APP level is seen after exposition to potentially toxic agents (*eg.* HAL), marked elevation in APP concentration is not present after treatment with the presumably non-toxic RISP.

In contrast with HAL, high potential atypical antipsychotic RISP did not have a significant effect on APP, nor does the literature have any information on its importance at any level in AD, be it clinical or molecular. The observed changes elicited by HAL are therefore better explained. Even though both drugs act as D_2-antagonists, involvement of σ_1-receptors, altered affinity to APP, direct cytotoxicity, or other unknown mechanisms may account for the differing impact of HAL from RISP on APP metabolism. However, the slight increase in APP levels as a result of RISP treatment (Figure 7) might also be regulated by gordian mechanisms; apart from D_2-receptors, the involvement of numerous factors beyond the first messenger level along with 5-HT_2 and NA-receptors, and above all Ca^{2+}, might be implicated.

Implication in AD

The pharmacological actions of HAL and RISP are important in the course of dementia. Antipsychotics did not affect intracellular Ca^{2+} levels in AD fibroblasts in the presence or absence of βAP treatment. In case of APP, however, their molecular impact is crucial in AD.

The major route of APP processing is via α-secretion, in which APP is cleaved by α-secretase at or adjacent to Lys_{16} in the βAP sequence [166] (Figures 1, 2). This route of processing has two important consequences. On one hand, C-terminally truncated form of APP (*ie.* sAPPα) is released from the membrane and can be subsequently secreted from the cell. Also, this pathway destroys the complete βAP sequence and therefore may mitigate amyloid formation. The amyloidogenic APP metabolic route produces βAP via the activation of β- and γ-secretases.

Intact cellular homeostasis, including relatively stable Ca^{2+}-level, leads to constitutive sAPPα-secretion. Because sAPPα is known to stabilize cytosolic Ca^{2+}-dynamics, normal processing of APP stimulates sAPPα production. Because HAL and also probably RISP stabilize Ca^{2+}, a protective circle is proposed in this study. By producing relatively constant Ca^{2+} concentrations, HAL and RISP stimulate sAPPα formation, which in turn stabilizes Ca^{2+} levels. Apart from constitutively secreted sAPPα, receptor-activated sAPPα release mediated by protein kinases – such as PKC – is a crucial pathway in APP metabolism [241,242]. As discussed above, activated PLC affects APP processing by either of two pathways, one involving an increase in PKC (PLC-DAG-PKC cascade) and the other an elevation in cytoplasmic Ca^{2+}-levels (PLC-IP_3-Ca^{2+} limb of the pathway). Both pathways, probably also activated by HAL and RISP, are known to stimulate sAPPα production and inhibit βAP formation [241].

Unlike usual conditions, however, cells in AD are exposed to neurotoxic βAP. The βAP-limb of the APP pathway has been shown to induce significantly altered Ca^{2+} levels both in this study and elsewhere. Increased or decreased levels of cytoplasmic Ca^{2+} plays a role in the abnormal proteolytic processing of APP resulting in increased βAP formation [522,523]. This feed-forward neurodegenerative pathway leads to cell death by sustained Ca^{2+}-levels and inhibited sAPPα secretion. By antagonizing βAP-induced Ca^{2+}-imbalance, stabilizing Ca^{2+}-levels, and inhibiting βAP formation, the protective circles of HAL and RISP are important events in slowing the progression of AD.

Protein kinases, such as PKC, are involved in receptor-mediated sAPPα production. These protein kinases can be activated by second messengers, such as Ca^{2+}, or in turn, these kinases may activate second messengers, including Ca^{2+}. Even though Ca^{2+}-cascade can regulate APP metabolism in a PKC-independent fashion, hyperactivation of kinases by incessant Ca^{2+}-imbalance can activate PKC-reliant amyloidogenic APP processing.

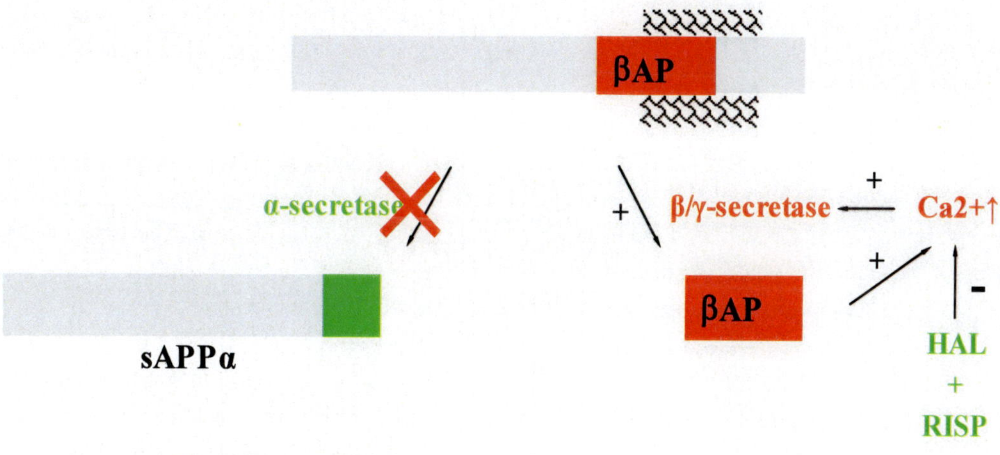

Figure 8: Proposed βAP-Ca^{2+} feed-forward

Because propeptides of β- and γ-secretases undergo phosphorylation in order to yield active proteases [524], PKC-mediated phosphorylation plays a critical role in the regulation of sAPPα and βAP secretion, and PKC-mediated pathways may help control amyloid formation. However, PKC appears to mediate APP secretory processing not directly but via additional intracellular messengers. Apart from Ca^{2+}, one reasonable candidate for the phosphoacceptor is the APP secretase itself. In the course of the disease, "mal-activation" of kinases leads to hyper-phosphorylation, which may stimulate not only amyloidogenic APP processing, but also hyper-phosphorylation of the axonal transport-associated protein *tau*, which ultimately yields PHF. By acting on Ca^{2+}, both HAL and RISP are proposed to alleviate these effects seen in AD.

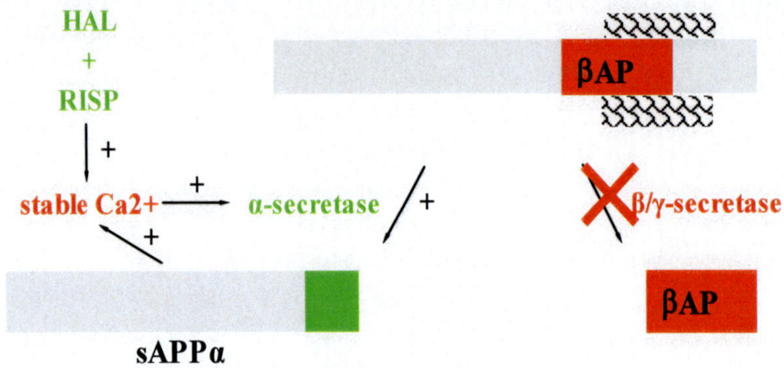

Figure 9: Proposed sAPPα-Ca^{2+} feed-back – "protective circle" of HAL

Taken together, βAP-Ca^{2+} feed-forward (Figure 8) and sAPPα-Ca^{2+} feed-back, *ie.* "protective circle" of HAL (Figure 9) is proposed in AD.

Gene Expression Profile of Antipsychotics in the Brain

It has been shown in this study that HAL and RISP interfere with Ca^{2+} and APP metabolism in AD. It is not yet fully understood whether this is a direct interaction taking place at the proteomic level, or their exact mode of action is interfering with genes and their expression – or perhaps both.

Recent advances in high-density cDNA microarray technique allow the possibility to analyze thousands of genes simultaneously for their differential gene expression patterns in various biologic processes, and determining mechanisms of drug action. The aim of this series of experiments was to gain experience in antipsychotic gene-expression profiling and characterize (in the parlance of genomics) the "antipsychotic transcriptome."

An important step to understanding the role of genes in the mechanism of drug action is to define gene expression profiles after drug manipulation. DNA expression array technology can monitor the expression of thousands of genes simultaneously and provide a format for identifying genes [525]. DNA microarrays allow genome-wide assessment of changes in relative messenger RNA abundance and thus can be used to monitor changes in gene expression as a result of various medications. Using this novel method we determined the effect of acute and chronic effect of HAL and RISP on the expression of approximately 8,000 gene targets in rat prefrontal/fronto-parieto-temporal cortex (Figure 10). This region of brain was chosen because this is an area in which abnormalities have been reported in SCH patients, where dopaminergic neurons largely project, and where antipsychotics have been shown to have numerous effects. Moreover, this is the area where AD-specific changes occur, and where HAL and RISP were shown in this study to have beneficial effects in dementia.

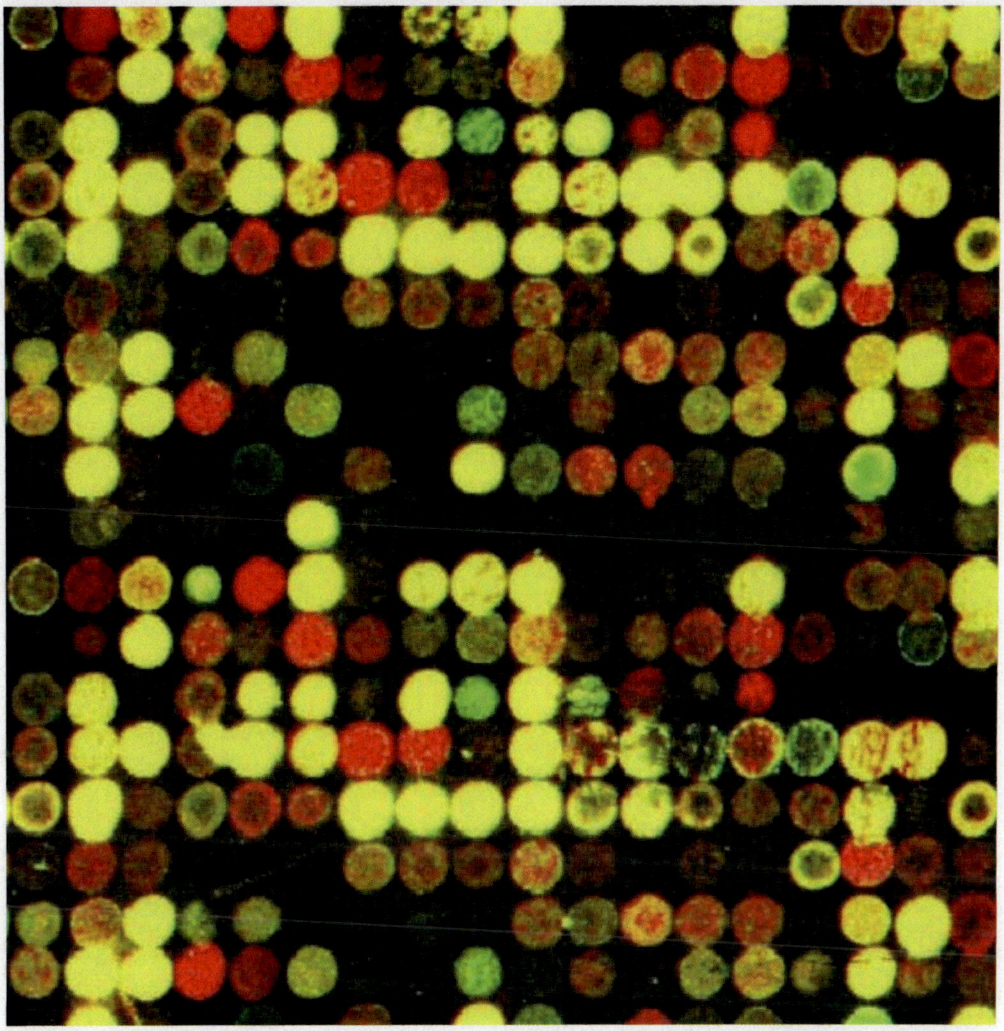

Figure 10: cDNA microarrays
Gene-chips hybridized with Cy5 labeled cDNA probe (obtained from untreated rat cortical total RNA) and Cy3 labeled cDNA probe (obtained from antipsychotic-treated cortical total RNA)

This study is very important, since the pharmacotherapy of psychiatric drugs continues to be based on trial and error, with psychiatric diagnosis serving as the principle guide for drug choice. Often drug selection within a therapeutic category (*eg*. antipsychotic) is based on side-effect profile (*eg*. sedation) to match a clinical need. Therapeutic response, even among diagnostically homogenous patient groups, is highly variable within classes of psychiatric drugs. Until now, little information is available on the molecular effect of antipsychotics.

Table XII: List of differentially expressed genes in rat fronto-temporo-parietal cortex after acute (96 hours) treatment with HAL. *A*: over-expressed, and *B*: repressed genes

A:

Clone name (mRNA)	Δ Fold (mean)	Accession number	Functional group
Follistatin-like protein	1.09	AA818445	Cell cycle and growth
Rat basic fibroblast growth factor (FGF) mRNA, complete cds	1.19	M22427	Cell cycle and growth
Protein O-mannosyltransferase 1	1.15	NM_053406	Development
Rattus norvegicus cDNA clone, similar to Human B12 protein mRNA	1.96	AA900186	Ion transport
S100 calcium-binding protein A8 (calgranulin A) (S100A8)	1.41	NM_053822	Ion transport
Glutamate receptor subunit GluR2-flip mRNA, complete cds	1.22	AF164344	Ion transport
Calmodulin 2, mRNA, complete cds	1.13	BC058485	Ion transport
Glutathione S-transferase A5 (Gsta5), mRNA	1.19	NM_031509	Metabolism
Rattus norvegicus similar to eosinophil peroxidase (LOC303414), mRNA	1.21	XM_220834	Metabolism
Adrenal mitochondrial protease	1.97	NM_153311	Protein metabolism
G protein-binding protein CRFG	1.06	NM_053689	Signal transduction
Ly-49 stimulatory receptor 3	1.20	NM_153726	Signal transduction
Rattus norvegicus APP770 mRNA, complete cds	1.76	AF513015	Other

B:

Clone name (mRNA)	Δ Fold (mean)	Accession number	Functional group
Pleiotrophin (heparine binding factor)	-1.11	NM_017066	Cell cycle and growth
Serotonin N-acetyltransferase	-2.15	NM_012818	Circadian rhythm
Activity regulated cytoskeletal-associated protein	-1.99	NM_019361	Development
α-prothymosin	-1.48	M86564	Immune response
Inwardly rectifying K^+-channel	-1.89	D61687	Ion transport
Voltage-dependent calcium channel γ-7 subunit (Cacng7) mRNA, complete cds	-1.21	AF361349	Ion transport
Tyrosine aminotransferase	-2.68	NM_012668	Metabolism
Rattus norvegicus mRNA for endothelial nitric oxide synthase, 3' region, partial	-1.57	AJ011116	Metabolism
Clathrin assembly protein long form	-1.04	AF041374	Protein metabolism
SH3 domain binding protein (CR16)	-1.82	U25281	Protein metabolism
Disintegrin and metalloprotease domain 2	-2.15	NM_020077	Protein metabolism
MIBP1 (c-myc intron-binding protein 1)	-0.99	D37951	Transcription regulator
Gap junction membrane channel β5	-1.24	NM_019241	Cell-cell interaction
Type I pro-α_2 collagen-like sequence	-3.25	AF050214	Other
TPCR09 protein	-2.07	X89698	Other

Analysis of DNA-chips

A number of transcripts were differentially expressed between control and treated samples. However, among the 8,000 genes examined in the present study the expression of only 36 genes (0.45%) was found to be significantly altered after treatment with HAL. In case of acute administration of HAL, 13 genes were more highly expressed and 15 transcripts were found to be significantly less abundant. After chronic HAL treatment, however, 9 transcripts were induced and none of them was considerably repressed.

The list of genes with significantly altered expression ($P<0.05$) is delineated in Tables XII-XIII. Values are presented as the expression ratio of treated / untreated cases for a certain gene. Apart from these known transcripts, a number of expressed sequence tags (ESTs) with unknown function were also differentially expressed in the antipsychotic-treated cortices; these are not listed.

Table XIII: List of differentially expressed genes in rat fronto-temporo-parietal cortex after chronic (4 weeks) treatment with HAL

Clone name (mRNA)	Δ Fold (mean)	Accession number	Functional group
Follistatin-like protein	1.58	AA818445	Cell cycle and growth
Transducin-like enhancer of split 4	1.51	NM_019141	Development
K^+-channel (erg2)	1.11	AF016192	Ion transport
Acyl-Coenzyme A dehydrogenase, C-4-12	0.97	NM_016986	Metabolism
Extracellular signal-related kinase (ERK2)	1.02	M64300	Signal transduction
G protein-binding protein CRFG	1.31	NM_053689	Signal transduction
Relaxin-like factor	1.37	AF139918	Signal transduction
Endothelin receptor	1.69	NM_017333	Signal transduction
U2 RNA 3' end	1.59	M10882	Others

In case of RISP, 89 genes (1.11%) were found to significantly differ in expression. Acutely, 43 genes were induced and 46 were down-regulated (Table XIV). Chronically, however, this number has decreased to 6 and 11, respectively (Table XV).

Table XIV: List of significantly A: induced, and B: down-regulated genes in rat fronto-temporo-parietal cortex after acute (96 hours) treatment with RISP

A:

Clone name (mRNA)	Δ Fold (mean)	Accession number	Functional group
Integrin α-subunit	1.13	S58528	Cell adhesion
Integrin β_1	1.70	NM_017022	Cell adhesion
Plakoglobin	2.20	U58858	Cell adhesion
Pleiotrophin (heparine binding factor)	1.11	NM_017066	Cell cycle and growth
KIAA1536 protein	1.28	NM_139190	Cell cycle and growth

Table XIV: List of significantly A: induced, and B: down-regulated genes in rat fronto-temporo-parietal cortex after acute (96 hours) treatment with RISP(Continued)

A:

Clone name (mRNA)	Δ Fold (mean)	Accession number	Functional group
Follistatin-like protein	1.96	AA818445	Cell cycle and growth
Protein O-mannosyltransferase 1	1.57	NM_053406	Development
HES-related repressor protein 1 (HERP1)	1.00	AY059382	Development
RING finger protein	2.07	AF036255	Development
Transducin-like enhancer of split 4	1.52	NM_019141	Development
Transmembrane receptor Robo1	1.71	AF041082	Development
T-cell receptor α-chain RT1L haplotype	1.57	L11027	Immune response
Small inducible cytokine subfamily A20	2.31	NM_019233	Immune response
Synaptotagmin VII	1.62	U20106	Protein metabolism
RAB15	2.24	M83679	Protein metabolism
ryk-tyrosine kinase-related protein	2.67	AB073721	Protein metabolism
Janus protein tyrosine kinase 1, JAK1	1.48	AJ000556	Protein metabolism
YME1 (*S. cerevisiae*)-like 1	1.14	NM_053682	Protein metabolism
Aminopeptidase PILS	1.15	AF148324	Protein metabolism
Adrenal mitochondrial protease	1.47	NM_153311	Protein metabolism
Cathepsin K	1.89	AF010306	Protein metabolism
CLN2 tripeptidyl peptidase I	2.35	AB043870	Protein metabolism
Similar to Human B12 protein	1.77	AA900186	Ion transport
K^+-channel (erg2)	2.08	AF016192	Ion transport
Na^+-channel, voltage-gated, type 10a	1.01	NM_017247	Ion transport
Vacuolar adenosine triphosphatase-β	1.40	Y12635	Ion transport
ATPase, Na^+/K^+ transporting, α_1	2.00	NM_012504	Ion transport
Acyl-Coenzyme A dehydrogenase, C-4-12	2.13	NM_016986	Metabolism
Kruppel-like factor LKLF	1.92	AA925780	Transcription regulator
Prophet of Pit1, paired-like homeodomain	1.96	NM_153627	Transcription regulator
Activating transcription factor 3	2.50	NM_012912	Transcription regulator
Splicing factor 1 homolog	1.75	AF079873	Signal transduction
SH-PTP2 protein tyrosine phosphatase 11	2.74	NM_013088	Signal transduction
Receptor activity modifying protein 2	1.00	AB042888	Signal transduction
Interleukin 6 receptor	3.12	NM_017020	Signal transduction
G protein-binding protein CRFG	1.89	NM_053689	Signal transduction
Endothelin receptor	2.04	NM_017333	Signal transduction
P-glycoprotein-like ATP cassette transporter	1.02	AF106563	Transport
Mg1	1.48	AY035343	Others
FSH-regulated protein	1.71	L26292	Others
Integrin-binding sialoprotein	2.14	NM_012587	Others
Ly-49 stimulatory receptor 3	1.84	NM_153726	Others
Neurexin IIIα (axon guidance)	2.84	L14851	Others

Table XIV: List of significantly A: induced, and B: down-regulated genes in rat fronto-temporo-parietal cortex after acute (96 hours) treatment with RISP(Continued)

B:

Clone name (mRNA)	Δ Fold (mean)	Accession number	Functional group
Neural adhesion molecule F3	-1.37	D38492	Cell adhesion
BIT	-1.12	D38468	Cell adhesion
CD5	-2.60	X78985	Cell adhesion
Growth accentuating protein 43	-1.01	NM_017195	Cell cycle and growth
Serotonin N-acetyltransferase	-1.31	NM_012818	Circadian rhythm
Activity regulated cytoskeletal-associated protein	-2.31	NM_019361	Development
α-actinin 4	-1.19	AF190909	Development
Protein phosphatase 1, regulatory 14a	-2.61	NM_130403	Immune response
α-prothymosin	-1.62	M86564	Immune response
EFA6 exchange factor for ARF6	-1.40	AB040468	Protein metabolism
ARL5 ARF-like protein 5	-1.31	X78604	Protein metabolism
Ubiquitin C (Ubc)	-1.54	NM_017314	Protein metabolism
Clathrin assembly protein long form	-1.61	AF041374	Protein metabolism
SH3 domain binding protein (CR16)	-1.59	U25281	Protein metabolism
Hippocalcin (Hpca)	-1.63	NM_017122	Protein metabolism
Disintegrin and metalloprotease domain 2	-2.15	NM_020077	Protein metabolism
Disintegrin and metalloproteinase domain 1	-1.09	NM_020078	Protein metabolism
Voltage-dependent calcium channel g8	-2.21	NM_080696	Ion transport
Plasma membrane CA^{2+}-ATPase 3	-1.47	M96626	Ion transport
Inwardly rectifying K^+-channel	-2.32	D61687	Ion transport
Mitochondrial H^+-ATP synthase-α	-1.16	J05266	Ion transport
Uncoupling protein 2, mitochondrial	-1.53	NM_019354	Ion transport
Alkaline phosphatase, tissue-nonspecific	-1.17	NM_013059	Metabolism
Tyrosine aminotransferase	-1.52	NM_012668	Metabolism
γ-glutamylcysteine synthetase	-2.12	NM_017305	Metabolism
MIBP1 (*c-myc* intron binding protein 1)	-1.13	D37951	Transcription regulator
Extracellular signal-related kinase (ERK2)	-1.95	M64300	Signal transduction
Interleukin 10 receptor-α	-2.68	NM_057193	Signal transduction
Basignin (Ox47 antigen or CE-9)	-2.02	NM_012783	Signal transduction
Relaxin-like factor	-1.95	AF139918	Signal transduction
Soluble adenylyl cyclase (SAC)	-2.37	AF081941	Transport
Messenger RNA for preproalbumin	-1.17	V01222	Transport
Vesicular GABA transporter	-1.41	AF030253	Transport
gcd-10S	-1.17	AB046592	Transport
Gap junction membrane channel β5	-1.11	NM_019241	Cell-cell interaction
Type I pro-α_2 collagen-like sequence	-3.77	AF050214	Others
Putative zinc-finger protein	-3.40	AJ007467	Others
U2 RNA 3' end	-2.63	M10882	Others
Glucagon (Gcg)	-2.17	NM_012707	Others

Table XIV: List of significantly A: induced, and B: down-regulated genes in rat fronto-temporo-parietal cortex after acute (96 hours) treatment with RISP(Continued)

B:

Clone name (mRNA)	Δ Fold (mean)	Accession number	Functional group
LIM homeodomain protein 3β	-1.61	AF370447	Others
Limkain β1 (Lkap)	-1.38	NM_133421	Others
mud-2	-1.28	U70266	Others
Zinc finger protein 2 (DZF2)	-1.11	U78130	Others
Axomer-8, transported in axons	-1.00	AB097858	Others
DnaJ-like protein (heat shock response)	-1.15	U53922	Others
TPCR09 protein	-1.21	X89698	Others

Table XV: List of significantly A: up-regulated, and B: repressed genes in rat fronto-temporo-parietal cortex after chronic (4 weeks) treatment with RISP

A:

Clone name (mRNA)	Δ Fold (mean)	Accession number	Functional group
Follistatin-like protein	1.09	AA818445	Cell cycle and growth
Protein phosphatase 1, regulatory 14a	2.30	NM_130403	Immune response
Acyl-Coenzyme A dehydrogenase, C-4-12	1.24	NM_016986	Metabolism
Prophet of Pit1, paired-like homeodomain	1.69	NM_153627	Transcription regulator
Extracellular signal-related kinase (ERK2)	1.35	M64300	Signal transduction
Receptor activity modifying protein 2	3.35	AB042888	Signal transduction

B:

Clone name (mRNA)	Δ Fold (mean)	Accession number	Functional group
EFA6 exchange factor for ARF6	-1.04	AB040468	Protein metabolism
Clathrin assembly protein long form	-1.49	AF041374	Protein metabolism
Janus protein tyrosine kinase 1, JAK1	-1.37	AJ000556	Protein metabolism
Inwardly rectifying K^+-channel	-1.65	D61687	Ion transport
Uncoupling protein 2, mitochondrial	-1.25	NM_019354	Ion transport
Alkaline phosphatase, tissue-nonspecific	-1.67	NM_013059	Metabolism
Endothelin receptor	-1.03	NM_017333	Signal transduction
P-glycoprotein-like ATP cassette transporter	-1.13	AF106563	Transport
U2 RNA 3' end	-1.56	M10882	Others
Limkain β1 (Lkap)	-1.16	NM_133421	Others
TPCR09 protein	-0.91	X89698	Others

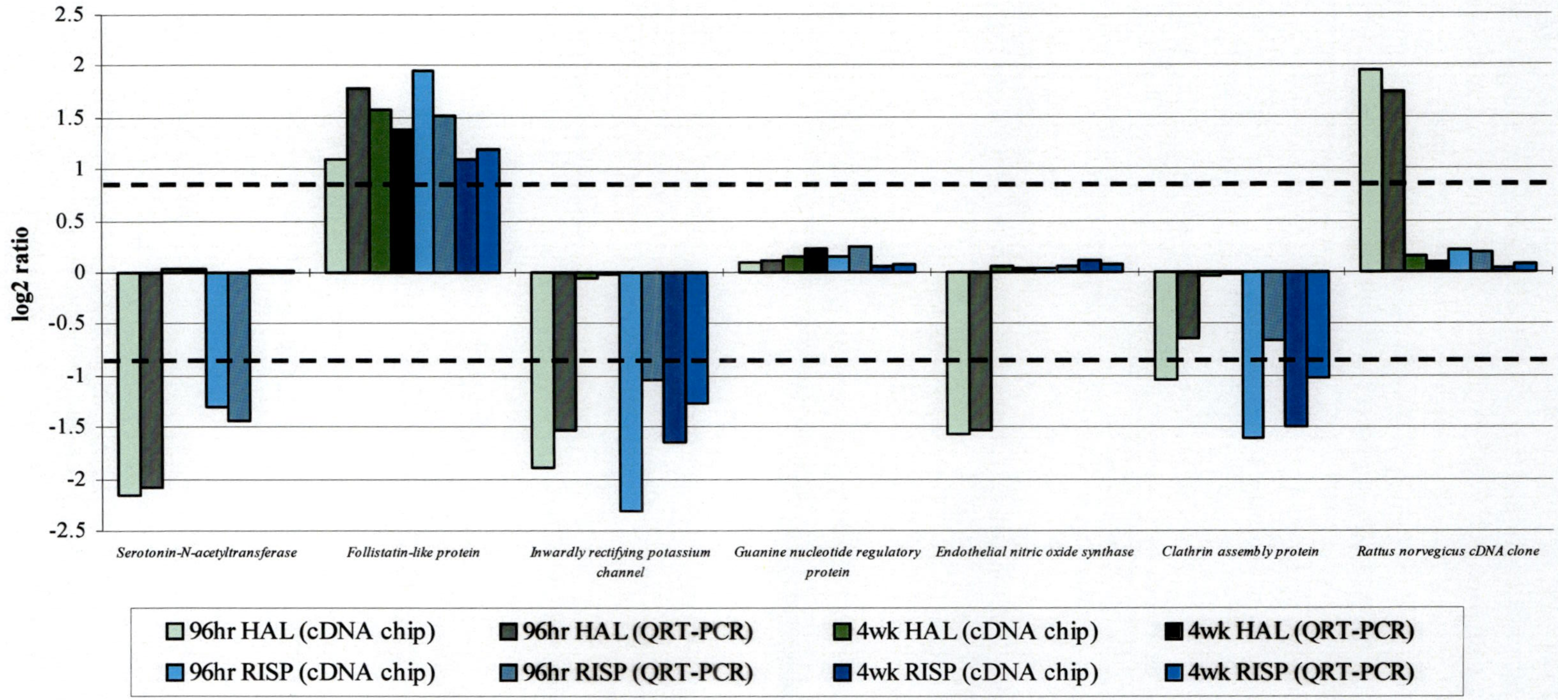

Figure 11: Quantitative determination of transcript levels by real time PCR
Changes in transcript levels in rat fronto-temporo-parietal cortex after acute or chronic intraperitoneal treatment with HAL (light and dark green box, respectively) or RISP (light and dark blue box, respectively) were confirmed by real time PCR (dashed light and dark green box for HAL, respectively; and dashed light and dark blue box for RISP, respectively). The expression of serotonin-N-acetyltransferase, follistatin-like protein, inwardly rectifying K^+-channel, guanine nucleotide regulatory protein, endothelial nitric oxide synthase, clathrin assembly protein, and a *Rattus norvegicus* cDNA clone was determined. Dashed lines indicate the interval -1.8 to 1.8-fold regulation (corresponding to $\log_2 = 0.85$) in which changes in expression were considered not significant.

Verification by QRT-PCR

To validate these semi-quantitative results, the expression of randomly selected 7 genes was analyzed by real-time quantitative PCR (QRT-PCR). The PCR primers used in this study are listed in Table XVI. The results for each sample were normalized to β-actin expression, then subjected to *t*-test analysis. All 7 transcripts were still identified as significantly more or less abundant, or unchanged, respectively (Figure 11). Comparison of the results derived from microarraying and the more accurate real-time PCR showed a qualitative agreement in that the differentially expressed genes in HAL or RISP-treated cortices were correctly assigned as up- or down-regulated, or significantly unchanged.

Table XVI: Primers used in QRT-PCR analysis

Gene product	Forward primer	Reverse primer
Serotonin N-acetyltransferase	ACCGCTTGCCTTCTGAACA	CCCGCTGCTCCCGTTT
Follistatin-like protein	TTTCGTGTATGGCATCAACCA	GCGTCCAATTCCCAAGGA
Inwardly rectifying K^+-channel	AGTGTGGCTGTGGCAAAGG	CCGCAACTCAGGACAAGGA
Guanine nucleotide regulatory protein	CCAGCCTGGCTCCTCAGA	TGTTGCAGTCTCGAATTAGCTTCT
Endothelial nitric oxide synthase	AATAAATTCCCCTTCACGACCTT	CCTGCGGCTGCATCAATAT
Clathrin assembly protein long form	GATTCTGGTGGTGGACTTCTCAA	GGCGGAAGTTTGGCAACA
Rattus norvegicus cDNA clone	GGCCCAGAATGATGAGCAA	TGATTGGCGGCGTCGTA

Significance of Genes with Altered Expression

In this study, we utilized the high-throughput gene chip, which contains 8,000 known genes and ESTs, to determine the alteration of gene expression profiles of rat cortex exposed to antipsychotics. Our results from microarray provided a genome-wide analysis of the cellular response to HAL and RISP treatment:

Neural Plasticity

Cellular and molecular responses of neurons to antipsychotic treatment are complex and are likely to be mediated by a variety of regulatory pathways. We found that the molecular response to these drugs in rat cortex involved inhibition or induction of genes, which have specific functions in signal transduction, transcriptional and translational regulation, protein turnover, and cellular metabolism (Tables XII-XV). All of these pathways are inevitable factors of neuronal plasticity. One of the central elements of this process is the activation of glutamatergic transmission.

HAL has been proved to mediate glutamate transmission by interfering with the expression of NMDA-receptor subunits [526] and glutamate transporters [527]. Glutamatergic transmission at either the N-methyl-D-aspartate (NMDA) or α-amino-3-hydroxy-5-methyl-4-isoxazole-propionic acid (AMPA) receptors is one mechanism for an increase in synaptic strength and the induction of central sensitization [528]. Activation of these ionotropic receptors causes a sustained elevation in postsynaptic Ca^{2+} that induces a cascade of persistent intracellular events contributing to synaptic plasticity [528,529]. Therefore, by acting on the glutamatergic system (*eg*. glutamate receptor subunit GluR2-flip), both HAL and RISP mediate Ca^{2+}-permeability of cells, and therefore mediates synaptic plasticity.

From the gene expression profiles of rat cortices exposed to HAL and RISP, it was also found that many genes are related to cell cycle control. Moreover, we also observed altered expression of genes encoding kinases including extracellular signal-related kinase (ERK2), *ryk*-tyrosine kinase-related protein, Janus protein tyrosine kinase 1, growth factors (*eg.* follistatin-like protein, basic fibroblast growth factor, pleiotrophin), and transcription factors, suggesting that HAL and RISP may also influence cell growth via mediation of multiple cell signal transduction pathways. These results confirm and provide mechanistic information about HAL- and RISP-induced neuro-plasticity.

Role in Neurometabolic Disorders

The results of our analysis implicate on the order of hundreds of genes as being involved in neurodegenerative diseases.

Several genes directly related to Ca^{2+} homeostasis were found to be differentially regulated by HAL and RISP, *eg.* S100 calcium-binding protein A8, calmodulin 2, voltage-dependent calcium channel γ-7 subunit, plasma membrane CA^{2+}-ATPase 3, *etc.* (Tables XII-XV). We have reported alterations of calcium homeostasis in neurometabolic conditions and other medical disorders in the present study. We have also shown that HAL and RISP are efficient in stabilizing $[Ca^{2+}]_i$, and therefore have a protective role in AD and various neuropsychiatric disorders. However, their mechanisms of action remained unknown. Broadly in line with the present findings on gene expression, the Ca^{2+}-stabilizing beneficial effects of these antipsychotics are based on the genomic level by interfering with Ca^{2+}-transporters, Ca^{2+}-channels, glutamatergic transmission, *etc.* (Tables XII-XV).

We have also reported above that HAL up-regulates the level of APP, a key precursor molecule in AD that yields neurotoxic βAP. Interestingly however, as aforementioned, APP might play a crucial role in neuroprotection by participating in cell-to-cell recognition, re-establishment of synaptic contacts, alleviation of glutamate-induced membrane depolarization and over-excitation, and stabilization of the intracellular Ca^{2+} homeostasis [484,513]. HAL therefore may stimulate cytoprotective pathways by inducing APP gene expression, as presented here and above. However, the direct neurotoxic effect of HAL [429-432] may also account for the elevated APP level as a defensive response seen in our experiments after HAL treatment.

The finding that HAL increases APP protein level, as discussed above, is now better explained, in that it is demonstrated here for the first time that HAL up-regulates APP mRNA. Ca^{2+} has been proposed to modify the production and processing of APP [59] By acting on Ca^{2+}-homeostasis, as we have presented above and herewith, HAL may also indirectly interfere with APP both at the genetic and at the proteome level.

Stress and Neuro-Toxicity

Documented consequences of HAL treatment are neurotoxicity and apoptosis [429-432], which apparently may be attributable to its impact on the central secondary messenger, Ca^{2+}. Despite its molecular and clinical adverse effects, HAL is widely used in the treatment of psychiatric symptoms associated with various medical conditions where the involvement of Ca^{2+} is implicated, such as in AD. Interestingly, we have demonstrated above that HAL efficiently attenuates βAP-induced Ca^{2+}-imbalance in AD, which is in controversy with the present findings regarding Ca^{2+}. Most, if not all, genes in mammalian cells are regulated by multiple mechanisms acting in concert to regulate their expression. As gene products change in response to cellular needs, it is likely that the repertoire of regulatory mechanisms influencing the expression of these genes also changes. It is likely, therefore, that HAL disrupts Ca^{2+}-levels to cause apoptosis, however in an already disturbed setting, *eg*. in AD, it acts as a cytoprotective agent by inflaming defensive mechanisms such as inducing APP expression and level.

Another major category of differentially regulated genes encode for cell defense proteins, which include anti-oxidative stress proteins and heat shock proteins. The majority of these transcripts were differentially expressed in HAL and RISP-treated animals.

Oxidative stress can occur as a result of a number of stimuli. During aerobic metabolism under stress, reactive oxygen species (ROS) are produced as a result of partial reduction of oxygen. ROS were originally considered to be detrimental to cells, but recently it has been shown that it is involved in redox regulation by adjusting cellular activities [530]. These ROS induce the expression of defense-related genes such as those encoding glutathione-S-transferase (GST). In this study, the GST- and peroxidase enzyme-encoding genes were both up-regulated after treatment with HAL (Table XIIA). Glutathione acts as a redox-sensor and is involved in the multiple regulatory systems coordinating the expression of defense genes [531], therefore the increasing glutathione biosynthetic capacity could enhance resistance to oxidative stress. Because HAL is known to be toxic to cells, up-regulation of defense genes might be a compensatory mechanism, just as discussed with the induction of APP.

Effect on the Dopaminergic System

Antipsychotics had been assumed to suppress DA-mediated neurotransmission either by decreasing brain DA-levels (*eg*. reserpin) or by antagonizing D_2-receptors. Contrary to the hyper-dopaminergic hypothesis, some previous data indicate a reduced central DA output associated with SCH, a phenomenon that was found to correlate with negative symptoms, such as anergia, emotional blunting, lack of drive, *etc* [532]. Consequently, both hypo- and hyperfunctioning of brain DA-systems in SCH have been proposed and, tentatively, both types of dysfunction might even occur simultaneously, albeit in different brain regions. Responsiveness of both negative and positive symptoms of SCH to antipsychotics has been documented in clinical settings. Importantly, however, no effect on DA metabolism and transmission were significantly altered by HAL or RISP in our experiments at the genetic level. Their antipsychotic effect therefore is not by modifying DA turnover and bioavailability or modifying genes encoding for D_2-receptors, but by fine-tuning signal transduction pathways. By acting on intracellular signaling, HAL and RISP are potent inductors of neuronal plasticity. As such, they are hypothesized to have an impact on remodeling pathological dopaminergic and glutamatergic neuronal circuits seen in SCH and other neuropsychiatric conditions as part of their modes of action.

Interestingly, however, both HAL and RISP significantly altered the expression of endothelin. There is increasing evidence that endothelins and endothelin receptors play a role in brain mechanisms associated with behavior and neuroendocrine regulation, as well as cardiovascular control. A growing mass of data suggests that they may be acting as true neurotransmitters or neuromodulators. They are known to stimulate DA-release in the CNS

[533]. The effects of endothelin in the brain could be a continuum from physiologically relevant neuromodulatory effects at low concentrations, through reversible hypermetabolism at higher doses, to permanent neurotoxic neuronal damage at very high concentrations.

Altered expression of the endothelin-system as a result of administration of either of the two antipsychotics underlying these neurochemical effects do not necessarily have to be different. While it is clear that physiological activation of endothelin receptors induces a rise in intracellular Ca^{2+} levels [534], supra-physiological levels of Ca^{2+}-influx have been associated with neurotoxic events [535]. This may add to the well-known toxic effect of HAL. Alternatively, it is possible that the low-dose, physiological action of endothelin is mediated by a direct effect on dopaminergic terminals, whereas the high-dose effect also involves other mechanisms, such as glutamate and NO release [536]. It is well-known that the high levels of glutamate may induce neurotoxic effects in the brain, further explaining the neurotoxic effect of HAL.

Taken together, the impact of HAL and RISP on the dopaminergic system might be accomplished not directly, as suggested above, but via the novel neuro-transmitter/modulator endothelin system.

Commonalities

The focus of this study was the identification of genes after treatment of rats with HAL and RISP. Even though there was an overlap of only a few genes at exposure to HAL and RISP, both drugs have interfered with the same cellular functions. RISP acutely had an impact on more genes than HAL. This might be attributable to the fact that the mode of action of RISP is more diverse: it is a high potential D_2-, $5HT_2$- and NA-antagonist, whereas HAL is a D_2-antagonist with some σ_1-activity. Taken together, this observation suggests that antipsychotics with different pharacological characteristics share targets at the molecular level.

Interestingly, some significantly enriched pathways differ after treatment with both drugs over time. Transcripts implicated in neuronal plasticity, survival, oxidative changes and metabolism in general are widely interfered with by HAL and RISP in our acute experiments, but these antipsychotic drugs had less diverse impact on these genes, or did not change chronically their expression profile (Table XII-XV). This finding is well established, since high potential agents like HAL and RISP are thought to have more profound impact on initiating therapy. Chronically, however, drastic changes subside.

Whereas conventional neurological assessment utilizes behavioral, functional, and/or morphological techniques to estimate overt effect of drugs, new methods are being developed that identify molecular changes on the gene expression level that are associated with medical treatment. Based on our findings, gene expression profiling is an invaluable method for delineating the neuro-psycho-pharmacological actions of antipsychotics by simultaneuously identifying thousands of genes modulated in response to drug exposure, and it allows the determination of genes responsible for treatment failure or development of drug resistance.

Additionally, cDNA arrays permit the measurement of early gene expression events in cells rather than cumulative systemic effects and, thus, may provide a means of early detection of therapeutic effects. Microarrays have considerable potential to define in molecular terms the progression of a disease and identify biochemical pathways that can be used as targets for future therapeutic intervention. The identification of unique gene expression patterns associated with cellular responses to a particular chemical or class of chemicals can be used in conjunction with a variety of pharmacological assessment platforms to identify the presence of both physiologically and medically useful substances. The feasibility of using DNA arrays for this assessment depends on the ability of this technology to identify reproducible gene expression signatures from relatively small quantities of RNA.

The use of QRT-PCR in combination with the microarray work ruled out the dissociation of protein translation from gene transcription. Therefore, gene expression profiling can be a powerful tool with which to compare the mechanisms of action of different agents, which elicit similar biological responses. As such, our study would serve as a platform for all further investigation of the molecular effects of HAL and RISP in rats and humans. Elucidation of the metabolic and signal transduction pathways and gene transcription factors that underlie the pathophysiology of neuropsychiatric disorders and the actions of various medications could also reveal targets for the development of novel therapeutic agents. The continued application of major advances in molecular and cellular neurobiology to psychiatric problems over the coming years holds a bright future for the treatment, and possibly prevention, of AD and other neurometabolic disorders.

Therapy of Alzheimer's Disease

AD is a chronically progressive neurodegenerative disorder and a major contributor to morbidity and mortality in the elderly. A number of emerging AD therapies now holds great promise. However, clinical introduction of new AD therapies is critically dependent on parallel development of technology for early diagnosis and disease monitoring. Ideally, such technology would provide a sensitive, specific and quantitative method to ascertain patients at risk, establish early diagnosis/prognosis, monitor disease progression, and assess treatment efficacy. The goal of this translational neuroscience proposal is to address this urgent need by developing and testing an innovative minimally-invasive diagnostic/monitoring technology for AD. The test on fibroblastic and/or lymphocytic Ca^{2+}-homeostasis described in this study might fulfill such a goal.

Current treatment approaches to dementia are based on variable degrees of scientific evidence, reflecting an incomplete understading of the basic pathophysiology of AD. The goals of treatment are to achieve improvement in cognition and to minimize behavioral disturbances, *ie.* depression, psychosis, agitation and insomnia. Prevention of other medical comorbidities can improve functioning of AD patients. Maintaining adequate hydration, nutrition, exercise and cleanliness, is also inevitable.

Psychosocial Treatment

Attempts to reverse cognitive impairment through reality orientation and memory retraining (*eg.* reminiscence therapy or life review) have become popular forms of treatment in individuals with dementia. These nonpharmacologic symptomatic approaches may yield some transient benefit, but also may cause frustration and depression in patients and caregivers. Patients with mild cognitive impairment may benefit from support groups that can provide a constructive environment that benefits cognition and behavioral disturbances.

Modulating the environment is helpful because patients with cognitive impairment are often sensitive to their environment and function best with moderate stimulation. Patient activities should have minimal changes. Overstimulation will worsen confusion and may even result in agitation, whereas understimulation may cause the patient to withdraw. Some patients benefit from increased exposure to the outside world through newspapers, radio, and television. Written daily reminders can be helpful in the performance of daily activities. Prominent clocks, calendars, and windows are important.

Working closely with family members and caregivers is critical, since they are important sources of information about cognitive and behavioral changes, and often implement and monitor treatment. However, family support is also essential, since many family members express concerns about their own memory loss, which should be addressed with counseling or neuropsychologic assessment. Depression, anxiety syndromes and insomnia in primary caregivers are common issues [537]. In some cases, support group participation will diminish caregiver distress; in others, antidepressant treatment is indicated. Community resources such as respite care may provide much-needed relief for caregivers. Information and emotional support may improve quality of life for patients and caregivers and even delay nursing-home placement. Psychotherapeutic interventions may diminish behavioral problems and improve mood in both patients and family members [538].

A variety of community supports is available. Home-care or family-service agencies and adult day-care centers are examples of resources that can help relieve the care burden. Food services can be arranged for patients who are homebound, and many senior citizens centers, church groups, community programs, and hospitals offer transportation options. Community support will help to maximize the patient's independent living and functioning, and bring relief to family members and caregivers.

Pharmacotherapy

The onset of symptoms in AD is preceded by months and years of biological and structural changes within the brain. Long-term therapeutic goals must be to slow, halt, reverse, or even prevent the disease process. Slowing or halting is likely to require intervention at an earlier stage, *eg.* MCI. Prevention (*ie.* delaying the onset by *eg.* utilizing hormone replacement therapy or NSAIDs) may likely to be feasible in those known to be at risk of FAD. Once the diagnosis has been made, however, it may be unduly optimistic to expect to reverse a process that has led to senile plaques, NFTs and neuronal loss.

Cholinergic Treatment Approaches

The well-established cholinergic defects in AD include cholinergic cell body loss in basal forebrain and nucleus basalis of Meynert, decline of cholinergic baso-cortical projections, reduced activity of cerebral cortical choline-acetyltransferase, and dramatic loss of choline uptake and Ach-level in the neocortex and hippocampus. The cholinergic hypothesis proposes that the cognitive deficits of AD are related to the observed profound decrease in central Ach-ergic activity, and that increasing intrasynaptic Ach could enhance cognitive function and clinical well-being. Therefore, AchEIs have been introduced as cognition-enhancing agents in the treatment of patients with mild to moderate AD. By using these compounds, there is an increase in the Ach concentration available for synaptic transmission by inhibiting enzymes responsible for its hydrolysis, *ie.* AchE. This action compensates at least partially for the effects of CNS cholinergic hypofunction in AD. AchEIs appear to be useful throughout the disease, but particularly in the middle stage [539]. The AchEIs available now worldwide for clinical use are donepezil, tacrine, galantamine, rivastigmine, and metrifonate. Also, Ach-precursors, such as lecithin and acethyl-L-carnitine increase Ach synthesis and therefore improve cognitive function [540,541].

Cholinergic disturbances present in the limbic and paralimbic cortices, which mediate functions critical to emotion [15,26], may also underlie the pathophysiological basis of neuropsychiatric manifestations of AD [542,543]. Therefore, demented patients with delusions and hallucinations may exhibit a beneficial neuropsychiatric response to therapy with AchEIs [544]. Visual hallucinations and apathy are the most predictably responsive symptoms. Anxiety, disinhibition, agitation, depression, delusions, dysphoria, irritability, vocal outbursts, suspiciousness, wandering, fearfulness, tearfulness, compulsiveness, mood swings, and aberrant motor behavior could also improve. Variability in the behavioral symptoms of AD and the responsiveness to cholinergic therapy may reflect dynamic interactions between the cholinergic changes and other transmitter systems involved in AD [542,23]. The cholinergic system interacts with a variety of other transmitters or neuromodulators, including norepinephrine, dopamine, serotonin, γ-aminobutyric acid, opioid peptides, galanin, substance P, and angiotensin II [545].

Cholinergic agents affect many aspects of cognition, which suggests that the primary effect may be on an attentional or executive system with a secondary, pan-intellectual modulating influence on memory, language, and visuospatial skills [546]. Improvement in attention may underlie the reductions in apathy that are commonly associated with AchEIs, possibly explaining why apathy is among the most responsive of neuropsychiatric symptoms to cholinergic therapy and why it correlates highly with cognitive improvement.

Although overall AchE activity is reduced [547], it is increased in neuritic plaque and neurofibrillary tangle at early stages in the brain of patients with AD [548]. It has also been suggested that AchE may promote aggregation of βAP into a more toxic amyloid form [549]. Therefore, the blocking of AchE activity by AchEIs might increase Ach neurotransmission in the synaptic cleft of the brain and diminish βAP burden, resulting in improving cognitive function and alleviating the process of amyloid deposition. Furthermore, because cholinergic agonists stimulate the non-amyloidogenic processing of APP [550], the administration of cholinergic agonist or AchEIs to patients with AD may produce the favorable clinical results in ameliorating not only Ach-dependent cognitive function but also AD pathogenesis by the modulation of APP metabolism (*ie.* increased production of neuroprotective sAPPα).

The Glutamatergic System

Treatment of patients with mild to moderate AD with AchEIs is only effective in about half of the patients for whom it is prescribed. While the cholinergic pathology is well documented, decline in other neurotransmitter systems is also present in AD. Glutamate-mediated excitotoxicity might be involved in this process [65].

Glutamate, a principal excitatory neurotransmitter, and its receptor of the NMDA-type have a clear role in long-term potentiation, a form of synaptic strengthening whereby brief, high-frequency stimulation leads to an increased response after subsequent activation. This process is considered to underlie learning and memory.

NMDA-receptors control calcium channels. At rest, Ca^{2+}-channels are controlled by a voltage-dependent Mg^{2+}-ion block. Upon activation, Mg^{2+} leaves the channel, giving way to calcium-influx. Excessive glutamate activity, however, is associated with excitotoxicity when receptors are over-stimulated. Chronically elevated glutamate levels displace Mg^{2+} and thereby remove the blockade of NMDA-receptor-channels in a tonic, rather than a phasic manner, causing continuous Ca^{2+}-flow into the neurons. This increased "background noise" will ultimately lead to cell death. Although it may not be the main feature of neuronal loss in AD, excitotoxicity involving glutamate contributes to the pathogenesis of this condition. Glutamatergic-cholinergic interactions are important, and damage to both of these systems is liable to exacerbate dementia.

Memantine is a voltage-dependent, moderate-affinity, uncompetitive NMDA-receptor antagonist. It acts as a surrogate for Mg^{2+}, reducing overstimulation of the NMDA-receptor caused by abnormally high concentrations of glutamate in its resting state and restoring normal receptor-signaling function. Memantine is less voltage-dependent than Mg^{2+}, implying that the neurons must be more depolarized to relieve the memantine block, suggesting that mementine in AD brains will reduce the background Ca^{2+}-influx (noise) through the NMDA-receptors.

The clinical effects of memantine on moderate to severe AD have also been justified [551]. Because NMDA-antagonists and AchEIs reduce dementia symptoms through distinct mechanisms of action, there is growing interest in co-prescribing memantine with AchEIs; because this appears to be promising, combination-treatment regimens seem to be on the horizon.

Other Treatment Options

Although much has been learned about AD, the only approved drug therapy is the use of AchEIs (*eg.*, tacrine, donepezil, rivastigmine, and galantamine) and NMDA-antagonists (*eg.* memantine). Other therapies are undergoing clinical trials. These include monoamine oxidase inhibitors (*eg.*, selegiline), antioxidants (*eg.*, vitamins E and C), estrogen replacement therapy, and anti-inflammatory agents (*eg.*, nonsteroidal anti-inflammatory drugs, or NSAIDs) [552]. Currently, such treatments are more palliative than preventative or curative.

Treatment strategies from the molecular knowledge of AD are currently in development. They include *tau*-phosphorylation-inhibitors, β- and/or γ-secretase inhibitors to limit βAP production, and immunization with βAP peptide to prevent βAP plaque formation [553,554], modulators of inflammation [555], and cholesterol-lowering drugs [556].

Interfering with APP Metabolism

βAP production can be reduced either by stimulating the normal secretory processing of APP, or as an alternative, by inhibiting the amyloidogenic metabolism.

The non-amyloidogenic APP processing yields neuroprotective sAPPα [166-176]. Increasing the normal metabolism of APP can be a goal in AD management. Factors stimulating α-secretase activity include Ach [550] (*ie.* cholinergic activation, decline of which is prominent in AD), electrical stimulation [557], metabotropic glutamate [558], 5-HT_{2A} and 5-HT_{2C} agonist [520], NGF and EGF [559].

Because BACE has been known as an enzyme converting APP into sAPPβ and CTF [560], the blocking of this enzyme would be expected to slow down the production of βAP. The optimal way to prevent the production of the toxic fragments of APP such as βAP and CTF might be the inhibition of BACE activity rather than γ-secretase or other factor that are able to use CTF as a substrate, because BACE can also produce toxic fragments of APP even though γ-secretase activity is blocked. The observation that β-secretase knockout (BACE -/-) animals are viable, show normal phenotype and display no gross anatomical or physiological abnormalities [561] suggests that BACE inhibition would be free of both major side effects and βAP generation.

Because $BACE_2$ is highly expressed in heart, kidney, and placenta, suggesting it may be important in highly vacuolarized systemic tissues [562], it would be critical to develop an AD drug that selectively inhibits $BACE_1$ but not $BACE_2$. Even though BACE is effectively inhibited, other enzymes such as caspases still may generate toxic CTFs, such as CTF_{31}. Additionally, molecular size, blood-brain barrier permeability, and activity in the neuron should be also considered. It is also believed that inhibitors that originate from natural products may be more superior in safety and stability than other peptide-based inhibitors.

γ-secretase may be an important enzyme for the processing of APP and other proteins (*eg.* Notch). Notch is involved in the regulation of neuronal differentiation, spermatogenesis, oogenesis, and myogenesis; its membrane-associated C-terminus is cut by a γ-secretase-like activity within the postulated transmembrane domain to release the Notch-intracellular-domain (NICD), which then translocates to the nucleus where it interacts with and activates transcription factors [563]. Therefore, a potential concern about γ-secretase inhibition was that it could inhibit Notch signaling, which could affect the bone marrow. However, the two effects may be dissociated with novel inhibitors [564]. Potential agents should be able to prevent γ-secretase cleavage of APP without affecting the processing of Notch.

In the reduction of βAP peptide formation, one area of research interest is in using antisense RNA to APP to reduce APP expression and subsequent production of βAP peptides. Recent evidence indicates that effective doses of βAP antisense phosphoro-thiolate oligonucleotides can cross the blood-brain barrier and serve to reduce learning and memory deficits in aged SAMP8 mice (a mouse model for AD characterized by the overexpression of APP with age) [565].

β-sheet Breakers, Metal Chelators and βAP-reducing Agents

The secondary structure determines several important properties of peptides that may be relevant to the pathogenesis of neurodegenerative diseases. It has been demonstrated that βAP-induced neurotoxicity is associated with the formation of β-sheet or amyloid fibrils [566]. The ability of βAP to form fibrils is directly correlated with the content of β-sheet structures adopted by the peptide [567]. Transition to the β-sheet conformation proceeds faster at a pH level lower than 6.5 and at higher βAP concentrations [568,569]. Because βAP-fibril disassemblers or β-sheet breakers could inhibit amyloid plaque deposition in the brain, this is one of possible therapeutic approaches for AD. Degradation by endogenous enzymes before reaching the affected region or absorbing into body; poor blood-brain barrier permeability; and putative side effects induced by conformational change of AD-unrelated peptides are some of the many barriers in developing an effective β-sheet breaker.

βAP can bind aluminum, iron, copper and zinc. The levels of these biometals are elevated in the neocortex in AD and particularly concentrated in neuritic plaques; they potentiate and promote βAP aggregation and neurotoxicity [570,571]. Solubilizing βAP from AD brains and inhibition of βAP accumulation by metal chelators may be one possible therapeutic approach for AD treatment [572].

Another direction focuses on the idea of hydrolyzing βAP with neprilysin and IDE to prevent its deposition into amyloid plaques [257]. Other research area focuses therapeutic efforts on preventing the toxic effects of βAP and CTF peptides. One area of interest is in the prevention of βAP-triggered complement proteins in the innate immune system response [573]. Because βAP and antibodies bind to different regions of C1q complement protein to initiate the inflammatory response, an ideal drug would inhibit β-amyloid binding (preventing autotoxicity and thus preventing neurodegeneration) while leaving the brain capable of responding to antibodies via the adaptive immune system [573]. Other research efforts have focused on the development of "decoy peptides" that coaggregate with βAPs and thus serve as antagonists/inhibitors of βAP toxicity [574]. Although these treatment strategies look exciting and promising, much work remains to be done to either prevent or affect a cure for AD.

Antioxidants and Anti-inflammatory Drugs

Accumulating evidence suggests that oxidative damage to neurons plays an important role in the AD pathogenesis. Thus, therapeutic strategies to reduce oxidative injury and increase antioxidant protection might retard and prevent the onset of the disease. βAP-induced neurotoxicity and cognitive impairments can be attenuated by antioxidants and free-radical scavengers, such as vitamin E and coenzyme Q_{10} [575]. Anti-AchEs can also attenuate βAP-induced oxidative damage [576]. Melatonin can also reduce neuronal oxidative damage, and it also has anti-amyloidogenic activities [577].

Increased activity of nNOS and iNOS in normal aging [578] is responsible for the increased NO and reactive nitrogen species in the aged brain; they all play important roles in the aging process [579]. All isoforms of NOS are aberrantly expressed in AD brains; most evidently, iNOS is present in high concentrations in neurons and astrocytes around senile plaques, directly associated with βAP-depositions [580]. βAP-directed elevated levels of iNOS, and βAP-derived Ca^{2+}-induced activation of nNOS has been shown to contribute to

AD pathology [581]. In view of the wide range of isoform-specific NOS inhibitors, they could be of therapeutic importance for the treatment of AD.

NSAIDs, corticosteroids, methotrexate, or other anti-inflammatory drugs might alleviate the neurodegenerative process of AD. Supporting this hypothesis, the occurrence of AD was reported to be significantly less frequent in patients with rheumatoid arthritis than in the rest of the population [582] because of administration of anti-inflammatory agents. The relative risk of developing AD is lower for NSAID users [583]. These drugs protect against AD by suppressing the inflammatory process associated with senile plaque [584]. NSAIDs reduce the numbers of activated microglia in brain [585]. NSAIDs also are known to antagonize some of the effects of interleukin-1, presumably mediated by suppression of prostaglandin synthesis by cyclo-oxygenase. NSAIDs also interfere with NO production; aspirin inhibits iNOS [586].

βAP-vaccines

Active immunization with βAP_{1-42} can reduce βAP levels, inhibit the deposition of amyloid onto existing plaques, and clear established senile plaques that are present in the brain [553]. The logic behind this scheme is that immunization will stimulate the immune system to fight the abnormal pathologies associated with βAP and thereby accelerate removal of the amyloid plaques. The appearance of βAP-containing microglial cells around remaining plaques indicates that antibody-mediated phagocytosis may be important for clearance. Anti-βAP antibodies cross the blood-brain-barrier, enter the CNS, bind to amyloid plaques, activate microglial cells, and induce the clearance of pre-existing amyloid.

Passive immunization with antibodies which recognize βAP_{1-42} injected into the peritoneum work as well as immunization with the βAP_{1-42} peptide. Monoclonal antibody (m266) directed against the central domain of βAP is able to bind and completely sequester plasma βAP [587], but does not bind to βAP deposits in the brain. Peripherally administered m266 can remove βAP deposits from brain by altering the dynamic equilibrium of βAP between brain, CSF, and plasma such that a reduction of plasma βAP can lead to an efflux of brain βAP to the CSF and into the circulation.

Vaccination is also accompanied by a corresponding improvement in cognitive and memory functions [554]. It would represent a novel therapeutic use of the immune response. Passive immunization might be able to circumvent any ineffectiveness of active immunization of elderly people whose immune responses are frequently blunted. However, some patients showed clinical signs consistent with inflammation in the CNS [588]. Antibodies reactive with βAP could potentially induce immune-mediated tissue damage: antibodies could also bind to the βAP sequence within APP and produce an autoimmune response in tissues expressing this protein in brain and in the periphery. However, this possibility is probably limited, because the relevant portion of APP is transmembranous. Few cases with adverse effects aside, vaccination has raised hope for a potentially important new therapeutic approach to the treatment of AD.

Inhibitors of Tau-hyperphosphorylation

The protein *tau* can be phosphorylated by several protein kinases. Among them, cdk-5 and GSK-3β are the major phosphorylators [589]. Hyperphosphorylated *tau* lacks normal *tau*'s activity to bind to microtubules and to stimulate their assembly. Instead, it sequesters normal *tau* and other microtubule-associated proteins and disassembles microtubules. The abnormal hyperphosphorylation of *tau* also promotes its self-assembly into PHFs and NFTs [590].

Synaptic, physiological, and behavioral abnormalities precede βAP plaque deposition in AD, supporting the possibility that plaque may not be the critical pathogenic entity, and soluble βAP levels are the βAP correlates of cognitive dysfunction in AD [591,592]. More authentic than soluble βAP levels, the number of NFTs correlates extremely well with both neuronal loss and degree of dementia. Therefore, inhibiting *tau*-hyperphosphorylation may be a reasonable therapeutical approach in AD. These may include cathepsin inhibitors, lithium, MAP-kinase inhibitors, protein phosphatase-2A, and estrogen replacement therapy.

Estrogen Replacement Therapy

AD is characterized mainly by a progressive loss of memory. Although the severity of this dementia is similar in both sexes, women with AD exhibit a greater difficulty in semantic memory tasks [593]. In men, testosterone-decrease is gradual, and neurons have the ability to convert this hormone into estradiol, which may explain a lesser incidence of cognitive disorders in elderly male patients. Besides its classic function as a sex steroid [594], estrogen modulates transmembrane receptor function [595], affects intracellular signal transduction cascades [596], and is regarded as a molecule that is important in providing neuroprotective and neurotrophic actions in mammal brain tissue [597]. Estrogen also participates in the process of synaptic morphogenesis and function [598]. It was shown that the ascendant cholinergic pathways and the hippocampus, seriously affected in AD, possess nuclear receptors for estrogen [599], which exert a modulatory action on cholinergic transmission [600].

Many clinical studies indicate that estrogen is closely associated with neurodegenerative diseases such as AD [601] and PD [602]. After menopause in women, plasma levels of the two principal estrogens, estradiol and estrone, plummet. It has been reported that the risk of developing AD and cognitive decline is lower in women with a history of estrogen-replacement therapy during the postmenopausal period [593].

Several mechanisms of estrogen have been proposed, including antioxidative action [603], blockade of βAP-induced neurotoxicity [604], and stimulation of α-secretory pathway of APP processing therefore reducing endogenous βAP production; it also prevents βAP aggregation [605]. Estrogen could also regulate the transcription of various genes, including apoptosis-related genes such as the bcl-2 family [606], and could modulate the expression of antiapoptotic protein bcl-xL [607] and NF-κB [608], and interacts with various intracellular signaling pathways such as MAPK. Estrogen modifies inflammation by acting on IL-6, a cytokine that contributes to the formation of senile plaques [609]. The phosphorylation of *tau* by GSK-3β is prevented by protein kinase A, involved in the neuroprotective effects of estrogens [610,611]. Estrogen also reduces plasma levels of ApoE [612], and is known to increase cerebral blood flow [613] and modulate cerebral glucose use [614]. Finally, estrogen interacts with and increases the anti-oxidative activity of glutathione [615] and decrease free-radical generation at the mitochondrial level. Because estrogen shows potent antioxidant effect and crosses the blood-brain barrier easily, it may have good therapeutic value as a free-radical scavenger in neurodegenerative diseases, such as in AD. However, estrogen does not show a beneficial effect on severe AD patients, suggesting that estrogen replacement therapy has a place in the prevention of AD but not in the treatment of severe AD [609].

Dietary Restriction

The lessons learned from the history of other age-related disorders, such as cardiovascular disease and cancer, should be applied to neurogenic conditions. To date, the approach that has the biggest impact in reducing risk of disease during aging is dietary restriction. Overeating is currently the most prominent contributor to age-related cardiovascular disease, stroke, type-2 diabetes, and cancers, and recent findings suggest that high caloric intake may also be a major risk factor for AD. Other approaches may include dietary supplementation with antioxidants such as vitamin E, folic acid, and low doses of anti-inflammatory drugs.

Management of Depression

The empirical evidence supporting various pharmacologic and nonpharmacologic interventions of depression has been sparse [616]. Several treatment options exist, including behavioral treatments focused on the patient, caregiver interventions, pharmacologic treatments, or electroconvulsive therapy (ECT) [392]. Nonpharmacologic interventions appear to be efficacious, particularly for milder depression and in situations in which the caregiver is depressed [617]. The major advantage of nonpharmacologic therapies is their greater safety.

With limited clinical trial data, the treatment of depression in AD remains empirical and consists in starting an antidepressant at a low dose and increasing it slowly. Sufficient dosage and duration of treatment are needed for clinical response in depressed patients without dementia. The depressed elderly may take up to 6 weeks to respond to antidepressant medication, and patients with AD should be expected to take as long.

No single class of antidepressants is clearly more effective than another; however, when compared with selective serotonin reuptake inhibitors (SSRIs), patients with dementia tolerate tricyclic antidepressant (TCAs) less well [618], since they have more anticholinergic properties. Once the decision has been made to initiate antidepressant therapy, initiation of an SSRI is believed to be the best first step, given the side-effect profile of these agents. The monoamine oxidase inhibitors (MAOI) also appear to have efficacy as treatments for major depression in dementia. Given consistent pathologic evidence suggesting that depression in dementia is associated with loss of norepinephrine-producing neurons of the locus ceruleus [393,394], preferred second-line agents are all newer antidepressants that produce norepinephrine reuptake inhibition, such as venlafaxine or mirtazapine. Antidepressant treatment may be associated with stabilization of activity of daily living decline [619].

For patients with severe, refractory depression, ECT might be considered, especially if there is risk of the patient doing harm to him or herself or to others. Interestingly, ECT has also been associated with improvement in cognition when used to treat depression in patients with degenerative dementia [620].

In addition to the side effects specific to individual agents or classes of agents, patients with dementia are vulnerable to the development of atypical or nonspecific medication side effects. These include falls, rashes, reduced oral intake, weight loss, constipation, increased confusion, or delirium. In addition, side effects seen rarely in other settings, such as bradycardia, hyponatremia, and syndrome of inappropriate antidiuretic hormone (SIADH).

For patients who have psychotic symptoms or agitation along with depression, an atypical antipsychotic (such as risperidone or olanzapine) might be indicated [621,622]. An anticonvulsant (such as carbamazepine) might be considered as additional therapy to an antidepressant if there is moderate to severe agitation [623].

Treatment of Psychiatric Symptoms

Reducing behavioral disturbances in patients with AD is an important treatment goal. Neuropsychiatric disturbances are distressing to the patient who experiences fear, sadness, or anxiety; they are a source of marked distress for the caregiver [624]; and they may precipitate institutionalization [625]. Thus, the treatment of behavioral disturbances in AD may ameliorate a patient's emotional distress and have secondary beneficial effects on the caregiver and on the opportunity for the patient to remain at home. Cholinergic agents may play an important role in these treatment goals. Whenever present, psychiatric symptoms should be treated in demented patients [626].

Classical (traditional) neuroleptics should not be considered anymore as first-choice drugs [627], since as strong D_2-receptor blockers they are associated with extrapyramidal symptoms, with an estimated incidence of tardive dyskinesia of ~30% per year. Also, they cause postural hypotension, a major cause of fall in the elderly; they have anticholinergic side effects and therefore cause worsening of dementia. This risk can be decreased by using atypical or novel antipsychotics, which represent the reference drugs for treating BPSD. Based on clinical experience and side-effect profiles, RISP is considered to be a drug of choice [628]. Its main characteristics are the efficacy at low doses (0.5-1.0 mg daily) and the absence of significant anticholinergic effects.

Another category of drugs which have proven useful in the treatment of BPSD are SSRIs. They have definitely replaced TCAs in this indication, clearly associated with anticholinergic effects.

The advent of AchEIs for the treatment of cognitive impairment in AD uncovered the potential use of these agents for the treatment of BPSD. AchEIs can have positive and independent effects on both cognitive and behavioral domains. Based on their cognitive-enhancing as well as psychotropic effects, AchEIs may play an important role in controlling BPSD [629]. The amelioration of apathy and reduction of visual hallucinations are the most reproducible effects on neuropsychiatric symptoms in AD, but other neuropsychiatric symptoms also respond. There may be variations among AchEIs in their psychotropic properties. The beneficial effects of AchEIs on emotion and behavior are most likely mediated through cholinergic influences on limbic and paralimbic brain structures. Patients who exhibit substantial cognitive improvement usually have a concomitant behavioral response, but behavioral and cognitive responses may be dissociated.

With the Last Word in Survival: Targeting Calcium

Effective clinical implementation of emerging AD treatments requires parallel development of early diagnostic/monitoring technology. Our interdisciplinary translational research team is developing novel AD diagnostic technology to meet this urgent need. Our approach capitalizes on our recent identification of βAP-induced biochemical changes both in the CNS and in the periphery, and pharmacological action of antipsychotics in AD with respect to Ca^{2+} homeostasis.

As presented in this study, cell death is lurking under Ca^{2+} overload. Because of its fundamental roles in regulating synaptic transmission, plasticity, and cell survival, Ca^{2+} homeostasis is undoubtedly altered, at some point in the pathogenesis of AD. The research described here suggests that altered Ca^{2+}-regulation contributes greatly to the pathogenic effects seen in neuropsychiatric conditions. Therefore, targeting Ca^{2+} for potential therapies is therefore worth consideration.

Proteolysis of APP affects cytosolic Ca^{2+}-dynamics by either of the two APP processing pathways, one involving an increase in sAPPα, and the other an elevation in βAP. Therefore, the regulation of formation of βAP is clearly of potential clinical relevance. As such, the apparent role of Ca^{2+} as a key regulator in this process is of importance. Targeting Ca^{2+} could prove advantageous over attempts to regulate other factors (*eg*. PKC activity) because, unlike other targets (*eg*. PKC), cytoplasmic Ca^{2+}-levels do not affect APP transcription [630]. The possible physiological relevance of the Ca^{2+}-regulation of APP processing is also supported by this study.

It is proposed in this study that relatively stable levels (*ie*. slight oscillations) of Ca^{2+} leads to normal or non-amyloidogenic APP processing (sAPPα-Ca^{2+} feed back, Figure 9), whereas pathological increase or decrease in Ca^{2+} triggers amyloidogenic APP metabolism (βAP-Ca^{2+} feed forward cascade, Figure 8), as discussed in this study. This provides a fine tuning mechanism in APP regulation. Both sAPPα and βAP have been demonstrated to modulate cytoplasmic Ca^{2+} concentration [58,173], raising the possibility that these compounds may regulate their own formation. Also, the biphasic regulation in AD fibroblasts presented in this study is well established. βAP results in aberrant regulation of $[Ca^{2+}]_i$ such that there is an enhancement of acute activation (*ie*. elevated levels), followed by a sustained depression of Ca^{2+}-concentration and signaling, when cells are exposed to βAP and other insults [631]. Subtle alterations in this metabolic balance over a long period of time are likely to influence cellular homeostasis, be it proteomic, ionic, genetic, or other. Therefore, Ca^{2+}-stabilizing interventions should be considered the foundation of treating dementing neurodegenerative conditions.

Because abnormalities in Ca^{2+} homeostasis are also present in psychosis [632,633,634] and depression [635,636], both associated with AD, this study points out that these psychiatric disturbances manifest in dementia because of the Ca^{2+}-imbalance present in the primary disease. It is not difficult to see how disruption of a single interaction (*ie.* Ca^{2+} dysregulation) can have multiple flow-on effects to other cellular processes. Based on the protective role of antipsychotics regarding Ca^{2+}-imbalance, antipsychotics may constitute future treatment strategies aimed at alleviating BPSD and the cognitive symptomatology of AD. The use of antipsychotic drugs to treat dementia and co-morbidities associated with AD is apparently justified at the molecular level with respect to Ca^{2+}.

The involvement of Ca^{2+} has been implicated in neurometabolic disorders other than AD and its corollary symptoms, such as in Parkinson's [637], Huntington's [638], amyotropic lateral sclerosis [639], SCH [632-634], muscular dystrophy [640], *etc.*, suggesting a broad-spectral role of Ca^{2+} in neuropsychiatric disorders. A great number of articles were devoted to Ca^{2+} in brain physiology and pathology, but literally no paper has ever studied the role for Ca^{2+} at the holistic level. Broadly in line with the above findings, this study proposes that disturbing the central secondary messenger, Ca^{2+}, plays pivotal roles in various, if not all neurometabolic disorders. Impaired Ca^{2+} homeostasis has different functional consequences that mainly involve mechanisms to compensate Ca^{2+} imbalance. Environmental and genetic factors may determine that these functional compensatory mechanisms involve which of the many tertiary and quaternary messenger systems, yielding differing brain pathology, *ie.* altered Ca^{2+} leads to β-amyloid deposition in Alzheimer's, α-synucleinopathy in Parkinson's, accumulation of Lewy bodies, *etc.*

The French philosopher, Rene Descartes, who crystallized the idea that mind and body are radically distinct and we are split into a brain-body dichotomy in which the brain occupies a hierarchically privileged place, has been dead for more than three and a half centuries. His ideas have been universally rejected by serious modern thinkers by democratizing the relationship between brain and body and positing a powerful interdependence, yet one would too rarely gain that impression in the offices of modern physicians. Would that this were just a dry academic matter; unfortunately it impacts many medical professionals in search of care.

Crosstalk between neuronal and systemic organs is inevitable, requiring a molecular unity of biochemical pathways. This holistic approach is signified in various neurometabolic disorders, in which primary changes in the brain are reflected in extraneural organs, such as in fibroblasts and lymphocytes. Not just cells of brain and body are interconnected, but brain cells also form an information network in the CNS. Conceivably, the break in any element of this harmonized system may result in a systemic malfunctioning. We therefore propose that brain pathology is reflected in not only the periphery, but various neuropsychiatric disorders might also share similarities with one another. This is supported by the existence of overlapping disorders (*eg.* AD with Lewy-bodies) and frequent co-morbid conditions (*eg.* depression in most neuropsychiatric maladies), as described earlier.

The effort to understand the molecular basis of degenerative neurological disorders has generated an enormous amount of data, however the biochemical studies failed to identify a narrow set of specific neurochemical abnormalities firmly linked to neurometabolic disorders. Instead, these studies identified possible changes in a seemingly perplexing array of molecules involved in almost all aspects of neuronal intracellular biochemistry. Based on our ongoing studies, the present work hypothesizes that neuropsychiatric diseases can be produced by alterations in various groups of intracellular molecules as long as these alterations lead to abnormal functioning of some central intracellular regulatory pathways, disregulation of which causes the affected neurons to express functional deficits underlying neurodegenerative symptomatology. It may also be profitable to look for common threads in different hypotheses of neurometabolic conditions. If patients with degenerative disorders of their brain display various neurochemical abnormalities, then the commonalities among these alterations should be able to point to the intracellular signaling pathway involved in generating neuropathology.

When the above discussion is taken together with the search for a commonality among the available biochemical data, one would have to assume as a conclusion that one of the unifying elements of the great majority of molecular changes reported in neurometabolic diseases is their association with potential modifications in Ca^{2+} signaling and/or homeostasis. Each of us is developed from a single cell: a fertilized oocyte. It is reasonable, therefore, to propose that our organism might be interconnected by the pathology of a single cell; this pathology might be equal to disturbed Ca^{2+} regulation.

Ca^{2+} is a ubiquitous molecule, with Ca^{2+} signaling being involved in most cellular events. The regulation of Ca^{2+} dynamics is very complex, involving proteins localized in essentially all major subcellular structures. The sheer diversity of molecular findings in neurodegenerative disorders predicts the ubiquitous nature of Ca^{2+} as the unifying factor.

The central role of Ca^{2+} in various diseases of the man is signified not only in neuropsychiatric disorders, but also in many medical conditions. Ca^{2+}-channel blockers, for example, in arrhythmias, cardiac failure and hypertension, have proved to efficiently restore Ca^{2+} homeostasis. Moreover, Ca^{2+}-channel blockers were reported to limit cellular degeneration [641]. This study, therefore, proposes a new approach to Ca^{2+} imbalance at the holistic level (Figure 12).

Therefore, stabilizing calcium levels with Ca^{2+}-channel blockers, calmodulin antagonists, *etc.* may slow or inhibit the progression of and may reverse several neurometabolic disorders, and conditions afflicting other organs, too. Because Ca^{2+} is a universal messenger in the body, alteration in its function seen in the central nervous system is ubiquitously reflected in the periphery, such as in AD. If further studies finally confirm this hypothesis in other disorders, too, and will gain disease-specificity, readily available tissues such as fibroblasts and lymphocytes would then be used for the diagnosis of a number of neuropsychiatric conditions. This may also be helpful in monitoring the efficiency of CNS treatment (*eg.* Ca^{2+}-antagonists) in the periphery.

A better understanding of the molecular mechanisms underlying the perturbations in Ca^{2+} regulation will likely lead to novel approaches to preventing neuronal dysfunction and death in individuals at risk for neurogenic diseases. It is important to understand how Ca^{2+}-regulating systems are altered in neurons early in the disease process. Such knowledge should then be used in studies that test the efficacy of drugs and probably dietary factors that stabilize neuronal Ca^{2+} homeostasis.

Obvious directions for future research include identification of the specific alterations that lead to altered Ca^{2+} regulation, and recognition of environmental factors that increase risk of diseases and how these risk factors might impact on cellular Ca^{2+} homeostasis. It will also be important to establish whether alterations outside the immediate environment of neurons in the brain contribute to disruption of neuronal Ca^{2+} homeostasis. Such indirect mechanisms might include vascular alterations [642] and disturbances in the immune system [643]; these alterations may disrupt cellular Ca^{2+} homeostasis and sensitize neurons to synaptic dysfunction and degeneration. In addition, drugs that buffer Ca^{2+} or affect downstream targets of Ca^{2+} should also be evaluated. These studies should focus on altered Ca^{2+} homeostasis and signaling both in the CNS and in the periphery, and should also be undertaken to prove this hypothesis in metabolic disorders affecting organs other than the brain.

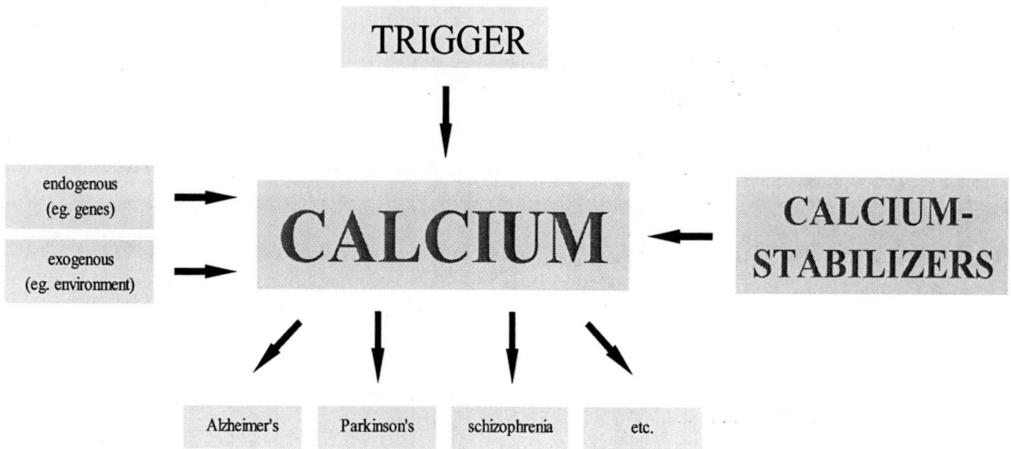

Figure 12: Proposed central role of Ca^{2+} in (neurometabolic) diseases

Afterword

AD is of great interest scientifically and important clinically, but the neurobiological process responsible for this illness is poorly understood. As the average age of the population increases, Alzheimer's dementia and other age-related neurodegenerative diseases are becoming more common.

Perhaps the most frequently posed question in current research in AD is whether βAP deposition is a cause or an effect of the disease. Based on the novel Ca^{2+} hypothesis proposed in this study, it appears increasingly likely that that answer is both. When an environmental factor, head injury, or a mutation in genes ultimately triggers Ca^{2+}-imbalance, this ionic dysregulation sets in train a series of inflammatory reactions, neuritic sprouting and other tissue repair responses which includes upregulation of APP expression; this can lead to increased βAP production, which furthers Ca^{2+} disruption. Thus, a biochemical cascade is generated, leading to further amyloid formation and further disturbances in APP metabolism.

The multiplicity of macromolecular and cellular alterations occurring in neurons and their processes, astrocytes, and microglia following the deposition of βAP and/or Ca^{2+}-imbalance suggests that βAP or molecules that become associated with it (*eg*. Ca^{2+}) may trigger a variety of complex molecular alterations, presumably followed by secondary feedback, or feed forward reactions that further aggravate the initial insult. In line with this, the present study has made major contributions on these novel findings, proposing that:

1. Fibroblasts and lymphocytes show biochemical differences in AD by exhibiting abnormal Ca^{2+} homeostasis.
2. Biphasic effect of βAP may shed new light on understanding cellular alterations in dementia and differences between tissue-specific pathologies in the same condition.
3. Biochemical alterations in the periphery may support the diagnosis of AD, may help monitoring the progression of the disease and responsiveness to treatments.
4. HAL and RISP are potent in antagonizing βAP-induced Ca^{2+}-imbalance.
5. These antipsychotics are safe with respect to APP metabolism, and acute HAL (ie. pulse-therapy) might be beneficial on APP processing in AD.
6. A "protective cycle" of antipsychotics is proposed in AD.
7. HAL and RISP are involved in interfering with the expression of a large array of genes involved in Ca^{2+}-signaling, survival and cellular plasticity.
8. Antipsychotics with different pharmacological characteristics share targets at the molecular level, including Ca^{2+}-homeostasis.
9. Ca^{2+} is central in the pathogenesis of AD, and agents targeting Ca^{2+} may slow or inhibit the progression of and might even reverse AD and its corollary symptoms.
10. Ca^{2+} might be a unifying molecule in most neurodegenerative disorders and also in metabolic diseases afflicting extraneuronal organs.

The fundamentals of neurometabolic disorders that have been reviewed here points to new therapeutic strategies aimed at one or more crucial steps in their molecular progression. Nevertheless, these pharmacological objectives will not be easy to reach. But the strength of accumulating data emerging from laboratories worldwide suggests that Ca^{2+} is a rational therapeutic target which offers the best hope of slowing or arresting the progression of the fundamental disease process early in its tragic course. Notwithstanding the elegant and fascinating cutting-edge science presented in this study, education of the general public concerning the steps they can take throughout their adult life to reduce risk of AD is likely to have a major impact on this devastating disease.

AD is characterised by memory loss associated with neuronal degeneration in the cortex and hippocampus, generation of amyloid plaques in many areas of the grey matter, death of neurons with formation of abnormal cytoskeletal structures, profound cholinergic deficiency, and striking Ca^{2+}-imbalance. However, our understanding of AD and other disorders of the central nervous system will only become clearer once we are able to study and understand the process at the molecular level.

Acknowledgements

First of all, the active participation of the *volunteers* in the ongoing study is highly appreciated.

I thank *Dr. Gábor Laskay* for always giving me a helping hand in understanding the pickiest details of fluorimetric studies. He taught me everything I will ever know about spectrofluorimetry. Sharing his teaching and research experiences with me has been a springboard for my scientific career.

I am grateful to *Prof. Dr. Lajos Kemény, Dr. Anna Kenderessy-Szabó* and their colleagues at the Department of Dermatology, University of Szeged, Hungary, for providing me with the skin samples and fibroblast cultures.

I would also like to thank my colleagues at the Department of Experimental Surgery, University of Szeged, Hungary, in granting me unlimited access to the Hitachi F-2000 spectrofluorimeter.

I acknowledge the assistance of each and every co-workers at the Alzheimer's Disease Research Center, University of Szeged, including *Prof. Dr. Péter Kása, Dr. Magdolna Pákáski, Marietta Hugyecz* and colleagues. The assistance and friendship of *László Kálmán* of the same institute is highly valued.

I am indebted to *Dr. László G. Puskás* and his colleagues of the Laboratory of Functional Genomics, Biological Research Center of the Hungarian Academy of Sciences, for assisting in the cDNA-micro-chips.

I am gratified that my clinical directors, *Prof. Dr. György Lázár* of the Department of Surgery and *Dr. Gábor Bogáts* of the Division of Cardiac Surgery, University of Szeged, Hungary, permitted me to conduct this research parallel with my daily routine in cardiac surgical patient care.

I am obliged for the co-operation of the staff of the Department of Psychiatry, University of Szeged, Hungary. I also appreciate *Dr. Anna Juhász* for her assistance in taking care of the biological samples and cultures. Predominantly, I owe gratitude with deepest respect to *Prof. Dr. Zoltán Janka* and *Dr. János Kálmán* for their scientific partnership, helping hand, encouragement, incessantly magnificent aid, lift, moral support, and friendship.

Above all, words cannot convey how thankful I am to *Prof. Dr. Botond Penke* of the Department of Medical Chemistry, University of Szeged, Hungary, for his contribution to this work in endlessly giving perpetual advice, hand, comfort, care, encouragement, and for being the most sincere counterpart & friend of mine I have ever worked with.

My admiration to my family and loved ones for closing ranks behind me all my life, and to my brother, *Dr. Miklós Palotás*, of the Department of Anesthesiology and Intensive Therapy, University of Szeged, Hungary, for collaborating with me throughout this study is self-explanatory.

I would also like to express my gratefulness to my fellow scientific colleagues and friends, *Dr. Yoo-Hun Suh* [644] of Seoul National University, South Korea, *Dr. Mark P. Mattson* [645] at the National Institute on Aging, USA, *Dr. Jefferey L. Cummings* [646] of UCLA School of Medicine, USA, and *Dr. Alexei Koudinov* [647] of the Russian Academy of Sciences, Russia, for permitting me to reproduce some of their theoretical work in this study, and for giving me future directions.

Finally, I highly value the assistance of all of my work-mates whom are not listed here by name.

May God bless all of you!

References

[1] Alzheimer A. Über eine eigenartige Erkrankung der Hirnrinde. *Allgemeine Zeitschrift für Psychiatrie und Psychisch-gerichtliche Medizin* 1907; 64: 146-148.

[2] Brookmeyer R, Gray S, Kawas C. Projections of Alzheimer's disease in the United States and the public health impact of delaying disease onset. *Am. J. Public Health* 1998; 88: 1337-1342.

[3] American Psychiatric Association, *Diagnostic and Statistical Manual of Mental Disorders*. 4th Ed. Washington, D.C., 1994.

[4] Small GW, Rabins PV, Barry PP, Buckholtz NS, DeKosky ST, Ferris SH, Finkel SI, Gwyther LP, Khachaturian ZS, Lebowitz BD, McRae TD, Morris JC, Oakley F, Schneider LS, Streim JE, Sunderland T, Teri LA, Tune LE. Diagnosis and treatment of Alzheimer's disease and related disorders. Consensus statement of the American Association for Geriatric Psychiatry, the Alzheimer's Association, and the American Geriatrics Society. *JAMA* 1997; 278: 1363-1371.

[5] Rabins PV, Lyketsos CG, Steele CD. *Practical Dementia Care*. New York: Oxford University Press.

[6] Selkoe DJ, Abraham CR, Podlisny MB, Duffy LK. Isolation of low-molecular-weight proteins from amyloid plaque fibers in Alzheimer's disease. *J. Neurochem*. 1986; 46: 1820-1834.

[7] Masters CL, Simms G, Weinman NA, Multhaup G, McDonald BL, Beyreuther K. Amyloid plaque core protein in Alzheimer disease and Down syndrome. *Proc. Natl. Acad. Sci. USA* 1985; 82: 4245-4249.

[8] Glenner GG, Wong CW. Alzheimer's disease: initial report of the purification and characterization of a novel cerebrovascular amyloid protein. *Biochem. Biophys. Res. Commun*. 1984; 120: 885-890.

[9] Perry EK, Tomlinson BE, Blessed G, Bergmann K, Gibson PH, Perry RH. Correlation of cholinergic abnormalities with senile plaques and mental test scores in senile dementia. *British Med. J.* 1978; 2: 1457-1459.

[10] Kidd M. Paired helical filaments in electron microscopy of Alzheimer's disease. *Nature* 1963; 197: 192-193.

[11] Terry RD. The fine structure of neurofibrillary tangles in Alzheimer's disease. J. Neuropathol. *Exp. Neurol.* 1963; 22: 629-642.

[12] Trojanowski JQ, Lee VM. Paired helical filament tau in Alzheimer's disease. The kinase connection. *Am. J. Pathol.* 1994; 144: 449-453.

[13] Perry EK, Court JA, Lloyd S, Johnson M, Griffiths MH, Spurden D, Piggott MA, Turner J, Perry RH. β-amyloidosis in normal aging and transmitter signalling in human temporal lobe. *Ann. New York Acad. Sci.* 1996; 777: 388-392.

[14] Storga D, Vrecko K, Birkmayer JG, Reibnegger JG. Monoaminergic neurotransmitters, their precursors and metabolites in brains of Alzhiemer patients. *Neurosci. Lett.* 1996; 203: 29-32.

[15] Mesulam MM. Structure and function of cholinergic pathways in the cerebral cortex, limbic system, basal ganglia, and thalamus of the human brain. In: Bloom FE, Kupfer DJ (Eds.) *Psychopharmacology: The Fourth Generation of Progress*. Raven Press, NY. 1995; 135-146.

[16] Xiang Z, Huguenard JR, Prince DA. Cholinergic switching within neocortical inhibitory networks. *Science* 1998; 281: 985-988.

[17] Guela C. Abnormalities of neural circuitry in Alzheimer's disease: hippocampus and cortical cholinergic innervation. *Neurology* 1998; 51: S18-S29.

[18] Richelson E. Cholinergic transduction. In: Bloom FE, Kupfer DJ (Eds.) *Psychopharmacology: The Fourth Generation of Progress*. Raven Press, NY. 1995; 125-134.

[19] Nicholls DG. *Proteins, transmitters and synapses*. Blackwell Science, London. 1994.

[20] Flynn DD, Ferrari-DiLeo G, Mash DC, Levery AL. Differential regulation of modular subtypes of muscarinic receptors in Alzheimer's disease. *J. Neurochem.* 1995; 64: 1888-1891.

[21] Schliebs R, Robner S. Distribution of muscarinic acetylcholine receptors in the CNS. In: Stone TW (Ed.) *CNS Neurotransmitters and Neuromodulators: Acetylcholine*. CRC Press, Boca Raton, FL. 1995; 67-83.

[22] Younkin SG, Goodridge B, Katz J, Lockett G, Nafziger D, Usiak MF, Younkin LH. *Molecular forms of acetylcholinesterase in Alzheimer's disease. Fed. Proc.* 1986; 45: 2982-2988.

[23] Cummings JL, Vinters HV, Cole GM, Khachaturian ZS. Alzheimer's disease: etiologies, pathophysiology, cognitive reserve, and treatment opportunities. *Neurology* 1998; 51: S2-S17.

[24] Whitehouse PJ, Price DL, Struble RG, Clark AW, Coyle JT. Alzheimer's disease and senile dementia: loss of neurons in the basal forebrain. *Science* 1982; 215: 1237-1239.

[25] Rasool CG, Svendsen CN, Selkoe DJ. Neurofibrillary degeneration of cholinergic and noncholinergic neurons of the basal forebrain in Alzheimer's disease. *Ann. Neurol.* 1986; 20: 482-488.

[26] Davies P. Studies on the neurochemistry of central cholinergic systems in Alzheimer's disease. In: Katzman R, Terry RD, Bick KL (Eds.) *Alzheimer's Disease: Senile Dementia and Related Disorders*. Raven Press, New York. 1978; 453-459.

[27] Guela C, Mesulam MM. Cholinergic systems and related neuropathological predilection patterns in Alzheimer disease. In: Terry RD, Katzman R, Bick KL (Eds.) *Alzheimer Disease*. Raven Press, New York. 1994; 263-291.

[28] Rodriguez-Puertas R, Pascual J, Vilaro T, Pazos A. Autoradiographic distribution of M_1, M_2, M_3, and M_4 muscarinic receptor subtypes in Alzheimer's disease. *Synapse* 1997; 26: 341-350.

[29] Perry EK, Perry RH, Smith CJ, Purohit D, Bonham J, Dick DJ, Candy JM, Edwardson JA, Fairbairn A. Cholinergic receptors in cognitive disorders. *Can. J. Neurol. Sci.* 1986; 13: 521-527.

[30] Fishman EB, Siek GC, MacCallum RD, Bird ED, Volicer L, Marquis JK. Distribution of the molecular forms of acetylcholinesterase in human brain: alterations in dementia of the Alzheimer type. *Ann. Neurol.* 1986; 19: 246-252.

[31] Poirier J, Delisle MC, Quirion R, Aubert I, Farlow M, Lahiri D, Hui S, Bertrand P, Nalbantoglu J, Gilfix BM, Gauthier S. Apolipoprotein E4 allele as a predictor of cholinergic deficits and treatment outcome in Alzheimer's disease. *Proc. Natl. Acad. Sci. USA* 1995; 92: 12260-12264.

[32] Allen SJ, MacGowan SH, Tyler S, Wilcock GK, Robertson AGS, Holden PH, Smith SKF, Dawbarn D. Reduced cholinergic function in normal and Alzheimer's disease brain is associated with apolipoprotein E4 genotype. *Neurosci. Lett.* 1997; 239: 33-36.

[33] Estus S, Golde TE, Kunishita T, Blades D, Lowery D, Eisen M, Usiak M, Qu XM, Tabira T, Greenberg BD. Potentially amyloidogenic, carboxyl-terminal derivatives of the amyloid protein precursor. *Science* (Wash DC) 1992; 255: 726-728.

[34] Golde TE, Estus S, Younkin LH, Selkoe DJ, and Younkin SG. Processing of the amyloid protein precursor to potentially amyloidogenic derivatives. *Science* (Wash DC) 1992; 255: 728-730.

[35] Seubert P, Vigo-Pelfrey C, Esch F, Lee M, Dovey H, Davis D, Sinha S, Schlossmacher M, Whaley J, Swindlehurst C, McCormack R, Wolfert R, Selkoe D, Lieberberg I, Schenk D. Isolation and quantification of soluble Alzheimer's β-peptide from biological fluids. *Nature* 1992; 359: 325-327.

[36] Schulz PE. Beta-peptides enhance the magnitude and probability of long-term potentiation. *Soc. Neurosci. Abstr.* 1996; 22: 2111.

[37] Wu J, Anwyl R, Rowan MJ. β-amyloid selectively augments NMDA receptor-mediated synaptic transmission in rat hippocampus. *NeuroReport* 1995; 6: 2409-2413.

[38] Huber G, Bailly Y, Martin JR, Mariani J, Brugg B. Synaptic β-amyloid precursor proteins increase with learning capacity in rats. *Neuroscience* 1997; 80: 313-320.

[39] Kamenetz FR, Tomita T, Borchelt DR, Sisodia SS, Iwatsubo T, Malinow R. Activity dependent secretion of β-amyloid: roles of β-amyloid in synaptic transmission. *Soc. Neurosci. Abstr.* 2000; 26: 491.

[40] Heese K, Nagai Y, Sawada T. Identification of a new synaptic vesicle protein 2B mRNA transcript which is up-regulated in neurons by amyloid-β peptide fragment (1-42). *Biochem Biophys. Res. Commun.* 2001; 289: 924-928.

[41] Pike CJ, Burdick D, Walencewicz AJ, Glabe CG, Cotman CW. Neurodegeneration induced by β-amyloid peptides *in vitro*: The role of peptide assembly state. *J. Neurosci.* 1993; 13: 1676-1687.

[42] Dumery L, Bourdel F, Soussan Y, Fialkowsky A, Viale S, Nicolas P, Reboud-Ravaux M. β-Amyloid protein aggregation: its implication in the physiopathology of Alzheimer's disease. *Pathol. Biol.* (Paris) 2001; 49: 72-85.

[43] Mehta PD, Pirttila T, Patrick BA, Barshatzky M, Mehta SP. Amyloid-β protein 1-40 and 1-42 levels in matched cerebrospinal fluid and plasma from patients with Alzheimer disease. *Neurosci. Lett.* 2001; 304: 102-106.

[44] Younkin SG. Evidence that Aβ42 is the real culprit in Alzheimer's disease. *Ann. Neurol.* 1995; 37: 287-288.

[45] Abraham CR, Selkoe DJ, Potter H. Immunochemical identification of the serine protease inhibitor, α_1-antichymotrypsin in the brain amyloid deposits of Alzheimer's disease. *Cell* 1988; 52: 487-501.

[46] Joachim CL, Mori H, Selkoe DJ. Amyloid β-protein deposition in tissues other than brain in Alzheimer's disease. *Nature* (Lond) 1989; 341: 226-230.

[47] Gearing M, Wilson RW, Unger ER, Shelton ER, Chan HW, Masters CL, Beyreuther K, Mirra SS. Amyloid precursor protein (APP) in the striatum in Alzheimer's disease: an immunohistochemical study. *J Neuropathol. Exp. Neurol.* 1993; 52: 22-30.

[48] Einstein G, Buranosky R, Crain BJ. Dendritic pathology of granule cells in Alzheimer's disease is unrelated to neuritic plaques. *J. Neurosci.* 1994; 14: 5077-5088.

[49] Cochran E, Bacci B, Chen Y, Patton A, Gambetti P, Autilio-Gambetti L. Amyloid precursor protein and ubiquitin immunoreactivity in dystrophic axons is not unique to Alzheimer's disease. *Am. J. Pathol.* 1991; 139: 485-489.

[50] Wilson CA, Doms RW, Lee VM. Intracellular APP processing and Aβ production in Alzheimer disease. *J. Neuropathol. Exp. Neurol.* 1999; 58: 787-794.

[51] Klein WL, Krafft GA, Finch CE. Targeting small Aβ oligomers: the solution to an Alzheimer's disease conundrum? *Trends. Neurosci.* 2001; 24: 219-224.

[52] Wilson D, Doms R, Lee V. Intracellular APP processing and Aβ production in Alzheimer's disease. *J. Neuropathol. Exp. Neurol.* 1999; 58: 787-794.

[53] Emre M, Geula C, Ransil BJ, Mesulam MM. The acute neurotoxicity and effects upon cholinergic axons of intracerebrally injected β amyloid in the rat brain. *Neurobiol. Aging* 1992; 13: 553-560.

[54] Kowall NW, McKee AC, Yankner BA, Beal MF. *In vivo* neurotoxicity of β-amyloid [β(1-40)] and the [β(25-35)] fragment. *Neurobiol. Aging* 1992; 13: 537-542.

[55] Pike CJ, Burdick D, Walencewicz AJ, Glabe CG, Cotman CW. Neurodegeneration induced by β-amyloid peptides *in vitro*: the role of peptide assembly state. *J. Neurosci.* 1993; 13: 1676-1687.

[56] Glenner GG, Wong CW. Alzheimer's disease and Down's syndrome: sharing of a unique cerebrovascular amyloid fibril protein. *Biochem. Biophys. Res. Commun.* 1984; 122: 1131-1135.

[57] Hardy JA, Higgins GA. Alzheimer's disease: the amyloid cascade hypothesis. *Science* (Wash DC) 1992; 256: 184-185.

[58] Mattson MP, Cheng D, Davis D, Bryant K, Lieberburg I, Rydel RE. β-amyloid peptides destabilize calcium homeostasis and render human cortical neurons vulnerable to excitotoxicity. *J. Neurosci.* 1992; 12: 376-389.

[59] Mattson MP, Barger SW, Cheng B, Lieberburg I, Smith-Swintosky VL, Rydel RE. β-Amyloid precursor protein metabolites and loss of neuronal Ca^{2+} homeostasis in Alzheimer's disease. *Trends Neurosci.* 1993; 16: 409-414.

[60] Ueda K, Shinohara S, Yagami T, Asakura K, Kawasaki K. Amyloid β protein potentiates Ca^{2+} influx through L-type voltage-sensitive Ca^{2+} channels: a possible involvement of free radicals. *J. Neurochem.* 1997; 68: 265-271.

[61] Mark RJ, Hensley K, Butterfield DA, Mattson MP. Amyloid β-peptide impairs ion-motive ATPase activities: evidence for a role in loss of neuronal Ca^{2+} homeostasis and cell death. *J. Neurosci.* 1995; 15: 6239-6249.

[62] Wu A, Derrico CA, Hatem L, Colvin RA. Alzheimer's amyloid-β peptide inhibits sodium/calcium exchange measured in rat and human brain plasma membrane vesicles. *Neuroscience* 1997; 80: 675-684.

[63] Good TA, Smith DO, Murphy RM. β-amyloid peptide blocks the fast-inactivating K^+ current in rat hippocampal neurons. *Biophys. J.* 1996; 70: 296-304.

[64] Arispe N, Pollard HB, Rojas E. β-Amyloid Ca^{2+}-channel hypothesis for neuronal death in Alzheimer disease. *Mol. Cell Biochem.* 1994; 140: 119-125.

[65] Harkány T, Ábrahám I, Timmerman W, Laskay G, Tóth B, Sasvári M, Kónya C, Sebens JB, Korf J, Nyakas C, Zarandi M, Soos K, Penke B, Luiten PG. β-Amyloid neurotoxicity is mediated by a glutamate-triggered excitotoxic cascade in rat nucleus basalis. *Eur. J. Neurosci.* 2000; 12: 2735-2745.

[66] Hensley K, Carney JM, Mattson MP, Aksenova M, Harris M, Wu JF, Floyd RA, Butterfield DA. A model for β-amyloid aggregation and neurotoxicity based on free radical generation by the peptide: relevance to Alzheimer's disease. *Proc. Natl. Acad. Sci. USA* 1994; 91: 3270-3274.

[67] Butterfield DA, Hensley K, Harris M, Mattson MP, Carney J. β-amyloid peptide free radical fragments initiate synaptosomal lipoperoxidation in a sequence-specific fashion: implications to Alzheimer's disease. *Biochem. Biophys. Res. Commun.* 1994; 200: 710-715.

[68] Lynch T, Cherny RA, Bush AI. Oxidative processes in Alzheimer's disease: the role of Aβ-metal interactions. *Exp. Gerontol.* 2000; 35: 445-451.

[69] Mark RJ, Lovell MA, Markesbery WR, Uchida K, Mattson MP. A role for 4-hydroxynonenal in disruption of ion homeostasis and neuronal death induced by amyloid β-peptide. *J. Neurochem.* 1997; 68: 255-264.

[70] Mark RJ, Pang Z, Geddes JW, Mattson MP. Amyloid β-peptide impairs glucose uptake in hippocampal and cortical neurons: involvement of membrane lipid peroxidation. J. Neurosci. 1997; 17: 1046-1054.

[71] Blanc EM, Keller JN, Fernandez S, Mattson MP. 4-hydroxynonenal, a lipid peroxidation product, inhibits glutamate transport in astrocytes. *Glia* 1998; 22: 149-160.

[72] Mattson MP, Lovell MA, Furukawa K, Markesbery WRK. Neurotrophic factors attenuate glutamate-induced accumulation of peroxides, elevation of $[Ca^{2+}]_i$ and neurotoxicity, and increase antioxidant enzyme activities in hippocampal neurons. *J. Neurochem.* 1995; 65: 1740-1751.

[73] Behl C, Davis JB, Lesley R, Schubert D. Hydrogen peroxide mediates amyloid protein toxicity. *Cell* 1994; 77: 817-827.

[74] Monji A, Utsumi H, Ueda T, Imoto T, Yoshida I, Hashioka S, Tashiro K, Tashiro N. The relationship between the aggregational state of the amyloid-β peptides and free radical generation by the peptides. *J. Neurochem.* 2001; 77: 1425-1432.

[75] Goodman Y, Mattson MP. Secreted forms of β-amyloid precursor protein protect hippocampal neurons against amyloid β-peptide-induced oxidative injury. *Exp. Neurol.* 1994; 128: 1-12.

[76] Keller JN, Kindy MS, Holtsberg FW, StClair DK, Yen HC, Germeyer A, SM, Bruce-Keller AJ, Hutchins JB, Mattson MP. Mitochondrial manganese superoxide dismutase prevents neural apoptosis and reduces ischemic brain injury: suppression of peroxynitrite production, lipid peroxidation, and mitochondrial dysfunction. *J. Neurosci.* 1998; 18: 687-697.

[77] Zou H, Henzel WJ, Liu X, Lutschg A, Wang X. Apaf-1, a human protein homologous to *C. elegans* CED-4, participates in cytochrome c-dependent activation of caspase-3. *Cell* 1997; 90: 405-413.

[78] Martinou JC, Desagher S, Antonsson B. Cytochrome c release from mitochondria: all or nothing. *Nat. Cell. Biol.* 2000; 2: 41-43.

[79] Green DR, Reed JC. Mitochondria and apoptosis. Science 1998; 281: 1309-1312.

[80] Akama KT, Albanese C, Pestell RG, van Eldik LJ. Amyloid β-peptide stimulates nitric oxide production in astrocytes through an NF-κB-dependent mechanism. *Proc. Natl. Acad. Sci. USA* 1998; 95: 5795-5800.

[81] Garthwaite J. Glutamate, nitric oxide and cell-cell signalling in the nervous system. *Trends Neurosci.* 1991; 14: 60-67.

[82] Holscher C. Nitric oxide, the enigmatic neuronal messenger: its role in synaptic plasticity. *Trends Neurosci.* 1997; 20: 298-303.

[83] Law A, Gauthier S, Quirion R. Say NO to Alzheimer's disease: the putative links between nitric oxide and dementia of Alzheimer's type. *Brain Res. Rev.* 2001; 35: 73-96.

[84] Zhang J, Snyder SH. Nitric oxide in the nervous system. *Annu. Rev. Pharmacol. Toxicol.* 1995; 35: 213-233.

[85] Hölscher C. Possible causes of Alzheimer's disease: amyloid fragments, free radicals, and calcium homoeostasis. *Neurobiology* 1998; 5: 129-141.

[86] Dumery L, Bourdel F, Soussan Y, Fialkowsky A, Viale S, Nicolas P, Reboud-Ravaux M. β-Amyloid protein aggregation: its implication in the physiopathology of Alzheimer's disease. *Pathol. Biol.* (Paris) 2001; 49: 72-85.

[87] Selkoe DJ. Alzheimer's disease: genes, proteins and therapy. *Physiol. Rev.* 2001; 81: 741-766.

[88] Tan J, Town T, Paris D, Mori T, Suo Z, Crawford F, Mattson MP, Flavell RA, Mullan M. Microglial activation resulting from CD40-CD40L interaction after β-amyloid stimulation. *Science* (Wash DC) 1999; 286: 2352-2355.

[89] Chen S, Frederickson RC, Brunden KR. Neuroglial-mediated immunoinflammatory responses in Alzheimer's disease: complement activation and therapeutic approaches. *Neurobiol. Aging* 1996; 17: 781-787.

[90] Hu J, Akama KT, Krafft GA, Chromy BA, Van Eldik LJ. Amyloid-β peptide activates cultured astrocytes: morphological alterations, cytokine induction and nitric oxide release. *Brain Res.* 1998; 785: 195-206.

[91] Lue LF, Rydel R, Brigham EF, Yang LB, Hampel H, Murphy GM Jr, Brachova L, Yan SD, Walker DG, Shen Y, Rogers J. Inflammatory repertoire of Alzheimer's disease and nondemented elderly microglia *in vitro*. *Glia* 2001; 35: 72-79.

[92] Patrick GN, Zukerberg L, Nikolic M, de la Monte S, Dikkes P, Tsai LH. Conversion of p35 to p25 deregulates Cdk5 activity and promotes neurodegeneration. *Nature* (Lond) 1999; 402: 615-622.

[93] Williamson R, Scales T, Clark BR, Gibb G, Reynolds CH, Kellie S, Bird IN, Varndell IM, Sheppard PW, Everall I, Anderton BH. Rapid tyrosine phosphorylation of neuronal proteins including tau and focal adhesion kinase in response to amyloid-β

peptide exposure: involvement of Src family protein kinases. *J. Neurosci.* 2002; 22: 10-20.

[94] Rapoport M, Ferreira A. PD98059 prevents neurite degeneration induced by fibrillar β-amyloid in mature hippocampal neurons. *J. Neurochem.* 2000; 74: 125-133.

[95] Yankner BA, Duffy LK, Kirschner DA. Neurotrophic and neurotoxic effects of amyloid β-protein: reversal by tachykinin neuropeptides. *Science* 1990; 250: 279-282.

[96] Varadarajan S, Kanski J, Aksenova M, Lauderback C, Butterfield DA. Different mechanisms of oxidative stress and neurotoxicity for Alzheimer's Aβ(1-42) and Aβ(25-35). *J. Am. Chem. Soc.* 2001; 123: 5625-5631.

[97] Qi JS, Qiao JT. Amyloid β-protein fragment 31-35 forms ion channels in membrane patches excised from rat hippocampal neurons. *Neuroscience* 2001; 105: 845-852.

[98] Vogt W. Oxidation of methionyl residues in proteins: tools, targets, and reversal. *Free Radic. Biol. Med.* 1995; 18: 93-105.

[99] Palmblad M, Westlind-Danielsson A, Bergquist J. Oxidation of methionine 35 attenuates formation of amyloid β-peptide 1-40 oligomers. *J. Biol. Chem.* 2002; 277: 19506-19510.

[100] Braak H, Braak E. Evolution of neuronal changes in the course of Alzheimer's disease. *J. Neural. Transm.* 1998; 105: S127-S140.

[101] Koudinov AR, Koudinova NV. Essential role for cholesterol in synaptic plasticity and neuronal degeneration. *FASEB J.* 2001; 15: 1858-1860.

[102] Koudinov AR, Matsubara E, Frangione B, Ghiso J. The soluble form of Alzheimer's amyloid β-protein is complexed to high density lipoprotein and very high density lipoprotein in normal human plasma. *Biochem. Biophys. Res. Commun.* 1994; 205: 1164-1171.

[103] Ladu MJ, Reardon C, Van Eldik L, Fagan AM, Bu G, Holtzman DM. Lipoproteins in the central nervous system. *Ann. N. Y. Acad. Sci.* 2000; 903: 167-175.

[104] Koudinov AR, Berezov TT, Koudinova NV. The levels of soluble Aβ in different HDL subfractions distinguish Alzheimer's and normal aging CSF: implication for brain cholesterol pathology? *Neurosci. Lett.* 2001; 314: 115-118.

[105] Cedazo-Minguez A, Huttinger M, Cowburn RF. β-VLDL protects against Aβ(1-42) and apoE toxicity in human SH-SY5Y neuroblastoma cells. NeuroReport. 2001; 12: 201-206.

[106] Montine TJ, Montine KS, Swift LL. Central nervous system lipoproteins in Alzheimer's-disease. *Am. J. Pathol.* 1997; 151: 1571-1575.

[107] Koudinov AR, Koudinova NV. Modulation of cholesterol metabolism initiates Alzheimer's amyloid deposition and neuronal dysfunction. *Soc. Neurosci. Abstr.* 2000; 26: 497.

[108] Markesbery WR. Oxidative stress hypothesis in Alzheimer's disease. *Free Radic. Biol. Med.* 1997; 23: 134-147.

[109] Miura T, Suzuki K, Kohata N, Takeuchi H. Metal binding modes of Alzheimer's amyloid β-peptide in insoluble aggregates and soluble complexes. *Biochemistry* 2000; 39: 7024-7031.

[110] Huang X, Atwood CS, Hartshorn MA, Multhaup G, Goldstein LE, Scarpa RC, Cuajungco MP, Gray DN, Lim J, Moir RD, Tanzi RE, Bush AI. The Aβ peptide of

Alzheimer's disease directly produces hydrogen peroxide through metal ion reduction. *Biochemistry* 1999; 38: 7609-7616.

[111] Varadarajan S, Yatin S, Aksenova M, Butterfield DA. Review: Alzheimer's amyloid β-peptide-associated free radical oxidative stress and neurotoxicity. *J. Struct. Biol.* 2000; 130: 184-208.

[112] Kontush A. Amyloid-β: an antioxidant that becomes a pro-oxidant and critically contributes to Alzheimer's disease. *Free Radic. Biol. Med.* 2001; 31: 1120-1131.

[113] Kontush A, Berndt C, Weber W, Akopyan V, Arlt S, Schippling S, Beisiegel U. Amyloid-β is an antioxidant for lipoproteins in cerebrospinal fluid and plasma. *Free Radic. Biol. Med.* 2001; 30: 119-128.

[114] Kontush A, Donarski N, Beisiegel U. Resistance of human cerebrospinal fluid to *in vitro* oxidation is directly related to its amyloid-β content. *Free Radic. Res.* 2001; 35: 507-517.

[115] Berndt C, Kontush A, Beisiegel U. Neuronal cell cultures protect low density lipoprotein from oxidation. *Neurobiol. Aging* 1998; 19: S284.

[116] Koudinov AR, Koudinova NV. Soluble amyloid-β protein is secreted by HepG2 cells as an apolipoprotein. *Cell Biol. Inter.* 1997; 25: 265-271.

[117] Whitson JS, Glabe CG, Shintani E, Abcar A, Cotman CW. β-amyloid protein promotes neuritic branching in hippocampal cultures. *Neurosci. Lett.* 1990; 110: 319-324.

[118] Yankner BA, Duffy LK, Kirschner DA. Neurotrophic and neurotoxic effects of amyloid-β protein: reversal by tachykinin neuropeptides. *Science* 1990; 250: 279-282.

[119] Koo EH, Park L, Selkoe DJ. Amyloid β-protein as a substrate interacts with extracellular matrix to promote neurite outgrowth. *Proc. Natl. Acad. Sci. USA.* 1993; 90: 4748-4752.

[120] Mattson MP. Calcium as sculptor and destroyer of neural circuitry. *Exp. Gerontol.* 1992; 27: 29-49.

[121] Bootman MD, Collins TJ, Peppiatt CM, Prothero LS, MacKenzie L, De Smet P, Travers M, Tovey SC, Seo JT, Berridge MJ, Ciccolini F, Lipp P. Calcium signalling: an overview. *Semin. Cell Dev. Biol.* 2001; 12: 3-10.

[122] Cameron AM, Steiner JP, Roskams AJ, Ali SM, Ronnett GV, Snyder SH. Calcineurin associated with the inositol 1,4,5-trisphosphate receptor-FKBP12 complex modulates Ca2+ flux. *Cell* 1996; 83: 463-472.

[123] Yamada M, Miyawaki A, Saito K, Nakajima T, Yamamoto-Hino M, Ryo Y, Furuichi T, Mikoshiba K. The calmodulin-binding domain in the mouse type 1 inositol 1,4,5-trisphosphate receptor. *Biochem. J.* 1995; 308: 83-88.

[124] Bourguignon LY, Jin H. Identification of the ankyrin-binding domain of the mouse T-lymphoma cell inositol 1,4,5-trisphosphate (IP3) receptor and its role in the regulation of IP3-mediated internal Ca2+ release. *J. Biol. Chem.* 1995; 270: 7257-7260.

[125] Babcock DF, Herrington J, Goodwin PC, Park YB, Hille B. Mitochondrial participation in the intracellular Ca2+ network. *J. Cell Biol.* 1997; 136: 833-844.

[126] Gunter TE, Buntinas L, Sparagna GC, Gunter KK. The Ca2+ transport mechanisms of mitochondria and Ca2+ uptake from physiological-type Ca2+ transients. *Biochim. Biophys. Acta* 1998; 1366: 5-15.

[127] Nicholls DG. A role for mitochondria in the protection of cells against calcium overload? *Prog. Brain Res.* 1985; 63: 97-106.

[128] Ichas F, Jouaville LS, Sidash SS, Mazat JP, Holmuhamedov EL. Mitochondrial calcium spiking: a transduction mechanism based on calcium-induced permeability transition involved in cell calcium signalling. *FEBS Lett.* 1994; 348: 211-215.

[129] Hajnóczky G, Robb-Gaspers LD, Seitz MB, Thomas AP. Decoding of cytosolic calcium oscillations in the mitochondria. *Cell* 1995; 82: 415-424.

[130] Dugan LL, Sensi SL, Canzoniero LM, Handran SD, Rothman SM, Goldberg MP, Choi DW. Mitochondrial production of reactive oxygen species in cortical neurons following exposure to N-methyl-D-aspartate. *J. Neurosci.* 1995; 15: 6377-6388.

[131] Keller JN, Guo Q, Holtsberg FW, Bruce-Keller AJ, Mattson MP. Increased sensitivity to mitochondrial toxin-induced apoptosis in neural cells expressing mutant presenilin-1 is linked to perturbed calcium homeostasis and enhanced oxyradical production. *J. Neurosci.* 1998; 18: 4439-4450.

[132] Watt F. Nuclear microscope analysis in Alzheimer's and Parkinson's disease: a review. *Cell. Mol. Biol.* 1996; 42: 17-26.

[133] Nixon RA, Saito KI, Grynspan F, Griffin WR, Katayama S, Honda T, Mohan PS, Shea TB, Beermann M. Calcium-activated neutral proteinase (calpain) system in aging and Alzheimer's disease. *Ann. NY Acad. Sci.* 1994; 747: 77-91.

[134] Selkoe DJ. Cell biology of the amyloid β-protein precursor and the mechanism of Alzheimer's disease. *Annu. Rev. Cell Biol.* 1994; 10: 373-403.

[135] Kang J, Lemaire HG, Unterbeck A, Salbaum JM, Masters CL, Grzeschik KH, Multhaup G, Beyreuther K, Muller-Hill B. The precursor of Alzheimer's disease amyloid A4 protein resembles a cell-surface receptor. *Nature* (Lond) 1987; 325: 733-736.

[136] Multhaup G, Bush AI, Pollwein P, Masters CL. Interaction between the zinc (II) and the heparin binding site of the Alzheimer's disease βA4 amyloid precursor protein (APP). *FEBS Lett.* 1994; 355: 151-154.

[137] Small DH, Nurcombe V, Reed G, Clarris H, Moir R, Beyreuther K, Masters CL. A heparin-binding domain in the amyloid protein precursor of Alzheimer's disease is involved in the regulation of neurite outgrowth. *J. Neurosci.* 1994; 14: 2117-2127.

[138] Goldgaber D, Lerman MI, McBride OW, Saffiotti U, Gajdusek DC. Characterization and chromosomal localization of a cDNA encoding brain amyloid of Alzheimer's disease. *Science* 1987; 235: 877-880.

[139] Johnstone EM, Chaney MO, Norris FH, Pascual R, Little SP. Conservation of the sequence of the Alzheimer's disease amyloid peptide in dog, polar bear and five other mammals by cross-species polymerase chain reaction analysis. *Mol. Brain Res.* 1991; 10: 299-305.

[140] Yoshikai S, Sasaki H, Dohura K, Furuya H, Sakaki Y. Genomic organization of the human amyloid β-protein precursor gene. *Gene* 1990; 87: 257-263.

[141] Zhong S, Wu K, Black IB, Schaar DG. Characterization of the genomic structure of the mouse APLP1 gene. *Genomics* 1996; 32: 159-162.

[142] Wasco W, Gurubhagavatula S, Tanzi RE. Molecular characterization of the members of the APP-like gene family. *Res. Adv. Alzh. Dis. Rel. Disord.* 1995; 71: 657-667.

[143] Rosen DR, Martin-Morris L, Luo LQ, White K. A *Drosophila* gene encoding a protein resembling the human β-amyloid protein precursor. *Proc. Natl. Acad. Sci. USA* 1989; 86: 2478-2482.

[144] Selkoe DJ, Podlisny MB, Joachim CL, Vickers EA, Lee G, Fritz LC, Oltersdorf T. β-amyloid precursor protein of Alzheimer disease occurs as 110- to 135-kilodalton membrane-associated proteins in neural and nonneural tissues. *Proc. Natl. Acad. Sci. USA* 1988; 85: 7341-7345.

[145] Manning RW, Reid CM, Lampe RA, Davis LG. Identification in rodents and other species of an mRNA homologous to the human β-amyloid precursor. *Brain Res.* 1988; 427: 293-297.

[146] Querfurth HW, Wijsman EM, St. George-Hyslop PH, Selkoe DJ. βAPP mRNA transcription is increased in cultured fibroblats from the familial Alzheimer's disease-1 family. *Mol. Brain Res.* 1995; 28: 319-337.

[147] Tanzi RE, McClatchey AI, Lamperti ED, Villakomaroff L, Gusella JF. Protease inhibitor domain encoded by an amyloid protein-precursor messenger-RNA associated with Alzheimer's-disease. *Nature* 1988; 331: 528-530.

[148] Ho L, Fukuchi K, Younkin SG. The alternatively spliced Kunitz protease inhibitor domain alters amyloid β-protein precursor processing and amyloid β-protein production in cultured cells. *J. Biol. Chem.* 1996; 271: 30929-30934.

[149] Goedert M. Neuronal localization of amyloid β-protein precursor mRNA in normal human brain and in Alzheimer's disease. *EMBO J.* 1987; 6: 3627-3632.

[150] Sandbrink R, Masters CL, Beyreuther K. βA4-amyloid protein precursor mRNA isoforms without exon 15 are ubiquitously expressed in rat tissues including brain, but not in neurons. *J. Biol. Chem.* 1994; 269: 1510-1517.

[151] Tanzi RE, Wenniger JJ, Hyman BT. Cellular specificity and regional distribution of amyloid-β protein precursor alternative transcripts are unaltered in Alzheimer's hippocampal formation. *Mol. Brain Res.* 1993; 18: 246-252.

[152] Li L, Ohman T, Deeb SS, Fukuchi KI. Analysis of mouse intron 7 DNA sequences of the APP gene: comparison with the human homologue. *DNA Seq.* 1999; 10: 219-228.

[153] Belandia B, Latasa MJ, Villa A, Pascual A. Thyroid hormone negatively regulates the transcriptional activity of the β-amyloid precursor protein gene. *J. Biol. Chem.* 1998; 273: 30366-30371.

[154] Genovesi G, Paolini P, Marcellini L, Vernillo E, Salvati G, Polidori G, Ricciardi D, drNuccio I, Re M. Relationship between autoimmune thyroid disease and Alzheimer's disease. *Panminerva Med.* 1996; 38: 61-63.

[155] Coulson EJ, Paliga K, Beyreuther K, Masters CL. What the evolution of the amyloid protein precursor supergene family tells us about its function? *Neurochem. Int.* 2000; 36: 175-184.

[156] Salbaum JM, Ruddle FH. Embryonic expression pattern of amyloid protein precursor suggests a role in differentiation of specific subsets of neurons. *J. Exp. Zool.* 1994; 269: 116-127.

[157] Schubert W, Prior R, Weidemann A, Dircksen H, Multhaup G, Masters CL, Beyreuther K. Localization of Alzheimer βA4 amyloid precursor protein at central and peripheral synaptic sites. *Brain Res.* 1991; 563: 184-194.

[158] Li QX, Berndt MC, Bush AI, Rumble B, Mackenzie I, Friedhuber A, Beyreuther K, Masters CL. Membrane-associated forms of the βA4 amyloid protein precursor of Alzheimer's disease in human platelets and brain: surface expression on the activated human platelet. *Blood* 1994; 84: 133-142.

[159] Sherriff FE, Bridges LR, Gentleman SM, Sivaloganathan S, Wilson S. Markers of axonal injury in post-mortem human brain. *Acta Neuropathol.* 1994; 88: 433-439.

[160] Furukawa K, Barger SW, Blalock EM, Mattson MP. Activation of K+ channels and suppression of neuronal activity by secreted β-amyloid precursor protein. *Nature* 1996; 379: 74-78.

[161] Mattson MP. Neurotransmitters in the regulation of neuronal cytoarchitecture. *Brain Res. Rev.* 1988; 13: 179-212.

[162] Kitaguchi N, Takahashi Y, Tokushima Y. Novel precursor of Alzheimer's disease amyloid protein shows protease inhibitory activity. *Nature* 1998; 331: 530-532.

[163] Nishimoto I, Okamoto T, Matsuura Y. Alzheimer amyloid protein precursor complexes with brain GTP-binding protein Go. *Nature* 1993; 362: 75-79.

[164] Tanzi RE, Gusella JF, Watkins PC, Bruns GAP, St.George-Hyslop P, van Keuren ML, Patterson D, Pagan S, Kurnit DM, Neve RL. Amyloid β protein gene: cDNA, mRNA distribution and genetic linkage near the Alzheimer's locus. *Science* 1987; 235: 880-884.

[165] Evin G, Beyreuther K, Masters CL. Alzheimer's disease amyloid precursor protein (AβPP): proteolytic processing, secretases and βA4 amyloid production. *Amyloid* 1994; 1: 263-280.

[166] Esch FS, Keim PS, Beattle EC, Blacher RW, Culwell AR, Oltersdorf T, McClure D, Ward PJ. Cleavage of amyloid-β peptide during constitutive processing of its precursor. *Science* 1990; 248: 1122-1124.

[167] Bowes MP, Masliah E, Otero DA, Zivin JA, Saitoh T. Reduction of neurological damage by a peptide segment of the amyloid β/A4 protein precursor in a rabbit spinal cord ischemia model. *Exp. Neurol.* 1994; 129: 112-119.

[168] Smith-Swintosky VL, Pettigrew LC, Craddock SD, Culwell AR, Rydel RE, Mattson MP. Secreted forms of β-amyloid precursor protein protect against ischemic brain injury. *J. Neurochem.* 1994; 63: 781-784.

[169] Fraser SP, Suh Y, Djamgoz MBA. Ionic effects of the Alzheimer's disease β-amyloid precursor protein and its metabolic fragments. *Trends Neurosci.* 1997; 20: 67-72.

[170] Milward EA, Papadopoulos R, Fuller SJ, Moir RD, Small D, Beyreuther K, Masters CL. The amyloid protein precursor of Alzheimer's disease is a mediator of the effects of nerve growth factor on neurite outgrowth. *Neuron* 1992; 9: 129-137.

[171] Saitoh T, Sundsmo M, Roch JM, Kimura N, Cole G, Schubert D, Oltersdorf T, Schenk DB. Secreted form of amyloid-β protein precursor is involved in the growth regulation of fibroblasts. *Cell* 1989; 58: 615-622.

[172] Araki W, Kitaguchi N, Tokushima Y, Ishii K, Aratake H, Shimohama S, Nakamura S, Kimura J. Trophic effect of β-amyloid precursor protein on cerebral cortical neurons in culture. *Biochem. Biophys. Res. Commun.* 1991; 181: 265-271.

[173] Mattson MP, Cheng B, Culwell AR, Esch FS, Lieberburg I, Rydel RE. Evidence for excitoprotective and intraneuronal calcium-regulating roles for secreted forms of the β-amyloid precursor protein. *Neuron* 1993; 10: 243-254.

[174] Kihara T, Shimohama S, Sawada H, Kimura J, Kume T, Kochiyama H, Maeda T, Akaike A. Nicotinic receptor stimulation protects neurons against β-amyloid toxicity. *Ann. Neurol.* 1997; 42: 159-163.

[175] Barger SW, Mattson MP. Induction of neuroprotective κB-dependent transcription by secreted forms of the Alzheimer's β-amyloid precursor. *Mol. Brain Res.* 1996; 40: 116-126.

[176] Ishiguro M, Ohsawa I, Takamura C, Morimoto T, Kohsaka S. Secreted from of β-amyloid precursor protein activates protein kinase C and phospholipase C-α1 in cultured embryonic rat neocortical cells. *Mol. Brain Res.* 1998; 53: 24-32.

[177] Mori F, Lai CC, Fusi F, Giacobini E. Cholinesterase inhibitors increase secretion of APPs in rat brain cortex. *Neuroreport* 1995; 6: 633-636.

[178] Rossner S, Ueberham U, Schliebs R, Perez-Polo JR, Bigl V. The regulation of amyloid precursor protein metabolism by cholinergic mechanisms and neurotrophin receptor signaling. *Prog. Neurobiology* 1998; 56: 541-569.

[179] Seo J, Kim S, Kim H, Park CH, Jeong S, Lee J, Choi SH, Chang K, Rah J, Koo J, Kim E, Suh Y. Effects of nicotine on APP secretion and Aβ- or CT105-induced toxicity. *Biol. Psychiatry* 2001; 49: 240-247.

[180] Moss ML, Jin SL, Becherer JD, Bickett DM, Burkhart W, Chen WJ, Hassler D, Leesnitzer MT, McGeehan G, Milla M, Moyer M, Rocque W, Seaton T, Schoenen F, Warner J, Willard D. Structural features and biochemical properties of TNF-α converting enzyme (TACE). *J. Neuroimmunol.* 1997; 72: 127-129.

[181] Buxbaum JD, Liu KN, Luo Y, Slack JL, Stocking KL, Peschon JJ, Johnson RS, Castner BJ, Cerretti DP, Black RA. Evidence that tumor necrosis factor-α converting enzyme is involved in regulated α-secretase cleavage of the Alzheimer amyloid protein precursor. *J. Biol. Chem.* 1998; 272: 27765-27767.

[182] Lammich S, Kojro E, Postina R, Gilbert S, Pfeiffer R, Jasionowski M, Haass C, Fahrenholz F. Constitutive and regulated α-secretase cleavage of Alzheimer's amyloid precursor protein by a disintegrin metalloprotease. *Proc. Natl. Acad. Sci. USA* 1999; 96: 3922-3927.

[183] Dingwall C. Spotlight on BACE: the secretases as targets for treatment in Alzheimer disease. *J. Clin. Invest.* 2001; 108: 1243-1246.

[184] Bennett BD, Denis P, Haniu M, Teplow DB, Kahn S, Louis JC, Citron M, Vassar R. A furin-like convertase mediates propeptide cleavage of BACE, the Alzheimer's β-secretase. *J. Biol. Chem.* 2000; 275: 37712-37717.

[185] Vassar R, Bennett BD, Babu-Khan S, Kahn S, Mendiaz EA, Denis P, Teplow DB, Ross S, Amarante P, Loeloff R, Luo Y, Fisher S, Fuller J, Edenson S, Lile J, Jarosinski MA, Biere AL, Curran E, Burgess T, Louis JC, Collins F, Treanor J, Rogers G, Citron M. β-secretase cleavage of Alzheimer's amyloid precursor protein by the transmembrane aspartic protease BACE. *Science* (Wash DC) 1999; 286: 735-741.

[186] Yan R, Bienkowski MJ, Shuck ME, Miao H, Tory MC, Pauley AM, Brashier JR, Stratman NC, Mathews WR, Buhl AE, Carter DB, Tomasselli AG, Parodi LA,

Heinrikson RL, Gurney ME. Membrane-anchored aspartyl protease with Alzheimer's disease β-secretase activity. *Nature* (Lond) 1999; 402: 533-537.

[187] Lin X, Koelsch G, Wu S, Downs D, Dashti A, Tang J. Human aspartic protease memapsin 2 cleaves the β-secretase site of β-amyloid precursor protein. *Proc. Natl. Acad. Sci. USA* 2000; 97: 1456-1460.

[188] Haass C, Schlossmacher MG, Hung AY, Vigo-Pelfrey C, Mellon A, Ostaszewski BL, Lieberburg I, Koo EH, Schenk D, Teplow DB. Amyloid β-peptide is produced by cultured cells during normal metabolism. *Nature* (Lond) 1992; 359: 322-325.

[189] Seubert P, Vigo-Pelfrey C, Esch F, Lee M, Dovey H, Davis D, Sinha S, Schlossmacher M, Whaley J, Swindlehurst C. Isolation and quantification of soluble Alzheimer's β-peptide from biological fluids. *Nature* (Lond) 1992; 359: 325-327.

[190] Shoji M, Golde TE, Ghiso J, Cheung TT, Estus S, Shaffer LM, Cai XD, McKay DM, Tintner R, Frangione B. Production of the Alzheimer amyloid β-protein by normal proteolytic processing. *Science* (Wash DC) 1992; 258: 126-129.

[191] Saunders AJ, Kim TW, Tanzi RE. BACE maps to chromosome 11 and a BACE homolog, BACE2, reside in the obligate Down syndrome region of chromosome 21. Science (Wash DC) 1999; 286: 1255.

[192] Farzan M, Schnitzler CE, Vasilieva N, Leung D, Choe H. BACE2, a β-secretase homolog, cleaves at the β-site and within the amyloid-β region of the amyloid-β precursor protein. *Proc. Natl. Acad. Sci. USA* 2000; 97: 9712-9717.

[193] Golde TE, Estus S, Younkin LH, Selkoe DJ, Younkin SG. Processing of the amyloid protein precursor to potentially amyloidogenic derivatives. *Science* 1992; 255: 728-730.

[194] Jarrett JT, Berger EP, Lansbury PT Jr. The carboxy terminus of the β-amyloid protein is critical for the seeding of amyloid formation: implications for the pathogenesis of Alzheimer's disease. *Biochemistry* 1993; 32: 4693-4697.

[195] Iwatsubo T, Odaka A, Suzuki N, Mizusawa H, Nukina N, Ihara Y. Visualization of Aβ42(43) and Aβ40 in senile plaques with end-specific Aβ monoclonals: evidence that an initially deposited species is Aβ 42(43). *Neuron* 1994; 13: 45-53.

[196] Kimberly WT, Zheng JB, Guenette SY, Selkoe DJ. The intracellular domain of the β-amyloid precursor protein is stabilized by Fe65 and translocates to the nucleus in a Notch-like manner. *J. Biol. Chem.* 2001; 276: 40288-40292.

[197] Ni CY, Murphy MP, Golde TE, Carpenter G. γ-Secretase cleavage and nuclear localization of ErbB-4 receptor tyrosine kinase. *Science* (Wash DC) 2001; 294: 2179-2181.

[198] Nunan J, Small DH. Regulation of APP cleavage by α-, β- and γ-secretases. FEBS Lett. 2000; 483: 6-10.

[199] Heldin CH, Ericsson J. Signal transduction. RIPping tyrosine kinase receptors apart. *Science* (Wash DC) 2001; 294: 2111-2113.

[200] Wolfe MS, Xia W, Ostaszewski BL, Diehl TS, Kimberly WT, Selkoe DJ. Two transmembrane aspartates in presenilin-1 required for presenilin endoproteolysis and γ-secretase activity. *Nature* 1999; 398: 513-517.

[201] Ancolio K, Dumanchin C, Barelli H, Warter JM, Brice A, Campion D, Frebourg T, Checler F. Unusual phenotypic alteration of β-amyloid precursor protein (βAPP)

maturation by a new Val-715 → Met βAPP-770 mutation responsible for probable early-onset Alzheimer's disease. *Proc. Natl. Acad. Sci. USA* 1999; 96: 4119-4124.

[202] Singh N, Talalayeva Y, Tsiper M, Romanov V, Dranovsky A, Colflesh D, Rudamen G, Vitek MP, Shen J, Yang X, Goldgaber D, Schwarzman AL. The role of Alzheimer's disease-related presenilin 1 in intercellular adhesion. *Exp. Cell Res.* 2001; 263: 1-13.

[203] Smine A, Xu X, Nishiyama K, Katada T, Gambetti P, Yadav SP, Wu X, Shi YC, Yasuhara S, Homburger V, Okamoto T. Regulation of brain G-protein by Alzheimer's disease gene presenilin-1. *J. Biol. Chem.* 1998; 273: 16281-16288.

[204] Naruse S, Thinakaran G, Luo JJ, Kusiak JW, Tomita T, Iwatsubo T, Qian X, Ginty DD, Price DL, Borchelt DR, Wong PC, Sisodia SS. Effects of PS1 deficiency on membrane protein trafficking in neurons. *Neuron* 1998; 21: 1213-1221.

[205] Schroeter EH, Kisslinger JA, Kopan R. Notch-1 signalling requires ligand-induced proteolytic release of intracellular domain. *Nature* 1998; 393: 382-386.

[206] Brown MS, Ye J, Rawson RB, Goldstein JL. Regulated intramembrane proteolysis: a control mechanism conserved from bacteria to humans. *Cell* 2000; 100: 391-398.

[207] [1]Ray WJ, Yao M, Mumm J, Schroeter EH, Saftig P, Wolfe M, Selkoe DJ, Kopan R, Goate AM. Cell surface presenilin-1 participates in the γ-secretase-like proteolysis of Notch. *J. Biol. Chem.* 1999; 274: 36801-36807.

[208] Mumm JS, Schroeter EH, Saxena MT, Griesemer A, Tian X, Pan DJ, Ray WJ, Kopan R. A ligand-induced extracellular cleavage regulates γ-secretase-like proteolytic activation of Notch 1. *Mol. Cell.* 2000; 5: 197-206.

[209] Wen C, Metzstein MM, Greenwald I. SUP-17, a *Caenorhabditis elegans* ADAM protein related to *Drosophila* KUZBANIAN and its role in LIN-12/NOTCH signalling. *Development* 1997; 124: 4759-4767.

[210] Edbauer D, Winkler E, Haass C, Steiner H. Presenilin and nicastrin regulate each other and determine amyloid β-peptide production via complex formation. *Proc. Natl. Acad. Sci. USA* 2002; 99: 8666-8671.

[211] Francis R, McGrath G, Zhang J, Ruddy DA, Sym M, Apfeld J, Nicoll M, Maxwell M, Hai B, Ellis MC, Parks AL, Xu W, Li J, Gurney M, Myers RL, Himes CS, Hiebsch R, Ruble C, Nye JS, Curtis D. Aph-1 and pen-2 are required for Notch pathway signaling, γ-secretase cleavage of βAPP, and presenilin protein accumulation. *Dev. Cell* 2002; 3: 85-97.

[212] Weidemann A, Eggert S, Reinhard FB, Vogel M, Paliga K, Baier G, Masters CL, Beyreuther K, Evin G. A novel epsilon-cleavage within the transmembrane domain of the Alzheimer amyloid precursor protein demonstrates homology with Notch processing. *Biochemistry* 2002; 41: 2825-2835.

[213] Morishima-Kawashima M, Ihara Y. γ(gamma)-cleavage and ε-(epsilon)-cleavage of β-amyloid precursor protein. *Int. Congr. Series* 2003; 1252: 355-361.

[214] Checler F. Processing of the β-amyloid precursor protein and its regulation in Alzheimer's disease. *J. Neurochem.* 1995; 65: 1431-1444.

[215] Suh YH. An etiological role of amyloidogenic carboxyl-terminal fragments of the β-amyloid precursor protein in Alzheimer's disease. *J. Neurochem.* 1997; 68: 1781-1791.

[216] Wolf D, Quon D, Wang Y, Cordell B. Identification and characterization of C-terminal fragments of the β-amyloid precursor produced in cell culture. *EMBO J.* 1990; 9: 2079-2084.

[217] Estus S, Golde TE, Kunishita T, Blades D, Lowery D, Eisen M, Usiak M, Qu XM, Tabira T, Greenberg BD. Potentially amyloidogenic, carboxylterminal derivatives of the amyloid protein precursor. *Science* (Wash DC) 1992; 255: 726-728.

[218] Caputo CB, Wischik C, Novak M, Scott CW, Brunner WF, De Garcini EM, Lo MM, Norris TE, Salama AI. Immunological characterization of the region of tau protein that is bound to Alzheimer paired helical filaments. *Neurobiol. Aging* 1992; 13: 267-274.

[219] Tamaoka A, Kalaria RN, Lieberburg I, Selkoe DJ. Identification of a stable fragment of the Alzheimer amyloid precursor containing the β-protein in brain microvessels. *Proc. Natl. Acad. Sci. USA* 1992; 89: 1345-1349.

[220] Tokuda T, Tanaka K, Kametani F, Ikeda S, Yanagisawa N. Secretory cleavage of β-amyloid precursor protein in the cerebral white matter produces amyloidogenic carboxyl-terminal fragments. *Neurosci. Lett.* 1995; 186: 149-152.

[221] Gardella JE, Gorgone GA, Newman P, Frangione B, Gorevic PD. Characterization of Alzheimer amyloid precursor protein transcripts in platelets and megakarocytes. *Neurosci. Lett.* 1992; 138: 229-232.

[222] Matsumoto A, Matsumoto R. Familial Alzheimer's disease cells abnormally accumulate β-amyloid-harbouring peptides preferentially in cytosol but not in extracellular fluid. *Eur. J. Biochem.* 1994; 225: 1055-1062.

[223] Kametani F, Tanaka K, Tokuda T, Ikeda S. Secretory cleavage site of Alzheimer amyloid precursor protein is heterogeneous in Down's syndrome brain. *FEBS Lett.* 1994; 351: 165-167.

[224] Kammesheidt A, Boyce FM, Spanoyannis AF, Cummings BJ, Ortegon M, Cotman C, Vaught JL, Neve RL. Deposition of β/A4 immunoreactivity and neuronal pathology in transgenic mice expressing the carboxyl-terminal fragment of the Alzheimer amyloid precursor in the brain. *Proc. Natl. Acad. Sci. USA* 1992; 89: 10857-10861.

[225] Oster-Granite ML, McPhie DL, Greenan J, Neve RL. Age-dependent neuronal and synaptic degeneration in mice transgenic for the C-terminus of the amyloid precursor protein. *J. Neurosci.* 1996; 16: 6732-6741.

[226] Hartell NA, Suh YH. Peptide fragments of β-amyloid precursor protein: effects on parallel fiber-Purkinje cell synaptic transmission in rat cerebellum. *J. Neurochem.* 2000; 74: 1112-1121.

[227] Choi SH, Park CH, Koo JW, Seo JH, Kim HS, Jeong SJ, Lee JH, Kim SS, Suh YH. Memory impairment and cholinergic dysfunction by centrally administered Aβ and carboxyl-terminal fragment of Alzheimer's APP in mice. *FASEB J.* 2001; 15: 1816-1818.

[228] Song DK, Won MH, Jung JS, Lee JC, Kang TC, Suh HW, Huh SO, Paek SH, Kim YH, Kim SH, Suh YH. Behavioral and neuropathologic changes induced by central injection of carboxyl-terminal fragment of β-amyloid precursor protein in mice. *J. Neurochem.* 1998; 71: 875-878.

[229] Fraser SP, Suh YH, Chong YH, Djamgoz MB. Membrane currents induced in *Xenopus* oocytes by the C-terminal fragment of the β-amyloid precursor protein. *J. Neurochem.* 1996; 66: 2034-2040.

[230] Fraser SP, Suh Y, Djamgoz MBA. Ionic effects of the Alzheimer's disease β-amyloid precursor protein and its metabolic fragments. *Trends Neurosci.* 1997; 20: 67-72.

[231] Kim JH, Rah JC, Fraser SP, Chang KA, Djamgoz MB, Suh YH. Carboxylterminal peptide of β-amyloid precursor protein blocks inositol 1,4,5-trisphosphate-sensitive Ca^{2+} release in *Xenopus laevis* oocytes. *J. Biol. Chem.* 2002; 277: 20256-20263.

[232] Kim HS, Lee JH, Suh YH. C-terminal fragment of Alzheimer's amyloid precursor protein inhibits sodium/calcium exchanger activity in SK-N-SH cell. *Neuroreport* 1999; 10: 113-116.

[233] Arispe N, Rojas E, Pollard HB. Alzheimer disease amyloid-β protein forms calcium channels in bilayer membranes: blockade by tromethamine and aluminum. *Proc. Natl. Acad. Sci. USA* 1993; 90: 567-571.

[234] Kim HS, Park CH, Cha SH, Lee JH, Lee S, Kim Y, Rah JC, Jeong SJ, Suh YH. Carboxyl-terminal fragment of Alzheimer's APP destabilizes calcium homeostasis and renders neuronal cells vulnerable to excitotoxicity. *FASEB J.* 2000; 14: 1508-1517.

[235] Duffy PE, Rapport M, Graf L. Glial fibrillary acidic protein and Alzheimer-type senile dementia. *Neurology* 1980; 30: 778-782.

[236] Wisniewski HM, Wegiel J. Spatial relationships between astrocytes and classical plaque components. *Neurobiol. Aging* 1991; 12: 593-600.

[237] Rah JC, Kim HS, Kim SS, Bach JH, Kim YS, Park CH, Seo JH, Jeong SJ, Suh YH. Effects of carboxyl-terminal fragment of Alzheimer's amyloid precursor protein and amyloid β-peptide on the production of cytokines and nitric oxide in glial cells. *FASEB J.* 2001; 15: 1463-1465.

[238] DeGiorgio LA, DeGiorgio N, Milner TA, Conti B, Volpe BT. Neurotoxic APP C-terminal and β-amyloid domains colocalize in the nuclei of substantia nigra pars reticulata neurons undergoing delayed degeneration. *Brain Res.* 2000; 874: 137-146.

[239] Gao Y, Pimplikar SW. The γ-secretase-cleaved C-terminal fragment of amyloid precursor protein mediates signaling to the nucleus. *Proc. Natl. Acad. Sci. USA* 2001; 98: 14979-14984.

[240] Buxbaum JD, Gandy SE, Cicchetti P. Processing of Alzheimer β/A4 amyloid precursor protein: modulation by agents that regulate protein phosphorylation. *Proc. Natl. Acad. Sci. USA* 1990; 87: 6003-6006.

[241] Buxbaum JD, Ruefli AA, Parker CA, Cypess AM, Greengard P. Calcium regulates processing of the Alzheimer amyloid protein precursor in a protein kinase C-independent manner. *Proc. Natl. Acad. Sci. USA* 1994; 91: 4489-4493.

[242] Checler F. Processing of the β-amyloid precursor protein and its regulation in Alzheimer's disease. *J. Neurochem.* 1995; 65: 1431-1444.

[243] Ivins KJ, Bui ET, Cotman CW. β-amyloid induces local neurite degeneration in cultured hippocampal neurons: evidence for neuritic apoptosis. *Neurobiol. Dis.* 1998; 5: 365-378.

[244] Forloni G, Bugiani O, Tagliavini F, Salmona M. Apoptosis-mediated neurotoxicity induced by β-amyloid and PrP fragments. *Mol. Chem. Neuropathol.* 1996; 28: 163-171.

[245] Earnshaw WC, Martins LM, Kaufmann SH. Mammalian caspases: structure, activation, substrates, and functions during apoptosis. *Annu. Rev. Biochem.* 1999; 68: 383-424.

[246] Jeon BS, Kholodilov NG, Oo F, Kim SY, Tomaselli KJ, Srinivasan A, Stefanis L, Burke RE. Activation of Caspase-3 in the developmental models of programmed cell death in neurons of the substantia nigra. *J. Neurochem.* 1999; 73: 322-333.

[247] Harada J, Sugimoto M. Activation of caspase-3 in β-amyloid-induced apoptosis of cultured rat cortical neurons. *Brain Res.* 1999; 842: 311-323.

[248] Gervais FG, Xu D, Robertson GS, Vaillancourt JP, Zhu Y, Huang J, LeBlanc A, Smith D, Rigby M, Shearman MS, Clarke EE, Zheng H, Van Der Ploeg LH, Ruffolo SC, Thornberry NA, Xanthoudakis S, Zamboni RJ, Roy S, Nicholson DW. Involvement of caspases in proteolytic cleavage of Alzheimer's amyloid-β precursor protein and amyloidogenic Aβ peptide formation. *Cell* 1999; 97: 395-406.

[249] Dumanchin-Njock C, Alves da Costa CA, Mercken L, Pradier L, Checler F. The caspase-derived C-terminal fragment of βAPP induces caspase-independent toxicity and triggers selective increase of Aβ42 in mammalian cells. *J. Neurochem.* 2001; 78: 1153-1161.

[250] de Boer RA, van Veldhuisen DJ, van der Wijk J, Brouwer RM, de Jonge N, Cole GM, Suurmeijer AJ. Additional use of immunostaining for active caspase 3 and cleaved actin and PARP fragments to detect apoptosis in patients with chronic heart failure. *J. Card. Fail.* 2000; 6: 330-337.

[251] Savage MJ, Trusko SP, Howland DS, Pinsker LR, Mistretta S, Reaume AG, Greenberg BD, Siman R, Scott RW. Turnover of amyloid β-protein in mouse brain and acute reduction of its level by phorbol ester. *J. Neurosci.* 1998; 18: 1743-1752.

[252] Saido T. Degradation of amyloid-β peptide: a key to Alzheimer pathogenesis, prevention and therapy. *Neurosci. News* 2000; 3: 52-62.

[253] Carson JA, Turner AJ. β-amyloid catabolism: Roles for neprilysin (NEP) and other metallopeptidases. *J. Neurochem.* 2002; 81: 1-8.

[254] Cohen AJ, Miller YE. Neuroendocrine differentiation, neuropeptides, and neprilysin. Am. J. Resp. *Cell Mol. Biol.* 1999; 21: 1-3.

[255] Yasojima K, Akiyama H, McGeer EG, McGeer PL. Reduced neprilysin in high plaque areas of Alzheimer brain: a possible relationship to deficient degradation of β-amyloid peptide. *Neurosci. Lett.* 2001; 297: 97-100.

[256] Selkoe DJ. Clearing the brain's amyloid cobwebs. *Neuron* 2001; 32: 177-180.

[257] Mukherjee A, Song E, Kihiko-Ehmann M, Goodman JP Jr, Pyrek JS, Estus S, Hersh LB. Insulysin hydrolyzes amyloid β-peptides to products that are neither neurotoxic nor deposit on amyloid plaques. *J. Neurosci.* 2000; 20: 8745-8749.

[258] Vekrellis K, Ye Z, Qiu WQ, Walsh D, Hartley D, Chesneau V, Rosner MR, Selkoe DJ. Neurons regulate extracellular levels of amyloid β-protein via proteolysis by insulin-degrading enzyme. *J. Neurosci.* 2000; 20: 1657-1665.

[259] Kurochkin IV. Amyloidogenic determinant as a substrate recognition motif of insulin-degrading enzyme. *FEBS Lett.* 1998; 427: 153-156.

[260] Edbauer D, Willem M, Lammich S, Steiner H, Haass C. Insulin-degrading enzyme rapidly removes the β-amyloid precursor protein intracellular domain (AICD). *J. Biol. Chem.* 2002; 277: 13389-13393.

[261] Reuning U, Magdolen V, Wilhelm O, Fischer K, Lutz V, Graeff H, Schmitt M. Multifunctional potential of the plasminogen activation system in tumor invasion and metastasis. *Int. J. Oncol.* 1998; 13: 893-906.

[262] Tucker HM, Kihiko M, Caldwell JN, Wright S, Kawarabayashi T, Price D, Walker D, Scheff S, McGillis JP, Rydel RE, Estus S. The plasmin system is induced by and degrades amyloid-β aggregates. *J. Neurosci.* 2000; 20: 3937-3946.

[263] vanNostrand WE, Porter M. Plasmin cleavage of the amyloid β-protein: Alteration of secondary structure and stimulation of tissue plasminogen activator activity. *Biochemistry* 1999; 38: 11570-11576.

[264] Kleiner DE Jr, Stetler-Stevenson WG. Structural biochemistry and activation of matrix metalloproteases. *Curr. Opin. Cell Biol.* 1993; 5: 891-897.

[265] Koike H, Tomioka S, Sorimachi H, Saido TC, Maruyama K, Okuyama A, Fujisawa-Sehara A, Ohno S, Suzuki K, Ishiura S. Membrane-anchored metalloprotease MDC9 has an α-secretase activity responsible for processing the amyloid precursor protein. *Biochem J.* 1999; 343: 371-375.

[266] Eckman EA, Reed DK, Eckman CB. Degradation of the Alzheimer's amyloid β-peptide by endothelin-converting enzyme. *J. Biol. Chem.* 2001; 276: 24540-24548.

[267] Mucke L, Yu GQ, McConlogue L, Rockenstein EM, Abraham CR, Masliah E. Astroglial expression of human α_1-antichymotrypsin enhances Alzheimer-like pathology in amyloid protein precursor transgenic mice. *Am. J. Pathol.* 2000; 157: 2003-2010.

[268] Nilsson LN, Bales KR, DiCarlo G, Gordon MN, Morgan D, Paul SM, Potter H. α_1-Antichymotrypsin promotes β-sheet amyloid plaque deposition in a transgenic mouse model of Alzheimer's disease. *J. Neurosci.* 2001; 21: 1444-1451.

[269] Kordula T, Bugno M, Rydel RE, Travis J. Mechanism of interleukin-1 and tumor necrosis factor α-dependent regulation of the α_1-antichymotrypsin gene in human astrocytes. *J. Neurosci.* 2000; 20: 7510-7516.

[270] Abraham CR, Selkoe DJ, Potter H. Immunochemical identification of the serine protease inhibitor α_1-antichymotrypsin in the brain amyloid deposits of Alzheimer's disease. *Cell* 1988; 52: 487-501.

[271] Fraser PE, Nguyen JT, McLachlan DR, Abraham CR, Kirschner DA. α_1-antichymotrypsin binding to Alzheimer Aβ peptides is sequence specific, and induces fibril disaggregation *in vitro*. *J. Neurochem.* 1993; 61: 298-305.

[272] Sun YX, Wright HT, Janciauskiene S. α_1-antichymotrypsin/Alzheimer's peptide Aβ(1-42) complex perturbs lipid metabolism and activates transcription factors PPARγ and NF-κB in human neuroblastoma (Kelly) cells. *J. Neurosci. Res.* 2002; 67: 511-522.

[273] Herz J, Strickland DK. LRP: A multifunctional scavenger and signalling receptor. *J. Clin. Invest.* 2001; 108: 779-784.

[274] Kang DE, Pietrzik CU, Baum L, Chevallier N, Merriam DE, Kounnas MZ, Wagner SL, Troncoso JC, Kawas CH, Katzman R, Koo EH. Modulation of amyloid β-protein clearance and Alzheimer's disease susceptibility by the LDL receptor-related protein pathway. *J. Clin. Invest.* 2000; 106: 1159-1166.

[275] Shibata M, Yamada S, Kumar SR, Calero M, Bading J, Frangione B, Holtzman DM, Miller CA, Strickland DK, Ghiso J, Zlokovic BV. Clearance of Alzheimer's amyloid-β (1-40) peptide from brain by LDL receptor-related protein-1 at the blood-brain barrier. *J. Clin. Invest.* 2000; 106: 1489-1499.

[276] Borth W. α_2-macroglobulin, a multifunctional binding protein with targeting characteristics. *FASEB J.* 1992; 6: 3345-3353.

[277] vanGool D, De Strooper B, vanLeuven F, Triau E, Dom R. Macroglobulin expression in neuritic-type plaques in patients with Alzheimer's disease. *Neurobiol. Aging* 1993; 14: 233-237.

[278] Du Y, Ni B, Glinn M, Dodel RC, Bales KR, Zhang Z, Hyslop PA, Paul SM. α2-Macroglobulin as a β-amyloid peptide-binding plasma protein. *J. Neurochem.* 1997; 69: 299-305.

[279] Qiu WQ, Borth W, Ye Z, Haass C, Teplow DB, Selkoe DJ. Degradation of amyloid-β protein by a serine protease-2-macroglobulin complex. J. Biol. Chem. 1996; 271: 8443-8451.

[280] Narita M, Holtzman DM, Schwartz AL, Bu G. α_2-macroglobulin complexes with and mediates the endocytosis of β-amyloid peptide via cell surface low-density lipoprotein receptor-related protein. *J. Neurochem.* 1997; 69: 1904-1911.

[281] Holtzman DM. Role of ApoE/Aβ interactions in the pathogenesis of Alzheimer's disease and cerebral amyloid angiopathy. *J. Mol. Neurosci.* 2001; 17: 147-155.

[282] Gearing M, Mori H, Mirra SS. Aβ-peptide length and apolipoprotein E genotype in Alzheimer's disease. *Ann. Neurol.* 1996; 39: 395-399.

[283] Evans DA, Hebert LE, Beckett LA, Scherr PA, Albert MS, Chown MJ, Pilgrim DM, Taylor JO. Education and other measures of socioeconomic status and risk of incident Alzheimer disease in a defined population of older persons. *Arch. Neurol.* 1997; 54: 1399-1405.

[284] Law A, Gauthier S, Quirion R. Say NO to Alzheimer's disease: the putative links between nitric oxide and dementia of the Alzheimer's type. *Brain Res. Rev.* 2001; 35: 73-96.

[285] Mayeux R, Costa R, Bell K, Merchant C, Tung MX, Jacobs D. Reduced risk of Alzheimer's disease among individuals with low calorie intake. *Neurology* 1999; 59: S296-S297.

[286] Weindruch R, Sohal RS. Seminars in medicine of the Beth Israel Deaconess Medical Center. Caloric intake and aging. *N. Engl. J. Med.* 1997; 337: 986-994.

[287] Lee CK, Weindruch R, Prolla TA. Gene-expression profile of the ageing brain in mice. *Nat. Genet.* 2000; 25: 294-297.

[288] Ingram DK, Weindruch R, Spangler EL, Freeman JR, Walford RL. Dietary restriction benefits learning and motor performance of aged mice. J. Gerontol. 1987; 42: 78-81.

[289] Lee J, Duan W, Long JM, Ingram DK, Mattson MP. Dietary restriction increases survival of newly-generated neural cells and induces BDNF expression in the dentate gyrus of rats. *J. Mol. Neurosci.* 2000; 15: 99-108.

[290] Duan W, Guo Z, Mattson MP. Brain-derived neurotrophic factor mediates an excitoprotective effect of dietary restriction in mice. *J. Neurochem.* 2001; 76: 619-626.

[291] Cheng B, Mattson MP. NT-3 and BDNF protect CNS neurons against metabolic/excitotoxic insults. *Brain Res.* 1994; 640: 56-67.

[292] Duan W, Mattson MP. Dietary restriction and 2-deoxyglucose administration improve behavioral outcome and reduce degeneration of dopaminergic neurons in models of Parkinson's disease. *J. Neurosci. Res.* 1999; 57: 195-206.

[293] Lowenstein DH, Chan P, Miles M. The stress protein response in cultured neurons: characterization and evidence for a protective role in excitotoxicity. *Neuron* 1991; 7: 1053-1060.

[294] Smith PJ, Hammar K, Tytell M. Effects of exogenous heat shock protein (hsp70) on neuronal calcium flux. *Biol. Bull.* 1995; 189: 209-210.

[295] Bruce-Keller AJ, Umberger G, McFall R, Mattson MP. Food restriction reduces brain damage and improves behavioral outcome following excitotoxic and metabolic insults. *Ann. Neurol.* 1999; 45: 8-15.

[296] Swain RA, St Clair L. The role of folic acid in deficiency states and prevention of disease. *J. Fam. Pract.* 1997; 44: 138-144.

[297] Choi SW, Mason JB. Folate and carcinogenesis: an integrated scheme. *J. Nutr.* 2000; 130: 129-132.

[298] Scott JM, Weir DG. Folic acid, homocysteine and one-carbon methabolism: a review of the essential biochemistry. *J. Cardiovasc. Risk* 1998; 5: 223-227.

[299] Kruman I, Chan SL, Culmsee C, Kruman Y, Penix L, Mattson MP. Homocysteine elicits a DNA damage response in neurons resulting in apoptosis and hypersensitivity to excitotoxicity. *J. Neurosci.* 2000; 20: 6920-6926.

[300] Kruman II, Kumaravel TS, Lohani A, Cutler RG, Kruman Y, Haughey N, Pedersen WA, Evans MK, Mattson MP. Folate deficiency and homocysteine enhance amyloid toxicity by impairing DNA repair. *Soc. Neurosci. Abstr.* 2001; 31: 962-969.

[301] Black JE, Sirevaag AM, Wallace CS, Savin MH, Greenough WT. Effects of complex experience on somatic growth and organ development in rats. *Dev. Psychobiol.* 1989; 22: 727-752.

[302] Kleim JA, Vij K, Ballard DH, Greenough WT. Learning-dependent synaptic modifications in the cerebellar cortex of the adult rat persist for at least four weeks. *J. Neurosci.* 1997; 17: 717-721.

[303] Saito S, Kobayashi S, Ohashi Y, Igarashi M, Komiya Y, Ando S. Decreased synaptic density in aged brains and its prevention by rearing under enriched environment as revealed by synaptophysin contents. *J. Neurosci. Res.* 1994; 39: 57-62.

[304] Kolb B, Gibb R. Environmental enrichment and cortical injury: behavioral and anatomical consequences of frontal cortex lesions. *Cereb. Cortex* 1991; 1: 189-198.

[305] Nilsson M, Perfilieva E, Johansson U, Orwar O, Eriksson P. Enriched environment increases neurogenesis in the adult rat dentate gyrus and improves spatial memory. *J. Neurobiol.* 1999; 39: 569-578.

[306] van Praag H, Kempermann G, Gage FH. Running increases cell proliferation and neurogenesis in the adult mouse dentate gyrus. *Nat. Neurosci.* 1999; 2: 266-270.

[307] Jones TA, Chu CJ, Grande LA, Gregory AD. Motor skills training enhances lesion-induced structural plasticity in the motor cortex of adult rats. *J. Neurosci.* 1999; 19: 10153-10163.

[308] Russo-Neustadt AA, Beard RC, Huang YM, Cotman CW. Physical activity and antidepressant treatment potentiate the expression of specific brain-derived neurotrophic factor transcripts in the rat hippocampus. *Neuroscience* 2000; 101: 305-312.

[309] Trejo JL, Carro E, Torres-Aleman I. Circulating insulin-like growth factor I mediates exercise-induced increases in the number of new neurons in the adult hippocampus. *J. Neurosci.* 2001; 21: 1628-1634.

[310] Tang MX, Jacobs D, Stern Y, Marder K, Schofield P, Gurland B, Andrews H, Mayeux R. Effect of oestrogen during menopause on risk and age at onset of Alzheimer's disease. *Lancet* 1996; 348: 429-432.

[311] Woolley CS, McEwen BS. Estradiol regulates hippocampal dendritic spine density via an N-methyl-D-aspartate receptor-dependent mechanism. *J. Neurosci.* 1994; 14: 7680-7687.

[312] Goodman Y, Bruce AJ, Cheng B, Mattson MP. Estrogens attenuate and corticosterone exacerbates excitotoxicity, oxidative injury and amyloid β-peptide toxicity in hippocampal neurons. *J. Neurochem.* 1996; 66: 1836-1844.

[313] Keller JN, Mattson MP. 17β-estradiol attenuates oxidative impairment of synaptic Na^+/K^+-ATPase activity, glucose transport and glutamate transport induced by amyloid β-peptide and iron. *J. Neurosci. Res.* 1997; 50: 522-530.

[314] Jaffe AB, Toran-Allerand CD, Greengard P, Gandy SE. Estrogen regulates metabolism of Alzheimer amyloid-β precursor protein. *J. Biol. Chem.* 1994; 269: 13065-13068.

[315] Hardy J. Amyloid, the presenilins and Alzheimer's disease. *Trends Neurosci.* 1997; 20: 154-159.

[316] Rumble B, Retallack R, Hilbich C, Simms G, Multhaup G, Martins R, Hockey A, Montgomery P, Beyreuther K, Masters CL. Amyloid A4 protein and its precursor in Down's syndrome and Alzheimer's disease. *N. Engl. J. Med.* 1989; 320: 1446-1452.

[317] Wisniewski KE, Dalton AJ, McLachlan C, Wen GY, Wisniewski HM. Alzheimer's disease in Down's syndrome: clinicopathologic studies. *Neurology* 1985; 35: 957-961.

[318] Mann DM, Esiri MM. The pattern of acquisition of plaques and tangles in the brains of patients under 50 years of age with Down's syndrome. *J. Neurol. Sci.* 1989; 89: 169-179.

[319] Chartier-Harlin MC, Crawford F, Houlden H, Warren A, Hughes D, Fidani L, Goate A, Rossor M, Roques P, Hardy J, Mullan M. Early-onset Alzheimer's disease caused by mutations at codon 717 of the β-amyloid precursor protein gene. *Nature* 1991; 353: 844-846.

[320] Tanzi R, Hyman BT. Alzheimer's mutation. *Nature* 1991; 350: 564.

[321] Cai XD, Golde TE, Younkin SG. Release of excess amyloid β protein from a mutant amyloid β protein precursor. *Science* 1993; 259: 514-516.

[322] Cruts M, Hendriks L, van Broeckhoven C. The presenilin genes: a new gene family involved in Alzheimer's disease pathology. *Hum. Mol. Genet.* 1996; 5: 1449-1455.

[323] Kovács DM, Fausett HJ, Page KJ, Kim TW, Moir RD, Merriam DE, Hollister RD, Hallmark OG, Mancini R, Felsenstein KM, Hyman BT, Tanzi RE, Wasco W. Alzheimer associated presenilins 1 and 2: neuronal expression in brain and localization to intracellular membranes in mammalian cells. *Nature Med.* 1996; 2: 224-229.

[324] Scheuner D, Eckman C, Jensen M, Song X, Citron M, Suzuki N, Bird TD, Hardy J, Hutton M, Kukull W, Larson E, Levy-Lehad E, Tiitanen M, Peskind E, Poorkaj P, Schellenberg G, Tanzi RE, Wasco W, Lannfelt L, Selkoe D, Younkin S. Aβ42(43) is increased *in vivo* by the PS1/2 and APP mutations linked to familial Alzheimer's disease. *Nature Med.* 1996; 2: 864-870.

[325] Rogaev EI, Sherrington R, Rogaeva EA, Levesque G, Ikeda M, Liang Y, Chi H, Lin C, Holman K, Tsuda T, Mar L, Sorbi S, Nacmias B, Piacentini S, Amaducci L, Chumakov I, Cohen D, Lannfelt L, Fraser PE, Rommens JM, St. George-Hyslop PH. Familial Alzheimer's disease in kindreds with missense mutations in a gene on chromosome 1 related to the Alzheimer's disease type 3 gene. *Nature* 1995; 376: 775-778.

[326] Huse JT, Doms RW. Closing in on the amyloid cascade: Recent insights into the cell biology of Alzheimer's disease. *Mol. Neurobiol.* 2000; 22: 81-98.

[327] Capell A, Grünberg J, Pesold B, Diehlmann A, Citron M, Nixon R, Beyreuther K, Selkoe DJ, Haass C. The proteolytic fragments of the Alzheimer's disease-associated presenilin-1 form heterodimers and occur as a 100-150-kDa molecular mass. *J. Biol. Chem.* 1998; 273: 3205-3211.

[328] Zhang J, Kang DE, Xia W, Okochi M, Mori H, Selkoe DJ, Koo EH. Subcellular distribution and turnover of presenilins in transfected cells. *J. Biol. Chem.* 1998; 273: 12436-12442.

[329] Smine A, Xu X, Nishiyama K, Katada T, Gambetti P, Yadav SP, Wu X, Shi YC, Yasuhara S, Homburger V, Okamoto T. Regulation of brain G-protein G_o by Alzheimer's disease gene presenilin-1. *J. Biol. Chem.* 1998; 273: 16281-16288.

[330] Dumanchin C, Czech C, Campion D, Cuif MH, Poyot T, Charbonnier F, Goud B, Pradier L, Frebourg T. Presenilins interact with Rab11, a small GTPase involved in the regulation of vesicular transport. *Hum. Mol. Genet.* 1999; 8: 1263-1269.

[331] Weidemann A, Paliga K, Dürrwang U, Czech C, Evin G, Masters C, Beyreuther K. Formation of stable complexes between two Alzheimer's disease gene products: presenilin-2 and β-amyloid precursor protein. *Nature Med.* 1997; 3: 328-332.

[332] Takashima A, Honda T, Yasutake K, Michel G, Murayama O, Murayama M, Ishiguro K, Yamaguchi H. Activation of tau protein kinase I/glycogen synthase kinase-3β by amyloid-β peptide (25-35) enhances phosphorylation of tau in hippocampal neurons. *Neurosci. Res.* 1998; 31: 317-323.

[333] Takashima A, Murayama M, Murayama O, Kohno T, Honda T, Yasutake K, Nihonmatsu N, Mercken M, Yamaguchi H, Sugihara S, Wolozin B. Presenilin1 associates with glycogen synthase kinase-3β and its substrate tau. *Proc. Natl. Acad. Sci. USA* 1998; 95: 9637-9641.

[334] Goedert M. Tau protein and the neurofibrillary pathology of Alzheimer's disease. *Trends Neurol. Sci.* 1993; 16: 460-465.

[335] Buxbaum JD, Choi EK, Luo Y, Lilliehook C, Crowley AC, Merriam DE, Wasco W. Calsenilin: a calcium-binding protein that interacts with the presenilins and regulates the levels of a presenilin fragment. *Nature Med.* 1998; 4: 1177-1181.

[336] Leissring MA, Paul BA, Parker I, Cotman CW, LaFerla FM. Alzheimer's presenilin-1 mutation potentiates inositol 1,4,5-trisphosphate-mediated calcium signaling in *Xenopus* oocytes. *J. Neurochem.* 1999; 72: 1061-1068.

[337] Mattson MP, LaFerla FM. Capacitative calcium entry deficits and elevated luminal calcium content in mutant presenilin-1 knockin mice. *J. Cell Biol.* 2000; 149: 793-798.

[338] Mattson MP, Duan W. Apoptotic biochemical cascades in synaptic compartments: roles in adaptive plasticity and neurodegenerative disorders. *J. Neurosci. Res.* 1999; 58: 152-166.

[339] Nabeshima T, Nitta A. Memory impairment and neuronal dysfunction induced by β-amyloid protein in rats. Tohoku *J. Exp. Med.* 1994; 174: 241-249.

[340] Ishida A, Furukawa K, Keller JN, Mattson MP. Secreted form of β-amyloid precursor protein shifts the frequency dependence for induction of LTD, and enhances LTP in hippocampal slices. *NeuroReport* 1997; 8: 2133-2137.

[341] Parent A, Linden DJ, Sisodia SS, Borchelt DR. Synaptic transmission and hippocampal long-term potentiation in transgenic mice expressing FAD-linked presenilin 1. *Neurobiol. Dis.* 1999; 6: 56-62.

[342] Barrow PA, Empson RM, Gladwell SJ, Anderson CM, Killick R, Yu X, Jefferys JG, Duff K. Functional phenotype in transgenic mice expressing mutant human presenilin-1. *Neurobiol. Dis.* 2000; 7: 119-126.

[343] van Gassen G, de Jonghe C, Pype S, van Criekinge W, Julliams A, Vanderhoeven I, Woodrow S, Beyaert R, Huylebroeck D, van Broeckhoven C. Alzheimer's disease associated presenilin 1 interacts with HC5 and Zeta, subunits of the catalytic 20S proteasome. *Neurobiol. Dis.* 1999; 6: 376-391.

[344] Marambaud P, Ancolio K, Lopez-Perez E, Checler F. Proteasome inhibitors prevent the degradation of familial Alzheimer's disease-linked presenilin 1 and trigger increased Aβ42 secretion by human cells. *Mol. Med.* 1998; 4: 147-157.

[345] Marambaud P, Chevallier N, Ancolio K, Checler F. Post-transcriptional contribution of a cAMP-dependent pathway to the formation of α and β/γ-secretases-derived products of βAPP maturation in human cells expressing wild-type and Swedish mutated βAPP. *Mol. Med.* 1998; 4: 715-723.

[346] Lee HJ, Jung KM, Huang YZ, Bennett LB, Lee JS, Mei L, Kim TW. Presenilin-dependent γ-secretase-like intramembrane cleavage of ErbB4. *J. Biol. Chem.* 2002; 277: 6318-6323.

[347] Baki L, Marambaud P, Efthimiopoulos S, Georgakopoulos A, Wen P, Cui W, Ozawa M, Friedrich VL Jr, Robakis NK. Presenilin-1 binds cytoplasmic epithelial cadherin, inhibits epithelial cadherin/p120 association and regulates stability and function of the cadherin adhesion complex. *Proc. Natl. Acad. Sci. USA* 2001; 98: 2381-2386.

[348] Annaert WG, Esselens C, Baert V, Boeve C, Snellings G, Cupers P, Craessaerts K, De Strooper B. Interaction with telencephalin and the amyloid precursor protein predicts a ring structure for presenilins. *Neuron* 2001; 32: 579-589.

[349] Ray WJ, Yao M, Nowotny P, Mumm J, Zhang W, Wu J, Kopan R, Goate A. Evidence for a physical interaction between presenilin and Notch. *Proc. Natl. Acad. Sci. USA* 1999; 96: 3263-3268.

[350] Kim TW, Pettingell WH, Jung YK, Kovács DM, Tanzi RE. Alternative cleavage of Alzheimer-associated presenilins during apoptosis by a caspase-3 family protease. *Science* (Wash DC) 1997; 277: 373-376.

[351] Alves da Costa C, Paitel E, Mattson MP, Amson R, Telerman A, Ancolio K, Checler F. Wild-type and mutated presenilin-2 trigger p53-dependent apoptosis and down-regulate presenilin-1 expression in HEK293 human cells and in murine neurons. *Proc. Natl. Acad. Sci. USA* 2002; 99: 4043-4048.

[352] Hong CS, Caromille L, Nomata Y, Mori H, Bredesen DE, Koo EH. Contrasting role of presenilin-1 and presenilin-2 in neuronal differentiation *in vitro*. *J. Neurosci.* 1999; 19: 637-643.

[353] Jeong SJ, Kim HS, Chang KA, Geum DH, Park CH, Seo JH, Rah JC, Lee JH, Choi SH, Lee SG, Kim K, Suh YH. Subcellular localization of presenilins during mouse preimplantation development. *FASEB J.* 2000; 14: 2171-2176.

[354] Strittmatter WJ, Saunders AM, Schmechel D, Pericak-Vance M, Enghild J, Salvesen GS, Roses AD. Apolipoprotein E: high-avidity binding to β-amyloid and increased frequency of type 4 allele in late-onset familial Alzheimer disease. *Proc. Natl. Acad. Sci. USA* 1993; 90: 1977-1981.

[355] Miyata M, Smith JD. Apolipoprotein E allele-specific antioxidant activity and effects on cytotoxicity by oxidative insults and β-amyloid peptides. *Nature Genet.* 1996; 14: 55-61.

[356] Pedersen WA, Chan SL, Mattson MP. A mechanism for the neuroprotective effect of apolipoprotein E: isoform-specific modification by the lipid peroxidation product 4-hydroxynonenal. J. Neurochem. 2000; 74: 1426-1433.

[357] Tolar M, Keller JN, Chan SL, Mattson MP, Marques MA, Crutcher KA. Truncated apolipoprotein E (ApoE) causes increased intracellular calcium and may mediate ApoE neurotoxicity. *J. Neurosci.* 1999; 19: 7100-7110.

[358] Strittmatter WJ, Saunders AM, Schmechel D, Pericak-Vance MA, Enghild J, Salvesen GS, Roses AD. Apolipoprotein E: high-avidity binding to β-amyloid and increased frequency of type 4 allele in late-onset familial Alzheimer's disease. *Proc. Natl. Acad. Sci. USA* 1993; 90: 1977-1981.

[359] Blacker D, Wilcox MA, Laird NM, Rodes L, Horváth SM, Go RC, Perry R, Watson B Jr, Bassett SS, McInnis MG, Albert MS, Hyman BT, Tanzi RE. α2-macroglobulin is genetically associated with Alzheimer disease. *Nature Genet.* 1998; 19: 357-360.

[360] Ali G, Wasco W, Cai X, Szabó P, Sheu KF, Cooper AJ, Gaston SM, Gusella JF, Tanzi RE, Blass JP. Isolation, characterization and mapping of gene encoding dihydrolipoyl succinyltransferase (E2k) of human α-ketoglutarate dehydrogenase complex. *Somat. Cell. Mol. Genet.* 1994; 20: 99-105.

[361] Lehmann DJ, Johnston C, Smith AD. Synergy between the genes for butyrylcholinesterase K variant and apolipoprotein E4 in late-onset confirmed Alzheimer's disease. *Hum. Mol. Genet.* 1997; 6: 1933-1936.

[362] Uéda K, Fukushima H, Masliah E, Xia Y, Iwai A, Yoshimoto M, Otero DAC, Kondo J, Ihara Y, Saitoh T. Molecular cloning of cDNA encoding an unrecognized component of amyloid in Alzheimer disease. *Proc. Natl. Acad. Sci. USA* 1993; 90: 11282-11286.

[363] Chen X, Rohan de Silva HA, Pettenati MJ, Rao PN, St. George-Hyslop P, Roses AD, Xia Y, Horsburgh K, Uéda K, Saitoh T. The human NACP/α-synuclein gene: chromosome assignment to 4q21.3-q22 and taqI RFLP analysis. *Genomics* 1995; 26: 425-427.

[364] Spillantini MG, Divane A, Goedert M. Assignment of human α-synuclein (SNCA) and β-synuclein (SNCB) genes to chromosomes 4q21 and 5q35. *Genomics* 1995; 27: 379-381.

[365] Lavedan C, Leroy E, Dehejia A, Biuchholtz S, Dutra A, Nussbaum RL, Polymeropoulos MH. Identification, localization and characterization of the human α-synuclein gene. *Hum. Mol. Genet.* 1998; 103: 106-112.

[366] Baba M, Nakajo S, Tu PH, Tomita T, Nakaya K, Lee VMY, Trojanowski JQ, Iwatsubo T. Aggregation of α-synuclein in Lewy bodies of sporadic Parkinson's disease and dementia with Lewy bodies. *Am. J. Pathol.* 1998; 152: 879-884.

[367] Ditter SM, Mirra SS. Neuropathologic and clinical features of Parkinson's disease in Alzheimer's disease patients. *Neurology* 1987; 37: 754-760.

[368] Hansen LA, Salmon D, Galasko D, Masliah E, Katzman R, DeTeresa R, Thal L, Pay MM, Hofstetter R, Klauber M. The Lewy body variant of Alzheimer's disease: a clinical and pathologic entity. *Neurology* 1990; 40: 1-8.

[369] Galasko D, Hansen L, Katzman R, Wiederholt W, Masliah E, Terry R, Hill LR, Lessin P, Thal LJ. Clinical-neuropathological correlations in Alzheimer's disease and related dementias. *Arch. Neurol.* 1994; 51: 888-895.

[370] Lippa CF, Schmidt ML, Lee VM, Trojanowski JQ. Antibodies to α-synuclein detect Lewy bodies in many Down's syndrome brains with Alzheimer's disease. Ann. Neurol. 1999; 45: 353-357.

[371] Hamilton RL. Lewy bodies in Alzheimer's disease: a neuropathological review of 145 cases using α-synuclein immunohistochemistry. *Brain Pathol.* 2000; 10: 378-384.

[372] Arai Y, Yamazaki M, Mori O, Muramatsu H, Asano G, Katayama Y. α-Synuclein-positive structures in cases with sporadic Alzheimer's disease: morphology and its relationship to tau aggregation. *Brain Res.* 2001; 888: 287-296.

[373] Hong L, Ko HW, Gwag BJ, Joe E, Lee S, Kim YT, Suh YH. The cDNA cloning and ontogeny of mouse α-synuclein. *NeuroReport* 1998; 9: 1239-1243.

[374] Maroteaux L, Campanelli JT, Scheller RH. Synuclein: a neuron-specific protein localized to the nucleus and presynaptic nerve terminal. *J. Neurosci.* 1988; 8: 2804-2815.

[375] El-Agnaf OMA, Jakes R, Curran MD, Middleton D, Ingenito R, Bianchi E, Pessi A, Neill D, Wallace A. Aggregates from mutant and wild-type α-synuclein proteins and NAC peptide induce apoptotic cell death in human neuroblastoma cells by formation of β-sheet and amyloid like filaments. *FEBS Lett.* 1998; 440: 71-75.

[376] Fujiwara H, Hasegawa M, Dohmae N, Kawashima A, Masliah E, Goldberg S, Shen J, Takio K, Iwatsubo T. α-synuclein is phosphorylated in synucleinopathy lesions. *Nat. Cell. Biol.* 2002; 4: 160-164.

[377] Iwata A, Maruyama M, Kanazawa I, Nukina N. α-Synuclein affects the MAP kinase pathway and accelerates cell death. *J. Biol. Chem.* 2001; 276: 45320-45329.

[378] Masters SC, Fu H. 14-3-3 proteins mediate an essential antiapoptotic signal. *J. Biol. Chem.* 2001; 276: 45193-45200.

[379] Alves da Costa C, Ancolio K, Checler F. Wild-type but not Parkinson's disease-related Ala53Thr-α-synuclein protect neuronal cells from apoptotic stimuli. *J. Biol. Chem.* 2000; 275: 24065-24069.

[380] Souza JM, Giasson BI, Lee VMY, Ischiropoulos H. Chaperone-like activity of synucleins. *FEBS Lett.* 2000; 474: 116-119.

[381] Henning-Jensen P, Hager H, Nielsen MS, Hojrup P, Gliemann J, Jakes R. α-Synuclein binds to tau and stimulates the protein kinase A-catalyzes tau phosphorylation of serine 262 and 356. *J. Biol. Chem.* 1999; 274: 25481-25485.

[382] Ghee M, Fournier A, Mallet J. Rat α-synuclein interacts with Tat binding protein 1, a component of the 26S proteasomal complex. *J. Neurochem.* 2000; 75: 2221-2224.

[383] Shimura H, Schlossmacher MG, Hattori N, Frosch MP, Trockenbacher A, Schneider R, Mizuno Y, Kosik KS, Selkoe DJ. Ubiquitination of a new form of α-synuclein by parkin from human brain: implications for Parkinson's disease. *Science* (Wash DC) 2001; 293: 263-269.

[384] Tofaris GK, Layfield R, Spillantini MG. α-synuclein metabolism and aggregation is linked to ubiquitin-independent degradation by the proteasome. *FEBS Lett.* 2001; 509: 22-26.

[385] Conway KA, Rochet JC, Bieganski RM, Lansbury PT Jr. Kinetic stabilization of the α-synuclein protofibril by a dopamine-α-synuclein adduct. *Science* (Wash DC) 2001; 294: 1346-1349.

[386] Hashimoto M, Rockenstein E, Mante M, Mallory M, Masliah E. β-Synuclein inhibits α-synuclein aggregation: a possible role as an anti-Parkinsonian factor. *Neuron* 2001; 32: 213-223.

[387] Drevets WC, Rubin EH.: Psychotic symptoms and the longitudinal course of Alzheimer disease. *Biol. Psychiatry* 1989; 25: 39-48.

[388] Mega MS, Cummings JL, Fiorello T, Gornbein J. The spectrum of behavioral changes in Alzheimer's disease. *Neurology* 1996; 46: 130-135.

[389] Lyketsos CG, Tune LE, Pearlson G, Steele C. Major depression in Alzheimer's disease: an interaction between gender and family history. *Psychosomatics* 1996; 37: 380-389.

[390] Migliorelli R, Tesona Sabe L, Petracchi M, Leiguarda R, Starkstein SE. Prevalence and correlates of dysthymia and major depression among patients with Alzheimer's disease. *Am. J. Psychiatry* 1995; 152: 37-44.

[391] Harwood DG, Barker WW, Ownby RL, Duara R. Association between premorbid history of depression and current depression in Alzheimer's disease. *J. Geriatr. Psychiatry Neurology* 1999; 12: 72-75.

[392] Olin JT, Katz IR, Meyers BS, Schneider LS, Lebowitz BD. Provisional diagnostic criteria for depression of Alzheimer disease: Rationale and background. *Am. J. Geriatr. Psychiatry* 2002; 10: 129-141.

[393] Förstl H, Burns A, Luthert P. Clinical and neuropathological correlates of depression in Alzheimer's disease. *Psychol. Med.* 1992; 22: 877-884.

[394] Zubenko GS, Moossy J, Martinez AJ, Rao G, Claassen D, Rosen J. Neuropathologic and neurochemical correlates of psychosis in primary dementia. *Arch. Neurol.* 1991; 48: 619-624.

[395] Lyketsos CG, Steinberg M, Tschantz J, Norton M, Steffens D, Breitner JCS. Mental and behavioral disturbances in dementia: Findings from the Cache County Study on Memory in Aging. *Am. J. Psychiatry* 2000; 157: 708-714.

[396] Gonzales-Salvador T, Lyketsos CG, Baker AS, Roques C, Hovanek L, Steele CD, Brandt J. Quality of life of patients with dementia in long-term care. *Int. J. Ger. Psychiatry* 2000; 15: 181-189.

[397] Kopetz S, Steele CD, Brandt J, Baker AS, Kronberg M, Galik E, Steinberg M, Warren A, Lyketsos CG. Characteristics and outcomes of dementia residents in an assisted living facility. *Int. J. Geriatr. Psychiatry* 2000; 15: 586-593.

[398] Gonzales-Salvador T, Aragano C, Lyketsos CG, Barba AC. The stress and psychological morbidity of the Alzheimer patient caregiver. *Int. J. Ger. Psychiatry* 1999; 14: 701-710.

[399] Finkel SI. The signs of the behavioural and psychological symptoms of dementia. *Clinician* 1998; 16: 33-42.

[400] Frisoni GB, Rozzini L, Gozzetti A, Binetti G, Zanetti O, Bianchetti A, Trabucchi M, Cummings JL. Behavioral syndromes in Alzheimer's disease: description and correlates. *Dement. Geriatr. Cogn. Disord.* 1999; 10: 130-138.

[401] Folstein MF. Differential diagnosis of dementia. The clinical process. *Psychiatr. Clin. North Am.* 1997; 20: 45-57.

[402] Förstl H, Burns A, Levy R, Cairns N. Neuropathological correlates of psychotic phenomena in confirmed Alzheimer's disease. *Br. J. Psychiatry* 1994; 165: 53-59.

[403] Mkaetova-Ladinska EB, Harrington CR, Xuereb J. In: Bergener M, Finkel SI (Eds.) *Treating Alzheimer's and other dementias.* Springer, New York, 1995.

[404] Kirby M, Lawlor BA. Biologic markers and neurochemical correlates of agitation and psychosis in dementia. *J. Geriatr. Psychiatry Neurol.* 1995; 8: S2-S7.

[405] Sultzer DL. Neuroimaging and the origin of psychiatric symptoms in dementia. *Int. Psychogeriatr.* 1996; 8: 239-243.

[406] Arnold SE, Franz BR, Trojanowski JQ. Elderly patients with schizophrenia exhibit infrequent neurodegenerative lesions. *Neurobiol. Aging* 1994; 15: 299-303.

[407] Casanova MF, Carosella NW, Gold JM, Kleinman JE, Weinberger DR, Powers RE. A topographical study of senile plaques and neurofibrillary tangles in the hippocampi of patients with Alzheimer's disease and congnitively impaired patients with schizophrenia. *Psychiatry Res.* 1993; 49: 41-62.

[408] Sisodia SS. Evidence that β-amyloid protein in Alzheimer's disease is not derived by normal processing. *Science* 1990; 248: 492-495.

[409] Soininen H, Syrjanen S, Heinonen O, Neittaanmaki H, Miettinen R, Paljarvi L, Syrjanen K, Beyreuther K, Riekkinen P. Amyloid-β protein deposition in skin of patients with dementia. *Lancet* 1992; 339: 245.

[410] Etcheberrigaray R, Bhagavan S. Ionic and signal transduction alterations in Alzheimer's disease: relevance of studies on peripheral cells. *Mol. Neurobiol.* 1999; 20: 93-109.

[411] Bergman M, Salman H, Beloosesky Y, Djaldetti M, Bessler H. Are peripheral blood cells from patients with Alzheimer's disease more sensitive to apoptotic stimuli? *Alzheimer Dis. Assoc. Dis.* 2002; 16: 156-160.

[412] Eckert A, Cotman CW, Zerfass R, Hennerici M, Muller WE. Lymphocytes as cell model to study apoptosis in Alzheimer's disease: vulnerability to programmed cell death appears to be altered. *J. Neural. Transm. Suppl.* 1998; 54: 259-67.

[413] Mórocz M, Kálmán J, Juhász A, Sinkó I, McGlynn AP, Downes CS, Janka Z, Raskó I. Elevated levels of oxidative DNA damage in lymphocytes from patients with Alzheimer's disease. *Neurobiol. Aging* 2002; 23: 47-53.

[414] Zubenko GS, Cohen BM, Growdon J, Corkin S. Cell membrane abnormality in Alzheimer's disease. *Lancet* 1984; 2: 235.

[415] Kálmán J, Dey I, Ilona SV, Matkovics B, Brown D, Janka Z, Farkas T, Joó F. Platelet membrane fluidity and plasma malondialdehyde levels in Alzheimer's demented patients with and without family history of dementia. *Biol. Psychiat.* 1994; 35: 190-194.

[416] Zubenko GS. Endoplasmic reticulum abnormality in Alzheimer's disease: selective alteration in platelet NADH-cytochrome C reductase activity. *J. Geriatr. Psychiat. Neurol.* 1989; 2: 3-10.

[417] Sevush S, Jy W, Horstman LL, Mao WW, Kolodny L, Ahn YS. Platelet activation in Alzheimer disease. *Arch. Neurol.* 1998; 55: 530-536.

[418] Perry RH, Wilson ID, Bober MJ, Atack J, Blessed G, Tomlinson BE, Perry EK. Plasma and erythrocyte acetylcholinesterase in senile dementia of Alzheimer type. *Lancet* 1982; 1: 174-175.

[419] Peterson C, Gibson GE, Blass JP. Altered calcium uptake in cultured skin fibroblasts from patients with Alzheimer's disease. *N. Engl. J. Med.* 1985; 312: 1063-1065.

[420] Peterson C, Goldman JE. Alterations in calcium content and biochemical processes in cultured skin fibroblasts from aged and Alzheimer donors. *Proc. Natl. Acad. Sci. USA* 1986; 83: 2758-2762.

[421] 1Sims NR, Finegan JM, Blass JP. Altered glucose metabolism in fibroblasts from patients with Alzheimer's disease. *N. Engl. J. Med.* 1985; 313: 683-684.

[422] Paoletti F, Tombaccini D. Proteolytic imbalance in Alzheimer fibroblasts as potential pathological trait of disease. *FASEB J.* 1998; 12: 925-927.

[423] Gibson G, Martins R, Blass J, Gandy S. Altered oxidation and signal transduction systems in fibroblasts from Alzheimer patients. In: Mailer D, Gispen W (Eds.) The current status of the calcium hypothesis of brain aging and Alzheimer's disease. *Life Science* 1996; 59: 477-490.

[424] Li JC, Kaminskas E. Deficient repair of DNA lesions in Alzheimer's disease fibroblasts. *Biochem. Biophys. Res. Commun.* 1985; 129: 733-738.

[425] Connolly GP. Fibroblast models of neurological disorders: fluorescence measurement studies. *Trends Pharmacol. Sci.* 1998; 19: 171-177.

[426] Dolman CL. Diagnosis of neurometabolic disorders by examination of skin biopsies and lymphocytes. *Semin. Diagn. Pathol.* 1984; 1: 82-97.

[427] Idoate-Gastearena MA, Vega-Vazquez F. Diagnosis of neurometabolic and neurodegenerative diseases by cutaneous biopsy. *Rev. Neurol.* 1997; 25: S269-S280.

[428] Huang HM, Martins R, Gandy S, Etcheberrigaray R, Ito E, Alkon DL, Blass JP, Gibson GE. The use of cultured fibroblasts in elucidating the pathophysiology and diagnosis of Alzheimer's disease. In: Disterhoft JF, Gispen WH, Traber J Khachaturian ZS (Eds.) *Calcium Hypothesis of Aging and Dementia* New York Academy of Science, New York, 1994; 225-244.

[429] 1Giambalvo CT. Protein kinase C and dopamine release – II: Effect of dopamine acting drugs *in vivo*. *Biochem. Pharmacol.* 1988; 37: 4009-4017.

[430] Walker JM, Bowen WD, Walker FO, Matsumoto RR, De Costa B, Rice KC. Sigma receptors: biology and function. *Pharmacol. Rev.* 1990; 42: 355-401.

[431] Behl C, Lezoualc'h F, Widmann M, Rupprecht R, Holsboer F. Oxidative stress-resistant cells are protected against haloperidol toxicity. *Brain Res.* 1996; 717: 193-195.

[432] Avent KM, Usuki E, Eyles DW, Keeve R, van der Schyf CJ, Castagnoli N, Pond SM. Haloperidol and its tetrahydropyridine derivative (HPTP) are metabolized to potentially neurotoxic pyridium species in the baboon. *Life Sci.* 1996; 59: 1473-1482.

[433] Prokopenko RA, Mogilevich SE, Luik AI, Naydyenova IY, Batrak GN, Hawryluk BR, Degtiar VY. Effects of haloperidol and chlorpromazine on smooth muscle contractility, platelet aggregation and neuronal calcium current. *Gen. Physiol. Biophys.* 1995; 14: 349-357.

[434] Higaki J, Murphy GM Jr, Cordell B. Inhibition of β-amyloid formation by haloperidol: a possible mechanism for reduced frequency of Alzheimer's disease pathology in schizophrenia. *J. Neurochem.* 1997; 68: 333-336.

[435] Lane HY, Chang YC, Su MH, Chiu CC, Huang MC, Chang WH. Shifting from haloperidol to risperidone for behavioral disturbances in dementia: safety, response predictors, and mood effects. *J. Clin. Psychopharmacol.* 2002; 22: 4-10.

[436] Chan WC, Lam LC, Choy CN, Leung VP, Li SW, Chiu HF. A double-blind randomised comparison of risperidone and haloperidol in the treatment of behavioural and psychological symptoms in Chinese dementia patients. *Int. J. Geriatr. Psychiatry* 2001; 16: 1156-1162.

[437] Borc JJ, Larson DB, Lyons JS, Beardsley RS. Expenditures for psychotropic medications in the United States. *Am. J. Psychiatry* 1991; 148: 644-647.

[438] Volavka J, Cooper TB. Review of haloperidol blood level and clinical response: looking through the window. *J. Clin. Psychopharmacol.* 1987; 7: 25-30.

[439] vanPutten T, Marder SR, Mintz J, Poland RE. Haloperidol plasma levels and clinical response: a therapeutic window relationship. *Psyhcopharmacol. Bull.* 1988; 24: 172-175.

[440] Aravagiri M, Marder SR, Wirshing D, Wirshing WC. Plasma concentrations of risperidone and its 9-hydroxy metabolite and their relationship to dose in schizophrenic patients: simultaneous determination by a high performance liquid chromatography with electrochemical detection. *Pharmacopsychiatry* 1998; 31: 102-109.

[441] Ereshefsky L, Lacombe S. Pharmacological profile of risperidone. *Can. J. Psychiatry* 1993; 38: S80-S88.

[442] Laskay G, Zarándi M, Varga J, Jost K, Fónagy A, Torday Cs, Latzkovits L, Penke B. A putative tetrapeptide antagonist prevents β-amyloid-induced long-term elevation of [Ca2+]i in rat astrocytes. *Biochem. Biophys. Res. Commun.* 1997; 235:479-481.

[443] Lowry OH, Rosenbrough NJ, Farr AL, Randall RJ. Protein measurement with the Folin-phenol reagents. *J. Biol. Chem.* 1951; 193: 265-275.

[444] Stewart J M, Young JD. *Solid Phase Peptide Synthesis*. 2nd Ed. Pierce, Rockford, IL. 1984.

[445] Kaiser E. Color test for detection of free terminal amino groups in the solid-phase synthesis of peptides. *Anal. Biochem.* 1970; 34: 595-598.

[446] Campbell A, Baldessarini RJ. Effects of maturation and aging on behavioral responses to haloperidol in the rat. *Psychopharmacology* (Berl) 1981; 73: 219-222.

[447] Dunn LA, Atwater GE, Kilts CD. Effects of antipsychotic drugs on latent inhibition: sensitivity and specificity of an animal behavioral model of clinical drug action. *Psychopharmacology* (Berl) 1993; 112: 315-323.

[448] Di Matteo V, Cacchio M, Di Giulio C, Di Giovanni G, Esposito E. Biochemical evidence that the atypical antipsychotic drugs clozapine and risperidone block 5-HT2C receptors *in vivo*. *Pharmacol. Biochem. Behav.* 2002; 71: 607-613.

[449] Broerse A, Timmerman W, Westerink BH, Den BJ. Antipsychotics and single-cell activity in the rat superior colliculus. *Prog. Neuropsychopharmacol. Biol. Psychiatry* 2002; 26: 327-33.

[450] Hess HH, Lees MB, Derr JE. A linear Lowry-Folin assay for both water-soluble and sodium dodecyl sulfate-solubilized proteins. *Ann. Biochem.* 1978; 85: 295-300.

[451] Puskás LG, Zvara Á, Hackler L Jr, Micsik T, van Hummelen P. Production of bulk amounts of universal RNA for DNA microarrays. *Biotechniques* 2002; 33: 898-904.

[452] Puskás LG, Hackler L Jr, Kovács G, Kupihár Z, Zvara Á, Micsik T, van Hummelen P. Recovery of cyanine-dye nucleotide triphosphates. *Anal. Biochem.* 2002; 305, 279-281.

[453] Onody A, Zvara Á, Hackler L Jr, Vígh L, Ferdinándy P, Puskás LG. Effect of classic preconditioning on the gene expression pattern of rat hearts: a DNA microarray study. *FEBS Lett.* 2003; 536: 35-40.

[454] Kitajka K, Puskás LG, Zvara Á, Hackler L Jr, Barcelo-Coblijn G, Yeo YK, Farkas T. The role of n-3 polyunsaturated fatty acids in brain: modulation of rat brain gene expression by dietary n-3 fatty acids. *Proc. Natl. Acad. Sci. USA* 2002; 99: 2619-2624.

[455] Cleveland WS. Robust locally weighted regression and smoothing scatterplots. *J. Am. Statist. Assoc.* 1974; 74: 829-836.

[456] Yang YH, Dudoit S, Luu P, Lin DM, Peng V, Ngai J, Speed TP. Normalization for cDNA microarray data: a robust composite method addressing single and multiple slide systematic variation. *Nucleic Acids Res.* 2002; 30: 15.

[457] Pfaffl MW. A new mathematical model for relative quantification in real-time RT-PCR. *Nucleic Acids Res.* 2001; 29: 45.

[458] Berridge MJ. Elementary and global aspects of calcium signalling. *J. Exp. Biol.* 1997; 200: 315-319.

[459] Bootman MD, Collins TJ, Peppiatt CM, Prothero LS, MacKenzie L, De Smet P, Travers M, Tovey SC, Seo JT, Berridge MJ. Calcium signaling – an overview. *Semin. Cell. Dev. Biol.* 2001; 12: 3-10.

[460] Lewis RS. Calcium signaling mechanisms in T lymphocytes. *Annu. Rev. Immunol.* 2001; 19: 497-521.

[461] Hickey WF, Hsu BL, Kimura H. T-lymphocyte entry into the central nervous system. *J. Neurosci. Res.* 1991; 28: 254-260.

[462] Fabry Z, Raine CS, Hart MN. Nervous tissue as an immune compartment: the dialect of the immune response in the CNS. *Immunol. Today* 1994; 15: 218-224.

[463] Nicoll JAR, Wilkinson D, Holmes C, Steart P, Markham H, Weller R. Neuropathology of human Alzheimer disease after immunization with amyloid-β peptide: a case report. *Nat. Med.* 2003; 9: 448-452.

[464] Etienne-Manneville S, Adamson P, Manneville JB, Wilbourn B, Greenwood J, Couraud PO. ICAM-1 coupled cytoskeletal rearrangements and lymphocyte migration across brain endothelium involves intracellular calcium signalling. *J. Immunol.* 2000; 165: 3375-3383.

[465] Schwarz MJ, Chiang S, Müller N, Ackenheil M. T-helper-1 and T-helper-2 responses in psychiatric disorders. *Brain Behav. Immun.* 2001; 15: 340-370.

[466] Meda L, Cassatella MA, Szendrei GI, Ötvös L Jr, Baron P, Villalba M, Ferrari D, Rossi F. Activation of microglial cells by β-amyloid protein and interferon-γ. *Nature* 1995; 374: 647-650.

[467] Brod SA. Unregulated inflammation shortens human functional longevity. *Inflamm. Res.* 2000; 49: 561-570.

[468] Marx F, Blasko I, Grubeck-Loebenstein B. Mechanisms of immune regulation in Alzheimer's disease: a viewpoint. *Arch. Immunol. Ther. Exp.* 1999; 47: 205-209.

[469] Wick G, Grubeck-Loebenstein B. Primary and secondary alterations of immune reactivity in the elderly: impact of dietary factors and disease. *Immunol. Rev.* 1997; 160: 171-184.

[470] Trieb K, Ransmayr G, Sgonc R, Lassmann H, Grubeck-Loebenstein B. APP peptides stimulate lymphocyte proliferation in normals, but not in patients with Alzheimer's disease. *Neurobiol. Aging* 1996; 17: 541-547.

[471] Hammarberg H, Lidman O, Lundberg C, Eltayeb SY, Gielen AW, Muhallab S, Svenningsson A, Linda H, van Der Meide PH, Cullheim S, Olsson T, Piehl F. Neuroprotection by encephalomyelitis: rescue of mechanically injured neurons and neurotrophin production by CNS-infiltrating T and natural killer cells. *J. Neurosci.* 2000; 20: 5283-5291.

[472] Schwartz M, Cohen I, Lazarov-Spiegler O, Moalem G, Yoles E. The remedy may lie in ourselves: prospects for immune cell therapy in central nervous system protection and repair. *J. Mol. Med.* 1999; 77: 713-717.

[473] Clayton A, Evans RA, Pettit E, Hallett M, Williams JD, Steadman R. Cellular activation through the ligation of intercellular adhesion molecule-1. *J. Cell. Sci.* 1998; 111: 443-453.

[474] Baldwin AS Jr. The NF-κB and IκB proteins: new discoveries and insights. *Annu. Rev. Immunol.* 1996; 14: 649-683.

[475] Bondy B, Hofmann M, Muller-Spahn F, Witzko M, Hock C. Reduced β-amyloid response in lymphocytes of patients with Alzheimer's disease. *Pharmacopsychiatry* 1995; 28: 143-146.

[476] Eckert A, Forstl H, Hartmann H, Muller WE. Decreased β-amyloid sensitivity in Alzheimer's disease. *Lancet* 1993; 342: 805-806.

[477] Samelson LE. Signal transduction mediated by the T cell antigen receptor: the role of adapter proteins. *Annu. Rev. Immunol.* 2002; 20: 371-394.

[478] Müller KM, Bickel M, Wiesmann UN, Spörri B. Natural killer cells activate human dermal fibroblasts. *Cytokine* 2000; 12: 1755-1762.

[479] Chirathaworn C, Kohlmeier JE, Tibbetts SA, Rumsey LM, Chan MA, Benedict SH. Stimulation through intercellular adhesion molecule-1 provides a second signal for T cell activation. *J. Immunol.* 2002; 168: 5530-5537.

[480] Greenwood J, Calder VL. Lymphocyte migration through cultured endothelial cell monolayers derived from the blood-retinal barrier. *Immunology* 1993; 80: 401-406.

[481] Pryce G, Male D, Campbell I, Greenwood J. Factors controlling T-cell migration across rat cerebral endothelium *in vitro*. *J. Neuroimmunol.* 1997; 75: 84-94.

[482] Muller WE, Koch S, Eckert A, Hartmann H, Scheuer K. β-Amyloid peptide decreases membrane fluidity. *Brain Res.* 1995; 674: 133-136.

[483] Harkány T, Ábraham I, Timmerman W, Laskay G, Tóth B, Sasvári M, Kónya C, Sebens JB, Korf J, Nyakas C, Zarándi M, Soós K, Penke B, Luiten PG. β-Amyloid neurotoxicity is mediated by a glutamate-triggered excitotoxic cascade in rat nucleus basalis. *Eur. J. Neurosci.* 2000; 12: 2735-2745.

[484] Mattson MP. Cellular actions of β-amyloid precursor protein and its soluble and fibrillogenic derivatives. *Physiol. Rev.* 1997; 77: 1081-1132.

[485] Mogensen HS, Beatty DM, Morris SJ, Jorgensen OS. Amyloid β-peptide (25-35) changes [Ca2+] in hippocampal neurons. *NeuroReport* 1998; 9: 1553-1558.

[486] Stix B, Reiser B. β-Amyloid peptide 25-35 regulates basal and hormone-stimulated Ca-levels in cultured rat astrocytes. *Neurosci. Lett.* 1998; 243: 121-124.

[487] Cowburn RF, Wiehager B, Sundstrom E. β-Amyloid peptides enhance binding of the calcium mobilising second messengers, inositol-(1,4,5)trisphosphate and inositol (1,3,4,5)tetrakis-phosphate to their receptor sites in rat cortical membranes. *Neurosci. Lett.* 1995; 191: 31-34.

[488] Pascale A, Etcheberrygaray R. Calcium alterations in Alzheimer's disease: pathophysiology, models and therapeutic opportunities. *Pharmacol. Res.* 1999; 39: 81-88.

[489] Ishikawa H, Ozawa H, Saito T, Takahata N, Takemura H. Calcium mobilization evoked by amyloid-β protein involves inositol 1,4,5-triphosphate production in human platelets. *Life Sci.* 1998; 62: 705-713.

[490] Crawford KW, Bowen WD. Sigma-2 receptor agonists activate a novel apoptotic pathway and potentiate antineoplastic drugs in breast tumor cell lines. *Cancer Res.* 2002; 62: 313-322.

[491] Mitchell IJ, Cooper AC, Griffiths MR, Cooper AJ. Acute administration of haloperidol induces apoptosis of neurones in the striatum and substantia nigra in the rat. *Neuroscience* 2002; 109: 89-99.

[492] Santi CM, Cayabyab FS, Sutton KG, McRory JE, Mezeyova J, Hamming KS, Parker D, Stea A, Snutch TP. Differential inhibition of T-type calcium channels by neuroleptics. *J. Neurosci.* 2002; 22: 396-403.

[493] Kontkanen O, Toronen P, Lakso M, Wong G, Castren E. Antipsychotic drug treatment induces differential gene expression in the rat cortex. *J. Neurochem.* 2002; 83: 1043-1053.

[494] Fletcher EJ, Church J, MacDonald JF. Haloperidol blocks voltage-activated Ca2+ channels in hippocampal neurones. *Eur. J. Pharmacol.* 1994; 267: 249-252.

[495] Kagan BL, Hirakura Y, Azimov R, Azimova R, Lin MC. The channel hypothesis of Alzheimer's disease: current status. *Peptides* 2002; 23: 1311-1315.

[496] Edell WS, Tunis SL. Antipsychotic treatment of behavioral and psychological symptoms of dementia in geropsychiatric inpatients. *Am. J. Geriatr. Psychiatry* 2001; 9: 289-297.

[497] Hermann N, Black SE. Behavioral disturbances in dementia: will the real treatment please stand up? *Neurology* 2000; 55: 1247-1248.

[498] Reuss B, Unsicker K. Atypical neuroleptic drugs downregulate dopamine sensitivity in rat cortical and striatal astrocytes. *Neurosci.* 2001; 18: 197-209.

[499] Fu W, Luo H, Parthasarathy S, Mattson MP. Catecholamines potentiate amyloid β-peptide neurotoxicity: involvement of oxidative stress, mitochondrial dysfunction, and perturbed calcium homeostasis. *Neurobiol. Dis.* 1998; 5: 229-243.

[500] DesJarlais RL, Scibel GL, Kuntz ID, Furth PS, Alvarez JC, Ortiz de Montellano PR, DeCamp DL, Babe LM, Craik CS. Structure-based design of nonpeptide inhibitors specific for the human immunodeficiency virus 1 protease. *Proc. Natl. Acad. Sci. USA* 1990; 87: 6644-6648.

[501] Paul SM, Purdy RH. Neuroactive steroids. *FASEB J.* 1992; 6: 2311-2322.

[502] French-Mullen JM, Spence KT. Neurosteroids block Ca2+ channel current in freshly isolated hippocampal CA1 neurons. *Eur. J. Pharmacol.* 1995; 202: 269-273.

[503] Vilner BJ, Bowen WD. Dual modulation of cellular calcium by sigma receptor ligands: release from intracellular stores and blockade of voltage-dependent influx. *Soc. Neurosci. Abstr.* 1995; 21: 1608.

[504] Chen M. Alzheimer's α-secretase may be a calcium-dependent protease. *FEBS Lett.* 1997; 417: 163-167.

[505] Matsuno K, Matsunaga K, Senda T, Mita S. Increase in extracellular acetylcholine level by sigma ligands in rat frontal cortex. *J. Pharmac. Exp. Ther.* 1993; 265: 851-859.

[506] Maurice T, Su TP, Privat A. Sigma-1 (σ1) receptor agonists and neurosteroids attenuate β25-35-amyloid peptide-induced amnesia in mice through a common mechanism. *Neuroscience* 1997; 83: 413-428.

[507] Cardounel A, Regelson W, Kalimi M. Dehydroepiandrosterone protects hippocampal neurons against neurotoxin-induced cell death: mechanism of action. *Proc. Soc. Exp. Biol. Med.* 1999; 222: 145-149.

[508] Dwivedi Y, Pandey GN. Effects of treatment with haloperidol, chlorpromazine, and clozapine on protein kinase C (PKC) and phosphoinositide-specific phospholipase C

(PI-PLC) activity and on mRNA and protein expression of PKC and PLC isozymes in rat brain. *J. Pharm. Exp. Ther.* 1999; 291: 688-704.
[509] Cedazo-Minguez A, Wiehager B, Winblad B, Huttinger M, Cowburn RF. Effects of apolipoprotein E (apoE) isoforms, β-amyloid (Aβ) and apoE/Aβ complexes on protein kinase C-α (PKC-α) translocation and amyloid precursor protein (APP) processing in human SH-SY5Y neuroblastoma cells and fibroblasts. *Neurochem. Int.* 2001; 38: 615-625.
[510] Pákáski M, Baláspiri L, Checler F, Kása P. Human amyloid-β causes changes in the levels of endothelial protein kinase C and its α-isoform *in vitro*. *Neurochem. Int.* 2002; 41: 409.
[511] Stea A, Soong TW, Snutch TP. Determinants of PKC-dependent modulation of a family of neuronal calcium channels. *Neuron* 1995; 15: 929-940.
[512] Westmark CJ, Malter JS. Extracellular-regulated kinase controls β-amyloid precursor protein mRNA decay. *Mol. Brain Res.* 2001; 90: 193-201.
[513] Masliah E, Westland CE, Rockenstein EM, Abraham CR, Mallory M, Veinberg I, Sheldon E, Mucke L. Amyloid precursor proteins protect neurons of transgenic mice against acute and chronic excitotoxic injuries *in vivo*. *Neuroscience* 1997; 78: 135-146.
[514] Papp H, Pákáski M, Kása P. Presenilin-1 and the amyloid precursor protein are transported bidirectionally in the sciatic nerve of adult rat. *Neurochem. Int.* 2002; 41: 429-435.
[515] Stephenson DT, Rash K, Clemens JA. Amyloid precursor protein accumulates in regions of neurodegeneration following focal cerebral ischemia in the rat. *Brain Res.* 1992; 593: 128-135.
[516] Kalaria RN, Bhatti SU, Palatinsky EA, Pennington DH, Shelton ER, Chan HW, Perry G, Lust WD. Accumulation of the β-amyloid precursor protein at sites of ischemic injury in rat brain. *NeuroReport* 1993; 4: 211-214.
[517] Post A, Rücker M, Ohl F, Uhr M, Holsboer F, Almeida OFX, Michaelidis TM. Mechanisms underlying the protective potential of α-tocopherol (vitamin E) against haloperidol-associated neurotoxicity. *Neuropsychopharmacology* 2002; 26: 397-407.
[518] Barger SW, Mattson MP. Induction of neuroprotective κB-dependent transcription by secreted forms of Alzheimer's β-amyloid precursor. *Mol. Brain Res.* 1996; 40: 116-126.
[519] Mattson MP, Camandola S. NF-κB in neuronal plasticity and neurodegenerative disorders. *J. Clin. Invest.* 2001; 107: 247-254.
[520] Nitsch RM, Deng A, Growdon JH, Wurtman RJ. Serotonin 5-HT2A and 5-HT2C receptors stimulate amyloid precursor protein ectodomain secretion. *J. Biol. Chem.* 1996; 271: 4188-4194.
[521] Wallace W, Haroutunian V. Using the subcortically lesioned rat cortex to understand the physiological role of amyloid precursor protein. *Behav. Brain Res.* 1993; 57: 199-206.
[522] Querfurth HW, Selkoe DJ. Calcium ionophore increases amyloid-β peptide production by cultured cells. *Biochemistry* 1994; 33: 4550-4561.
[523] Sennvik K, Benedikz E, Fastbom J, Sundstrom E, Winblad B, Ankarcrona M. Calcium ionophore A23187 specificially decreases the secretion of β-secretase cleaved amyloid

precursor protein during apoptosis in primary rat cortical cultures. *J. Neurosci. Res.* 2001; 63: 429-437.

[524] Walter J, Fluhrer R, Hartung B, Willem M, Kaether C, Capell A, Lammich S, Multhaup G, Haass C. Phosphorylation regulates intracellular trafficking of β-secretase. *J. Biol. Chem.* 2001; 276: 14634-14641.

[525] Lockhart DJ, Winzeler EA. Genomics, gene expression and DNA arrays. *Nature* 2000; 405: 827-836.

[526] McCullumsmith RE, Stincic TL, Agrawal SM, Meador-Woodruff JH. Differential effects of antipsychotics on haloperidol-induced vacuous chewing movements and subcortical gene expression in the rat. *Eur. J. Pharmacol.* 2003; 477: 101-112.

[527] Schmitt A, Zink M, Petroianu G, May B, Braus DF, Henn FA. Decreased gene expression of glial and neuronal glutamate transporters after chronic antipsychotic treatment in rat brain. *Neurosci. Lett.* 2003; 347: 81-84.

[528] Coderre TJ, Katz J, Vaccarino AL, Melzack R. Contribution of central neuroplasticity to pathological pain: review of clinical and experimental evidence. *Pain* 1993; 52: 259-285.

[529] Moore KA, Baba H, Woolf CJ. Synaptic transmission and plasticity in the superficial dorsal horn. *Prog. Brain Res.* 2000; 129: 63-80.

[530] Desikan R, Mackerness SAH, Hancock JT, Neill SJ. Regulation of the Arabidopsis transcriptome by oxidative stress. *Plant Physiol.* 2001; 127: 159-172.

[531] Wagner U, Edwards R, Dixon DP, Mauch F. Probing the diversity of the Arabidopsis glutathione S-transferase gene family. *Plant Mol. Biol.* 2002; 49: 515-532.

[532] Karoum F, Karson CN, Bigelow LB, Lawson WB, Wyatt RJ. Preliminary evidence of reduced combined output of dopamine and its metabolites in chronic schizophrenia. *Arch. Gen. Psychiat.* 1987; 44: 604-607.

[533] van den Buuse M, Webber KM. Endothelin and dopamine release. *Prog. Neurobiol.* 2000; 60: 385-405.

[534] Reiser G, Donié F. Endothelin induces a rise of inositol 1,4,5-triphosphate, inositol 1,3,4,5-tetrakisphosphate levels and of cytosolic Ca2+ activity in neural cell lines. *Eur. J. Neurosci.* 1990; 2: 769-775.

[535] Choi DW. Calcium-mediated neurotoxicity: relationship to specific channel types and role in ischemic damage. *Trends Neurosci.* 1988; 11: 465-469.

[536] Kataoka Y, Koizumi S, Kohzuma M, Shibaguchi H, Shigematsu K, Niwa M, Taniyama K. NMDA receptor involvement in endothelin neurotoxicity in rat striatal slices. *Eur. J. Pharmacol.* 1995; 273: 285-289.

[537] Stern Y, Tang MX, Albert M. Predicting time to nursing home care and death in individuals with Alzheimer's disease. *JAMA* 1997; 277: 806-812.

[538] Mintzer JE, Lewis L, Pennypaker L, Simpson W, Bachman D, Wohlreich G, Meeks A, Hunt S, Sampson R. Behavioral Intensive Care Unit (BICU): a new concept in the management of acute agitated behavior in elderly demented patients. *Clin. Gerontologist* 1993; 33: 801-806.

[539] Schachter AS, Davis KL. Guidelines for the appropriate use of cholinestrase inhibitors in patients with Alzheimer's disease. *CNS Drugs* 1999; 11: 281-288.

[540] Piovesan P, Pacifici L, Taglialatela G, Ramacci MT, Angelucci L. Acetyl-L-carnitine treatment increases choline acetyltransferase activity and NGF levels in the CNS of adult rats following total fimbria-fornix transection. *Brain Res.* 1994; 633: 77-82.

[541] Levy R, Little A, Chuaqui P, Reith M. Early results from double blind, placebo controlled trial of high dose phosphatidylcholine in Alzheimer's Disease. *Lancet* 1982; 1: 474-476.

[542] Cummings JL, Kaufer D. Neuropsychiatric aspects of Alzheimer's disease: the cholinergic hypothesis revisited. *Neurology* 1996; 47: 876-883.

[543] Cummings JL, Back C. The cholinergic hypothesis of neuropsychiatric symptoms in Alzheimer's disease. *Am. J. Geriatr. Psychiatry* 1998; 6: S64-S78.

[544] Hutchinson M, Fazzini E. Cholinesterase inhibition in Parkinson's disease. *J. Neurol. Neurosurg. Psychiatry* 1996; 61: 324-325.

[545] Decker MW, McGaugh JL. The role of interactions between the cholinergic system and other neuromodulatory systems in learning and memory. *Synapse* 1991; 7: 151-168.

[546] Lawrence AD, Sahakian BJ. Alzheimer disease, attention, and the cholinergic system. *Alzheimer Dis. Assoc. Disord.* 1995; 9: S43-S49.

[547] Fishman EB, Siek GC, MacCallum RD, Bird ED, Volicer L, Marquis JK. Distribution of the molecular forms of acetylcholinesterase in human brain: alterations in dementia of the Alzheimer type. *Ann. Neurol.* 1986; 19: 246-252.

[548] Martzen MR, Nagy A, Coleman PD, Zwiers H. Altered phosphorylation of growth-associated protein B50/GAP-43 in Alzheimer disease with high neurofibrillary tangle density. *Proc. Natl. Acad. Sci. USA* 1993; 90: 11187-11191.

[549] Inestrosa NC, Alvarez A, Perez CA, Moreno RD, Vicente M, Linker C, Casanueva OI, Soto C, Garrido J. Acetylcholinesterase accelerates assembly of amyloid β-peptides into Alzheimer's fibrils: possible role of the peripheral site of the enzyme. *Neuron* 1996; 16: 881-891.

[550] Buxbaum JD, Oishi M, Chen HI, Pinkas-Kramarski R, Jaffe EA, Gandy SE, Greengard P. Cholinergic agonists and interleukin-1 regulate processing and secretion of the Alzheimer β/A4 amyloid protein precursor. *Proc. Natl. Acad. Sci. USA* 1992; 89: 10075-10078.

[551] Reisberg B, Doody R, Stoffler A, Schmitt F, Ferris S, Mobius HJ. Memantine in moderate-to-severe Alzheimer's disease. N. Engl. J. Med. 2003; 348: 1333-1341.

[552] Cutler NR, Sramek JJ. Review of the next generation of Alzheimer's disease therapeutics: challenges for drug development. *Prog. Neuropsychopharmacol. Biol. Psychiatry* 2001; 25: 27-57.

[553] Schenk D, Barbour R, Dunn W, Gordon G, Grajeda H, Guido T, Hu K, Huang J, Johnson-Wood K, Khan K, Kholodenko D, Lee M, Liao Z, Lieberburg I, Motter R, Mutter L, Soriano F, Shopp G, Vasquez N, Vandevert C, Walker S, Wogulis M, Yednock T, Games D, Seubert P. Immunization with amyloid-β attenuates Alzheimer-disease-like pathology in the PDAPP mouse. *Nature* (Lond) 1999; 400: 173-177.

[554] Janus C, Pearson J, McLaurin J, Mathews PM, Jiang Y, Schmidt SD, Chishti MA, Horne P, Heslin D, French J, Mount HT, Nixon RA, Mercken M, Bergeron C, Fraser PE, St George-Hyslop P, Westaway D. Aβ peptide immunization reduces behavioural

impairment and plaques in a model of Alzheimer's disease. *Nature* (Lond) 2000; 408: 979-982.

[555] Lim GP, Yang F, Chu T, Chen P, Beech W, Teter B, Tran T, Ubeda O, Ashe KH, Frautschy SA, Cole GM. Ibuprofen suppresses plaque pathology and inflammation in a mouse model for Alzheimer's disease. *J. Neurosci.* 2000; 20: 5709-5714.

[556] Fassbender K, Simons M, Bergmann C, Stroick M, Lutjohann D, Keller P, Runz H, Kuhl S, Bertsch T, von Bergmann K, Hennerici M, Beyreuther K, Hartmann T. Simvastatin strongly reduces levels of Alzheimer's disease β-amyloid peptides Aβ42 and Aβ40 *in vitro* and *in vivo*. *Proc. Natl. Acad. Sci. USA* 2001; 98: 5856-5861.

[557] Nitsch RM, Farber SA, Growdon JH, Wurtman RJ. Release of amyloid β-protein precursor derivatives by electrical depolarization of rat hippocampal slices. *Proc. Natl. Acad. Sci. USA* 1993; 90: 5191-5193.

[558] Lee RK, Wurtman RJ, Cox AJ, Nitsch RM. Amyloid precursor protein processing is stimulated by metabotropic glutamate receptors. *Proc. Natl. Acad. Sci. USA* 1995; 92: 8083-8087.

[559] Refolo LM, Salton SR, Anderson JP, Mehta P, Robakis NK. Nerve and epidermal growth factors induce the release of the Alzheimer amyloid precursor from PC 12 cell cultures. *Biochem. Biophys. Res. Commun.* 1989; 164: 664-670.

[560] Sinha S, Anderson JP, Barbour R, Basi GS, Caccavello R, Davis D, Doan M, Dovey HF, Frigon N, Hong J, Jacobson-Croak K, Jewett N, Keim P, Knops J, Lieberburg I, Power M, Tan H, Tatsuno G, Tung J, Schenk D, Seubert P, Suomensaari SM, Wang S, Walker D, Zhao J, McConlogue L, John V. Purification and cloning of amyloid precursor protein β-secretase from human brain. *Nature* (Lond) 1999; 402: 537-540.

[561] Cai H, Wang Y, McCarthy D, Wen H, Borchelt DR, Price DL, Wong PC. BACE1 is the major β-secretase for generation of Aβ peptides by neurons. *Nat. Neurosci.* 2001; 4: 233-234.

[562] Farzan M, Schnitzler CE, Vasilieva N, Leung D, Choe H. BACE2, a β-secretase homolog, cleaves at the β-site and within the amyloid-β region of the amyloid-β precursor protein. *Proc. Natl. Acad. Sci. USA* 2000; 97: 9712-9717.

[563] Artavanis-Tsakonas S, Rand MD, Lake RJ. Notch signaling: cell fate control and signal integration in development. *Science* (Wash DC) 1999; 284: 770-776.

[564] Petit A, Bihel F, Alves da Costa C, Pourquié O, Kraus JL, Checler F. New protease inhibitors prevent γ-secretase-mediated Aβ40/42 production without affecting Notch cleavage. *Nat. Cell Biol.* 2001; 3: 507-511.

[565] Banks WA, Farr SA, Butt W, Kumar VB, Franko MW, Morley JE. Delivery across the blood-brain barrier of antisense directed against amyloid-β: reversal of learning and memory deficits in mice overexpressing amyloid precursor protein. *J. Pharmacol. Exp. Ther.* 2001; 297: 1113-1121.

[566] Pike CJ, Burdick D, Walencewicz AJ, Glabe CG, Cotman CW. Neurodegeneration induced by β-amyloid peptide *in vitro*: the role of peptide assembly state. *J. Neurosci.* 1993; 13: 1676-1687.

[567] Soto C, Casatano EM, Kumar RA, Beavis RC, Frangione B. Fibrillogenesis of synthetic amyloid-β peptides is dependent on their initial secondary structure. *Neurosci. Lett.* 1995; 200: 105-108.

[568] Barrow CJ, Zagorski MG. Solution structures of β-peptide and its constituent fragments: relation to amyloid deposition. *Science* (Wash DC) 1991; 253: 179-182.

[569] Burdick D, Soreghan B, Kwon M, Kosmoski J, Knauer M, Henschen A, Yates J, Cotman C, Glabe C. Assembly and aggregation properties of synthetic Alzheimer's A4/β amyloid peptide analogs. *J. Biol. Chem.* 1992; 267: 546-554.

[570] Mantyh PW, Ghilardi JR, Rogers S, DeMaster E, Allen CJ, Stimson ER, Maggio JE. Aluminum, iron and zinc ions promote aggregation of physiological concentrations of β-amyloid peptide. *J. Neurochem.* 1993; 61: 1171-1174.

[571] Lovell MA, Teesdale WJ, Campbell JL, Markesbery WR. Copper, iron and zinc in Alzheimer's disease senile plaques. *J. Neurol. Sci.* 1998; 158: 47-52.

[572] Cherny RA, Legg JT, Mclean CA, Fairlie D, Huang X, Atwood CS, Beyreuther K, Tanzi RE, Masters CL, Bush AI. Aqueous dissolution of Alzheimer's disease Aβ amyloid deposits by biometal depletion. *J. Biol. Chem.* 1999; 274: 23223-23228.

[573] Akiyama H, Barger S, Barnum S, Bradt B, Bauer J, Cole GM, Cooper NR, Eikelenboom P, Emmerling M, Fiebich BL, Finch CE, Frautschy S, Griffin WS, Hampel H, Hull M, Landreth G, Lue L, Mrak R, Mackenzie IR, McGeer PL, O'Banion MK, Pachter J, Pasinetti G, Plata-Salaman C, Rogers J, Rydel R, Shen Y, Streit W, Strohmeyer R, Tooyoma I, Van Muiswinkel FL, Veerhuis R, Walker D, Webster S, Wegrzyniak B, Wenk G, Wyss-Coray T. Inflammation and Alzheimer's disease. *Neurobiol. Aging* 2000; 21: 383-421.

[574] Blanchard V, Czech C, Bonici B, Clavel N, Gohin M, Dalet K, Revah F, Pradier L, Imperato A, Moussaoui S. Immunohistochemical analysis of presenilin 2 expression in the mouse brain: distribution pattern and co-localization with presenilin 1 protein. *Brain Res.* 1997; 758: 209-217.

[575] Behl C, Davis J, Cole GM, Schubert D. Vitamin E protects nerve cells from amyloid-β protein toxicity. *Biochem. Biophys. Res. Commun.* 1992; 186: 944-950.

[576] Xiao XQ, Wang R, Tang XC. Huperzine-A and tacrine attenuate β-amyloid peptide-induced oxidative injury. *J. Neurosci. Res.* 2000; 61: 564-569.

[577] Pappolla MA, Chyan YJ, Poeggeler B, Frangione B, Wilson G, Ghiso J, Reiter RJ. An assessment of the antioxidant and the antiamyloidogenic properties of melatonin: implications for Alzheimer's disease. *J. Neural. Transm.* 2000; 107: 203-231.

[578] Law A, Dore S, Blackshaw S, Gauthier S, Quirion R. Alteration of expression levels of neuronal nitric oxide synthase and haem oxygenase-2 messenger RNA in the hippocampi and cortices of young adult and aged cognitively unimpaired and impaired Long-Evans rats. *Neuroscience* 2000; 100: 769-775.

[579] McCann SM, Licinio J, Wong ML, Yu WH, Karanth S, Rettorri V. The nitric oxide hypothesis of aging. *Exp. Gerontol.* 1998; 33: 813-826.

[580] Sohn YK, Ganju N, Bloch KD, Wands JR, de la Monte SM. Neuritic sprouting with aberrant expression of the nitric oxide synthase III gene in neurodegenerative disease. *J. Neurol. Sci.* 1999; 162: 133-151.

[581] Lüth HJ, Holzer M, Gärtner U, Staufenbiel M, Arendt T. Expression of endothelial and inducible NOS-isoforms is increased in Alzheimer's disease, in APP23 transgenic mice and after experimental brain lesion in rat: evidence for an induction by amyloid pathology. *Brain Res.* 2001; 913: 57-67.

[582] Myllykangas-Luosujarvi R, Isomaki H. Alzheimer's disease and rheumatoid arthritis. *Br. J. Rheumatol.* 1994; 33: 501-502.

[583] Stewart WF, Kawas C, Corrada M, Metter EJ. Risk of Alzheimer's disease and duration of NSAID use. *Neurology* 1997; 48: 626-632.

[584] Netland EE, Newton JL, Majocha RE, Tate BA. Indomethacin reverses the microglial response to amyloid β-protein. *Neurobiol. Aging* 1998; 19: 201-204.

[585] Mackenzie IR, Munoz DG. Nonsteroidal anti-inflammatory drug use and Alzheimer-type pathology in aging. *Neurology* 1998; 50: 986-990.

[586] Amin AR, Attur MG, Thakker GD, Patel PD, Vyas PR, Patel RN, Patel IR, Abramson SB. A novel mechanism of action of tetracyclines: effects on nitric oxide synthases. *Proc. Natl. Acad. Sci. USA* 1996; 93: 14014-14019.

[587] DeMattos RB, Bales KR, Cummins DJ, Dodart JC, Paul SM, Holtzman DM. Peripheral anti-Aβ antibody alters CNS and plasma Aβ clearance and decreases brain Aβ burden in a mouse model of Alzheimer's disease. *Proc. Natl. Acad. Sci. USA* 2001; 98: 8850-8855.

[588] Check E. Nerve inflammation halts trial for Alzheimer's drug. *Nature* (Lond) 2002; 415: 462.

[589] Lovestone S, Reynolds CH. The phosphorylation of *tau*: a critical stage in neurodevelopment and neurodegenerative processes. *Neuroscience* 1997; 78: 309-324.

[590] Alonso AC, Zaidi T, Novak M, Barra HS, Grundke-Iqbal I, Iqbal K. Interaction of *tau* isoforms with Alzheimer's disease abnormally hyperphosphorylated *tau* and *in vitro* phosphorylation into the disease-like protein. *J. Biol. Chem.* 2001; 276: 37967-37973.

[591] McLean CA, Cherny RA, Fraser FW, Fuller SJ, Smith MJ, Beyreuther K, Bush AI, Masters CL. Soluble pool of Aβ amyloid as a determinant of severity of neurodegeneration in Alzheimer's disease. *Ann. Neurol.* 1999; 46: 860-866.

[592] Wilson CA, Doms RW, Lee VM. Intracellular APP processing and Aβ production in Alzheimer disease. *J. Neuropathol. Exp. Neurol.* 1999; 58: 787-794.

[593] Henderson VW, Paganini-Hill A, Emanuel CK, Dunn ME, Buckwalter JG. Estrogen replacement therapy in older women. Comparisons between Alzheimer's disease cases and nondemented control subjects. *Arch. Neurol.* 1994; 51: 896-900.

[594] Evans MI, Silva R, Burch JB. Isolation of chicken vitellogenin I and III cDNAs and the developmental regulation of five estrogen-responsive genes in the embryonic liver. *Genes Dev.* 1988; 2: 116-124.

[595] Wetzel CH, Hermann B, Behl C, Pestel E, Rammes G, Zieglgansberger W, Holsboer F, Rupprecht R. Functional antagonism of gonadal steroids at the 5-hydroxytryptamine type 3 receptor. *Mol. Endocrinol.* 1998; 12: 1441-1451.

[596] Migliaccio A, Di Domenico M, Castoria G, de Falco A, Bontempo P, Nola E, Auricchio F. Tyrosine kinase/p21ras/MAP-kinase pathway activation by estradiol-receptor complex in MCF-7 cells. *EMBO (Eur. Mol. Biol. Organ.) J.* 1996; 15: 1292-1300.

[597] Wickelgren I. Estrogen stakes claim to cognition. *Science* (Wash DC) 1997; 276: 675-678.

[598] Matsumoto A. Synaptogenic action of sex steroids in developing and adult neuroendocrine brain. *Psychoneuroendocrinology* 1991; 16: 25-40.

[599] Shughrue PJ, Scrimo PJ, Merchenthaler I. Evidence for the colocalization of estrogen receptor-β mRNA and estrogen receptor-α immunoreactivity in neurons of the rat forebrain. *Endocrinology* 1998; 139: 5267-5270.

[600] Kritzer MF, Kohama SG. Ovarian hormones differentially influence immunoreactivity for dopamine β-hydroxylase, choline acetyltransferase and serotonin in the dorsolateral prefrontal cortex of adult *Rhesus* monkeys. *J. Comp. Neurol.* 1999; 409: 438-451.

[601] Tang MX, Jacobs D, Stern Y, Marder K, Schofield P, Gurland B, Andrews H, Mayeux R. Effect of oestrogen during menopause on risk and age at onset of Alzheimer's disease. *Lancet* 1996; 348: 429-432.

[602] Marder K, Tang MX, Alfaro B, Mejia H, Cote L, Jacobs D, Stern Y, Sano M, Mayeux R. Postmenopausal estrogen use and Parkinson's disease with and without dementia. *Neurology* 1998; 50: 1141-1143.

[603] Smalheiser NR, Swanson DR. Linking estrogen to Alzheimer's disease: an informatics approach. *Neurology* 1996; 47: 809-810.

[604] Green PS, Gridley KE, Simpkins JW. Estradiol protects against β-amyloid (25-35)-induced toxicity in SK-N-SH human neuroblastoma cells. *Neurosci. Lett.* 1996; 218: 165-168.

[605] Jaffe AB, Toran-Allerand CD, Greengard P, Gandy SE. Estrogen regulates metabolism of Alzheimer amyloid-β precursor protein. *J. Biol. Chem.* 1994; 269: 13065-13068.

[606] Dena BD, Melinda EW, Phyllis MW. Estradiol: a protective and trophic factor in the brain. *Alzheimers Dis. Rev.* 1999; 4: 1-9.

[607] Pike CJ. Estrogen modulates neuronal Bcl-xL expression and β-amyloid induced apoptosis: relevance to Alzheimer's disease. *J. Neurochem.* 1999; 72: 1552-1563.

[608] Dodel RC, Du Y, Bales KR, Gao F, Paul SM. Sodium salicylate and 17β-estradiol attenuate nuclear transcription factor NF-κB translocation in cultured rat astroglial cultures following exposure to amyloid Aβ(1-40) and lipopolysaccharides. *J. Neurochem.* 1999; 73: 1453-1460.

[609] Paganini-Hill A, Henderson VW. Estrogen replacement therapy and risk of Alzheimer disease. *Arch. Intern. Med.* 1996; 156: 2213-2217.

[610] Imahori K, Uchida T. Physiology and pathology of *tau* protein kinases in relation to Alzheimer's disease. *J. Biochem.* (Tokyo) 1997; 121: 179-188.

[611] Pap M, Cooper GM. Role of glycogen synthase kinase-3 in the phosphatidylinositol 3-Kinase/Akt cell survival pathway. *J. Biol. Chem.* 1998; 273: 19929-19932.

[612] Applebaum-Bowden D, McLean P, Steinmetz A, Fontana D, Matthys C, Warnick GR, Cheung M, Albers JJ, Hazzard WR. Lipoprotein, apolipoprotein and lipolytic enzyme changes following estrogen administration in postmenopausal women. *J. Lipid Res.* 1989; 30: 1895-1906.

[613] Belfort MA, Saade GR, Snabes M, Dunn R, Moise KJ Jr, Cruz A, Young R. Hormonal status affects the reactivity of the cerebral vasculature. *Am. J. Obstet. Gynecol.* 1995; 172: 1273-1278.

[614] Bishop J, Simpkins JW. Role of estrogens in peripheral and cerebral glucose utilization. *Rev. Neurosci.* 1992; 3: 121-137.

[615] Gridley KE, Green PS, Simpkins JW. A novel, synergistic interaction between 17 β-estradiol and glutathione in the protection of neurons against β-amyloid 25-35-induced toxicity *in vitro*. *Mol. Pharmacol.* 1998; 54: 874-880.

[616] Doody RS, Stevens JC, Beck C, Dubinsky RM, Kaye JA, Gwyther L, Mohs RC, Thal LJ, Whitehouse PJ, DeKosky ST, Cummings JL. Practice parameter: Management of dementia (an evidence based review). Report of the Quality Standards Subcommittee of the American Academy of Neurology. *Neurology* 2001; 56: 1154-1166.

[617] Teri L, Logsdon RG, Uomoto J, McCurry SM. Behavioral treatment of depression in dementia patients: A controlled clinical trial. *J. Gerontol. B. Psychol. Sci. Soc. Sci.* 1997; 52: 159-166.

[618] Katona CL, Hunter BN, Bray J. A double blind comparison of the efficacy and safety of paroxetine and imipramine in the treatment of depression with dementia. *Int. J. Geriar. Psychiatry* 1998; 13: 100-108.

[619] Lyketsos CG, Sheppard JM, Steele CS, Kopunek S, Steinberg M, Baker AS, Brandt J, Rabins PV. A randomized, placebo-controlled, double-blind, clinical trial of sertraline in the treatment of depression complicating Alzheimer disease: initial results from the Depression in Alzheimer Disease Study (DIADS). *Am. J. Psychiatry* 2000; 157: 1686-1689.

[620] Rao V, Lyketsos CG. The benefits and risks of ECT for patients with degenerative dementia who also suffer from depression. *Int. J. Ger. Psychiatry* 2000; 15: 729-735.

[621] Katz IR, Jeste DV, Mintzer JE, Clyde C, Napolitano J, Brecher M. Comparison of risperidone and placebo for psychosis and behavioral disturbances associated with dementia: A randomized, double-blind trial. Risperidone Study Group. *J. Clin. Psychiatry* 1999; 60: 107-115.

[622] Street JS, Clark WS, Gannon KS, Cummings JL, Bymatser FP, Tamura RN, Mitan SJ, Kadam DL, Sanger TM, Feldman PD, Tollefson GD, Breier A. Olanzapine in the treatment of psychotic and behavioral symptoms in patients with Alzheimer's disease in nursing care facilities. The HGEU Study Group. *Arch. Gen. Psychiatry* 2000; 57: 968-976.

[623] Tariot PN, Erb R, Podgorski CA, Cox C, Patel S, Jakimovich L, Irvine C. Efficacy and tolerability of carbamazepine for agitation and aggression in dementia. *Am. J. Psychiatry* 1998; 155: 54-61.

[624] Kaufer DI, Cummings JL, Christine D, Bray T, Castellon S, Masterman D, MacMillan A, Ketchel P, DeKosky ST: Assessing the impact of neuropsychiatric symptoms in Alzheimer's disease: the Neuropsychiatric Inventory Caregiver Distress Scale. *J Am Geriatr Soc* 1998; 46:210-215

[625] Steele C, Rovner B, Chase GA, Folstein M: Psychiatric symptoms and nursing home placement of patients with Alzheimer's disease. *Am J Psychiatry* 1990; 147:1049-1051

[626] American Psychiatric Association. Practice guideline for the treatment of patients with Alzheimer's disease and other dementias of late life. *Am. J. Psychiatry* 1997; 154: 1-39.

[627] Devanand DP. Role of neuroleptics in treatment of behavioral complications. In: Lawlor BA (Ed.) *Behavioral Complications in Alzheimer's Disease*. APA Press Inc, Washington, DC. 1995.

[628] Defilippi JL, Crismon ML. Antipsychotic agents in patients with dementia. *Pharmacotherapy* 2000; 20: 23-33.

[629] Levy ML, Cummings JL, Kahn-Rose R. Neuropsychiatric symptoms and cholinergic therapy for Alzheimer's disease. *Gerontology* 1999; 45: 15-22.

[630] Goldgaber D, Harris HW, Hla T, Maciag T, Donnelly RJ, Jacobsen JS, Vitek MP, Gajdusek DC. Interleukin-1 regulates synthesis of amyloid β-protein precursor mRNA in human endothelial cells. *Proc. Natl. Acad. Sci. USA* 1989; 86: 7606-7610.

[631] Guo Q, Robinson N, Mattson MP. Secreted β-amyloid precursor protein counteracts the proapoptotic action of mutant presenilin-1 by activation of NF-κB and stabilization of calcium homeostasis. *J. Biol. Chem.* 1998; 273: 12341-12351.

[632] Jimerson DC, Post RM, Carman JS, van Kammen JS, Wood JH, Goodwin FK, Bunney WE. CSF calcium: clinical correlates in affective illness and schizophrenia. *Biol. Psychiatry* 1979; 14: 37-51.

[633] Riley DM, Watt DC. Hypercalcemia in the etiology of puerperal psychosis. *Biol. Psychiatry* 1985; 20: 479-488.

[634] Yarlagadda A. Role of calcium regulation in pathophysiology model of schizophrenia and possible interventions. *Med. Hypotheses* 2002; 58: 182-186.

[635] Xing G, Chavko M, Zhang LX, Yang S, Post RM. Decreased calcium-dependent constitutive nitric oxide synthase (cNOS) activity in prefrontal cortex in schizophrenia and depression. *Schizophr. Res.* 2002; 58: 21-30.

[636] Suzuki K, Kusumi I, Sasaki Y, Koyama T. Serotonin-induced platelet intracellular calcium mobilization in various psychiatric disorders: is it specific to bipolar disorder? *J. Affect. Disord.* 2001; 64: 291-296.

[637] Koller CW. Neuroprotective Therapy for Parkinson's Disease. *Exp. Neurol.* 1997; 144: 24-28.

[638] Panov AV, Burke JR, Strittmatter WJ, Greenamyre JT. *In vitro* effects of polyglutamine tracts on Ca2+-dependent depolarization of rat and human mitochondria: relevance to Huntington's disease. Arch. Biochem. Biophys. 2003; 410: 1-6.

[639] Yan HD, Lim W, Lee KW, Kim J. Sera from amyotrophic lateral sclerosis patients reduce high-voltage activated Ca2+ currents in mice dorsal root ganglion neurons. *Neurosci. Lett.* 1997; 235: 69-72.

[640] Gailly P. New aspects of calcium signaling in skeletal muscle cells: implications in Duchenne muscular dystrophy. *Biochim. Biophys. Acta* 2002; 1600: 38-44.

[641] Frasson M, Sahel JA, Fabre M, Simonutti M, Dreyfus H, Picaud S. Retinitis pigmentosa: rod photoreceptor rescue by a calcium-channel blocker in the rd mouse. *Nat. Med.* 1999; 5: 1183-1187.

[642] Farkas E, Luiten PG. Cerebral microvascular pathology in aging and Alzheimer's disease. *Prog. Neurobiol.* 2001; 64: 575-611.

[643] Marx F, Blasko I, Grubeck-Loebenstein B. Mechanisms of immune regulation in Alzheimer's disease: a viewpoint. *Arch. Immunol. Ther. Exp.* (Warsz) 1999; 47: 205-209.

[644] Suh YH, Checler F. Amyloid precursor protein, presenilins, and α-synuclein: molecular pathogenesis and pharmacological applications in Alzheimer's disease. *Pharmacol. Rev.* 2002; 54: 469-525.

[645] Mattson MP, Chan SL. Dysregulation of cellular calcium homeostasis in Alzheimer's disease: bad genes and bad habits. *J. Mol. Neurosci.* 2001; 17: 205-224.

[646] Cummings JL. Cholinesterase inhibitors: a new class of psychotropic compounds. *Am. J. Psychiatry* 2000; 157: 4-15.

[647] Koudinova NA, Kontush A, Berezov TT, Koudinov AR. Amyloid-β, neural lipids, cholesterol & Alzheimer's disease. *Neurobiol. Lipids* 2003; 1: 6.

In: Trends In Neurochemistry Research
Editor: Robert M. Coleman, pp. 131-144
ISBN: 1-59454-418-2

Chapter 2

Amyloid Neurodegeneration and the Role of the Wnt Signaling Pathway

Nibaldo C. Inestrosa,* Ginny Farías and Marcela Colombres
Centro de Regulación Celular y Patología "Joaquín V. Luco" (CRCP), MIFAB,
Facultad de Ciencias Biológicas,
Pontificia Universidad Católica de Chile, Santiago, Chile

Abstract

Amyloid-β-peptide (Aβ) deposition is an important event in the neurodegenerative processes that occur in Alzheimer's disease (AD). The cholinergic pathway plays a pivotal role in modulating learning and memory. AChE and the M1 muscarinic receptor (mAChR) has been the focus for a vast diversity of drugs for AD treatment. Cholinesterase inhibitors and muscarinic receptor agonists have been shown to increase the secretion of non-amyloidogenic soluble derivates of amyloid precursor protein (APP). In this context, IBU-PO, a bifunctional compound, which combines a non-steroidal anti-inflammatory drug (Ibuprofen) and a cholinesterase inhibitor (Octyl-Pyridostigmine), acts as a neuroprotective factor against Aβ neurotoxicity. Furthermore, its activity is associated to Wnt signaling components in rat hippocampal and mouse cortical neurons reducing the amount of $A\beta_{1-40}$ and increasing APP soluble levels.
The Wnt signaling pathway has been characterized as an essential pathway in development and oncogenesis. Recent evidence supports a role of this pathway in neurodegenerative disorders such as autism, squizofrenia and AD. A relationship between Aβ-induced neurotoxicity and a decrease in cytoplasmatic levels of β-catenin has been proposed. Also, the inhibition of glycogen synthase kinase (GSK-3β) a central modulator of the pathway, protects rat neurons from Aβ-induced damage. Interestingly,

* *Corresponding author:* Biomedical Center (CRCP), P. Catholic University of Chile, P.O. Box 114-D, Santiago, Chile. Fax: 56-2-686 2717 ; e-mail: ninestr@genes.bio.puc.cl

during the progression of AD, it has been described that GSK-3β is found in neuronal cell bodies and neurites, co-localizing with pre-neurofibrillary tangles observed in AD brains. These evidences suggest that a sustained loss of Wnt signaling function may be involved in the Aβ dependent neurodegeneration observed in Alzheimer's brain.
Progress in the development of more specific drugs targeted to cholinergic pathway would offer the possibility of improved therapeutic strategies for AD.

Introduction

The Wnt signaling pathway has been described as essential in cell adhesion and regulation of cell fate determination during animal development [1]. Recently, it has been implicated in oncogenic processes and in neurodegenerative disorders such as AD [2]. Wnt genes were identified by sequence homology with Wnt-1 in mouse and wingless in *Drosophila*; which encode for secreted glycoproteins (350-400 aminoacids). From the classical view of the Wnt signal transduction pathway, in the presence of an extracellular Wnt ligand (such as Wnt-1, -2, -3, and -8), membrane-anchored receptors of the Frizzled protein family transduce its signal to the intracellular space and activate Dishevelled (Dsh), which in turn inactivates GSK-3β, a key modulator of this signaling pathway, through its binding to a high-molecular weight complex that includes the protein Axin, the adenomatous poliposis coli (APC) and β-catenin [3]. As a net result, intracellular levels of β-catenin increase, allowing its import to the nucleus where it binds to components of the high mobility group family of transcription factors T-cell factor/lymphoid enhancer-binding factor (TCF/LEF) and activate the expression of Wnt target genes (Figure 1). Conversely, in the absence of a Wnt ligand, the activity of GSK-3β is switched on and phosphorylating β-catenin for ubiquitin-proteosome-mediated degradation [4]. β-catenin levels are then decreased within the cytosol and therefore the expression of Wnt-target genes such as cyclin-D and engrailed decrease. Recent studies, have shown that familial AD linked presenilin-1 (PS1) proteins form high molecular weight multi-protein complexes with the cell adhesion/signaling β-catenin and GSK-3β. Further studies have shown that β-catenin levels are markedly reduced in AD patients carrying PS1-inherited mutations and that these mutations may also disturb β-catenin translocation to the nucleus, thus affecting the Wnt signaling pathway cascade.

The function of the Wnt signaling pathway in the central nervous system (CNS) has been extensively studied during neuronal development; however in adult CNS it is less characterized. It has been described that an increase in Wnt-7a signaling mediates the maturation of synaptic contacts between cerebellar granule cells and mossy fibers in the cerebellum cortex *in vivo* by remodeling axonal microtubules, without the need of transcriptional activity of β-catenin. Other studies have shown that incomplete Wnt signaling occurs during depolarization-induced β-catenin-mediated dendritic arborization. In cultured hippocampal neurons, Wnt ligands with β-catenin promoting activity are modulated by depolarizating-inducing stimulus and hence β-catenin stabilization regulation and increased n-cadherin levels mediate dendritic morphogenesis. Recent evidence shows that the Wnt signaling pathway is required for both self-renewal of neuronal precursors and neuronal differentiation [1].

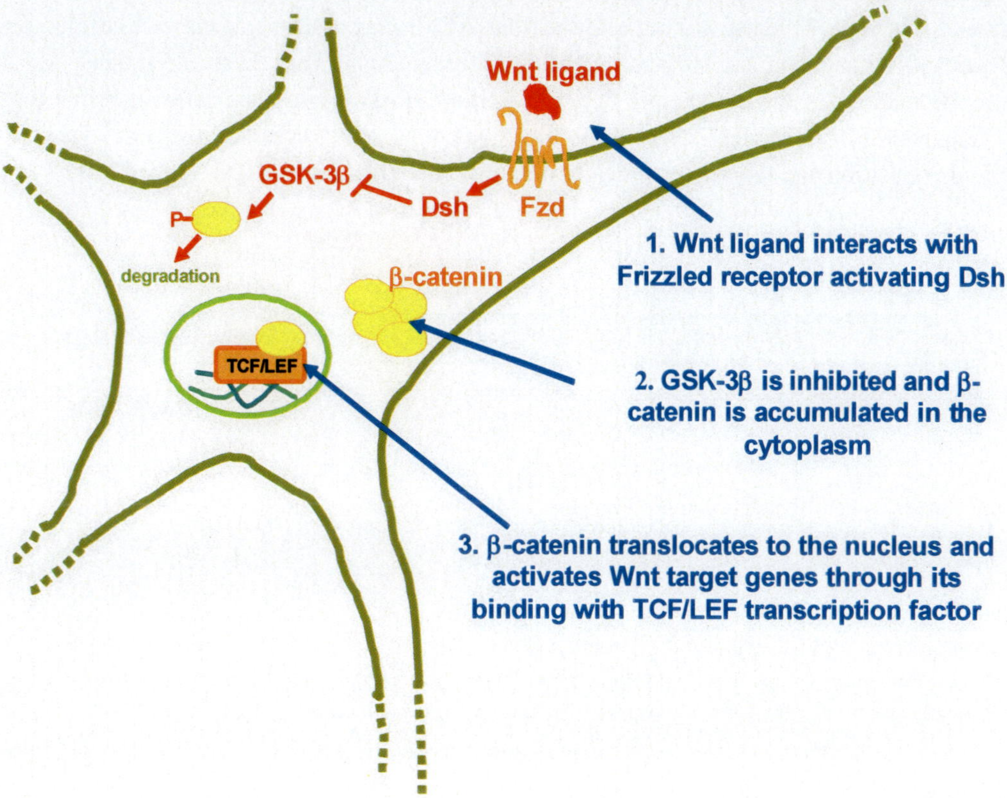

Figure 1: The Wnt signaling pathway.
The Wnt ligand interacts with its receptor Frizzled (Fzd). As a consequence GSK-3β is inactivated and β-catenin is not longer targeted to degradation. β-catenin is stabilized and accumulated in the cytoplasm followed by its translocation to the nucleus where Wnt target genes are activated. This neuroprotective signaling pathway is down-regulated by the toxic properties of Aβ, which activates GSK-3β.

The Role of AChE in Amyloid Neurotoxicity

Aβ is produced from the processing of amyloid precursor protein (APP) through the action of β- and γ-secretases (Figure 2). The detailed mechanism for the physiological transition of a soluble non-amyloidogenic form of Aβ to an amyloidogenic form remains not fully determined; however, the role of some senile plaque companion molecules has been proposed in this process. The list of these chaperone molecules includes acute phase proteins, serum proteins, extracellular matrix proteins, lipids, metals and apolipoproteins [5].

One of the proteins associated with amyloid plaque deposits is AChE, which is found bound to the amyloid core of mature senile plaques, pre-amyloid diffuse deposits and cerebral blood vessels. Interestingly, the areas where senile plaques are present are strongly AChE positive. It has been observed under these conditions the Aβ-AChE complexes are more toxic than the amyloid fibrils alone. In fact, AChE promotes Aβ fibril formation forming macromolecular complexes with the growing fibrils, and it is incorporated into senile-like plaques *in vitro* [6,7]. Early studies showed that AChE was associated with senile plaques and neurofibrillary tangles *in vivo*. Considering that AChE has been designated as a molecular marker for the cholinergic deficit observed in AD brains; this data suggested that the "companion" molecules detected in senile plaques may represent the link between amyloid deposition and the selective death of neuronal populations in AD brains.

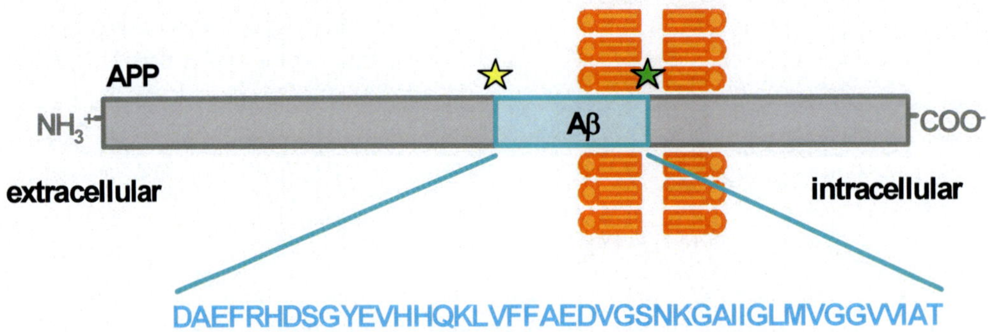

Figure 2: Diagram of APP and cleavage sites to produce Aβ.
When the integral protein APP (~700 aa) is processed by β- (BACE1: β-site APP-cleaving enzyme) and γ-secretase (shown in yellow and green stars, respectively), Aβ is produced. When this peptide accumulates as an insoluble form it forms senile plaques; known as one of the hallmarks of AD.

Previously, it was found that bovine brain AChE, as well as the human and mouse recombinant enzyme, were able to accelerate Aβ fibril formation, forming a high stable molecular complex with Aβ fibrils, which was resistant to the action of detergent, to low pH, high substrate concentrations, and present a lower sensitivity to AChE inhibitors and high ionic strength conditions. The action of AChE on fibril formation was independent of the subunit composition of the enzyme, it was not affected by edrophonium, (active site inhibitor), but it was affected by propidium, (a peripheral anionic binding site ligand). Butyrylcholinesterase, which lacks the peripheral site, did not show an effect on amyloid formation. Aβ aggregation process in the presence of AChE has been shown to be a thermodynamically favored process, as only a small quantity of AChE (stoichiometric ratio of AChE/Aβ 1/1000) is needed to promote the aggregation process. Moreover, under some conditions, it is possible to show cross-linking between Aβ peptide and AChE. Besides anti-cholinesterase drugs, the effect of AChE on amyloid formation was affected by a monoclonal antibody (mAb 25B1) directed towards the peripheral anionic site of the enzyme, which was able to completely abolish the effect of AChE on fibrillogenesis [8].

The effects of AChE on cultured cell lines such as neuronal (Neuro 2a) and glial-like (B12) cells were also analyzed. LDH release and MTT reduction assays have shown that the AChE increases the neurotoxicity produced by the Aβ peptide. This effect was dependent on the enzyme concentration, time of incubation and cellular density. The Aβ-increased neurotoxicity induced by AChE was not related to its catalytic activity, because the anti-cholinesterase drug BW284C51 was not able to block the effects of the enzyme. Upon incubation to Aβ-AChE complexes, cytoplasmic shrinkage, condensation and fragmentation of neuronal nuclei were detected indicating that apoptotic cell death is involved in the effect of the enzyme.

Further *in vivo* studies indicated that morphological and behavioral changes occur in astrocytes and hippocampal neurons when exposed to Aβ-AChE complexes. Neuronal cell loss was found in the upper leaf of the dentate gyrus in rats injected with AChE together with an astrocytic hypertrophy. Thus, a local increase in hippocampal AChE may play a role in triggering neuropathological and behavioral changes such as those observed in AD brains. Moreover, Aβ-AChE complexes show an increased dose-dependent neurotoxicity than the Aβ fibrils alone both in PC12 cells and chick primary cultures of retina cells.

Aβ-AChE complexes trigger dramatic neuropathological changes in comparison with Aβ fibrils alone as recently characterized *in vivo*. In fact, amyloid deposits were larger than those produced in the absence of AChE. Aβ-AChE complexes induced extensive astrocytosis and a stronger neuronal cell death in comparison with Aβ injected animals. This data is consistent with the notion that Aβ-AChE complexes are more toxic than Aβ fibrils and that AChE may trigger some of the neurodegenerative changes observed in AD brains [9].

The Activation of the Wnt/β-Catenin Signaling Pathway Prevents Aβ-induced Neurotoxicity

Aβ neurotoxicity has been shown to be associated with morphological alterations in rat and in human neurons including neuronal shrinkage plus severe axonal and dendritic dystrophy. Similar dystrophic changes were also seen in AD neurons that developed neurofibrillar pathology. During AD, protein levels of GSK-3β are mainly found in neuronal cell bodies and neurites, which co-localize with neurofibrillary changes present in AD brains. Remarkably, the activation of the tau protein kinase I/GSK-3β enzyme, the hyperphosphorylation of tau protein and the loss of the microtubular network, have all been observed in primary cultures of fetal rat hippocampal and human cortical neurons exposed to the Aβ peptide.

The hyperphosphorylation of tau in AD brains occurs primarily at (Ser or Thr)-Pro motifs, which suggests that proline-directed kinases may be responsible for this activity. Among this group of kinases, the stress-activated kinase Jun N-terminal kinase (JNK) and the GSK-3β enzyme are both known to phosphorylate tau protein and to generate AD-like antibody epitopes. It is presently accepted that GSK-3β plays a major role in maintaining the stability of tau protein and other MAPs such as MAP1B. In fact, it has been observed that blocking GSK-3β activity prevents tau hyperphosphorylation and thus enhances its binding to the microtubular network.

Interestingly, lithium, which inhibits GSK-3β in a reversible manner, protects rat neurons from Aβ insults suggesting that a sustained loss of the Wnt signaling function may be involved in the Aβ dependent neurodegeneration observed in AD. In fact, a direct activation of the Wnt pathway by its endogenous Wnt-3a ligand overcomes the neurotoxic effects induced by Aβ in rat hippocampal neurons [10] and this effect is prevented by sFRP, a secreted Wnt-antagonist protein [11]. Also, when rat hippocampal neurons exposed to Aβ fibrils induce the destabilization of endogenous levels of β-catenin and this effect is prevented by pre-treating neurons with lithium or by Wnt-3a conditioned media [12]. Wnt/β-catenin signaling, thus, might be playing an important role in neuroprotection against Aβ toxicity and this event is probably a crucial process involved in development of Aβ neurodegeneration observed early in the onset of AD.

The emerging role of Wnt/β-catenin signaling as a therapeutically target for treatment of AD has led to the study of potential Wnt signaling interacting pathways that might improve cell survival against Aβ toxicity.

Activation of M1 Muscarinic Receptor Protects From Aβ-neurotoxicity Through the Wnt Signaling Pathway

In AD, Aβ plays a critical role in neurodegeneration, primarily affecting certain types of neurons, particulary those in the cortex and hippocampus. It is known that cholinergic neurotransmission pathway plays a key role in modulating learning and memory processes [13] and is found strongly affected in AD patients, including a reduction in cholinergic neurotransmission and certain cholinergic markers [14]. Notably, brain presynaptic cholinergic cells in AD are depleted progressively in the basal forebrain nucleus of Meynert affecting the signaling in post-synaptic cholinergic neurons of AD patients [15].

A selective vulnerability of cholinergic neurons has been the focus of intensive studies aimed to elucidate the biochemical mechanism that lead to AD. Despite an improved understanding of the critical role of the cholinergic system in normal cognition and dementia, it has been unclear how cholinergic deficits and Aβ accumulation might be related to each other.

Two major strategies for AD treatments have been extensively employed so far via activation of muscarinic receptors by specific muscarinic agonists and inhibition of ACh hydrolysis by AChE inhibitors [16], respectively. The acetylcholine muscarinic receptor (mAChR) presents established roles in central cholinergic transmission and functioning in learning and memory. The m1 (type-1) mAChR is expressed in the cerebral cortex and hippocampus and its major role is in cognitive processing including short-term memory [17,18]. Thus, a loss of pre-synaptic cholinergic neurons would affect the mAChR-mediated signaling in post-synaptic cholinergic neurons in AD, however, m1 mAChR levels are relatively not changed in the pathology, and therefore may serve as an excellent target for an anti-demential drug treatment [19].

Recent advances in understanding the molecular pharmacology of the vulnerable transmitter systems in AD have provided the opportunity to develop improved pharmacological agents. A vast diversity of m1 mAChR agonists has been synthesized, however, agonists of m1 mAChR have not been effective in Alzheimer's treatment due to the lack of subtype selectivity, very low bioavailability and poor tolerability [19,20].

The series of AF compounds synthesized by Fisher [19] are specific agonists for m1 mAChR. The specificity of these compounds is associated to their structures. While acetylcholine is a highly flexible molecule capable of attaining numerous conformations allowing its interaction with various mAChR subtypes, the AF series are rigid analogs of ACh, whose rigidity limits their interaction with m1 mAChR subtype [19]. Specific agonists such as AF series have shown to promote the nonamyloidogenic APP processing pathway, decrease Aβ and increase APP soluble production. M1 mAChR stimulation leads to nonamyloidogenic APP processing at least by two transduction pathways: protein kinase C (PKC)-dependent and mitogen-activated protein kinase (MAPK)-dependent pathways [20]. On the other hand, M1 mAChR cross-talks with neuronal cytoskeleton regulating the phosphorylation of tau microtubule-associated protein through GSK-3β inhibition [21].

It has not been demonstrated how cholinergic activation would lead to neuroprotection against Aβ-toxicity. Recently it was shown that M1 mAChR activation protects rat hippocampal neurons from Aβ toxicity where the activation of M1 mAChR enhances the survival of hippocampal neurons and protects them against neurite dystrophy and nuclear condensation induced by Aβ-toxicity [22].

It is known that the stimulation of M1 mAChR induces PKC activation through G-protein signaling pathways. Interestingly, PKC has been related to the processing of APP [23], and to protection of the neuron against the apoptosis induced by Aβ [24]. Also, it has been shown that PKC inhibits by Ser-9 phosphorylation the activity of GSK-3β in neurons exposed to Aβ [25] inducing tau hyper-phosphorylation [26]. Conversely, its inactivation increases the non-amyloidogenic processing of APP [27]. More important is its function in the Wnt pathway where the phosphorylation of β-catenin by GSK-3β targets β-catenin to the ubiquitin-proteosome degradation [28].

It has been recently demostrated that M1 mAChR activation leads to neuroprotection against Aβ-toxicity through its cross talk with the Wnt pathway. In this context, it has been shown that the stimulation of M1 mAChR by AF267B, an specific agonist of M1, inhibits GSK-3β activity elicited by Aβ, rescues β-catenin from its reduction induced by Aβ and Wnt target genes are activated, finally leading to neuroprotection from Aβ-toxicity in hippocampal neurons cultures (Table 1). The modulation of tau phosphorylation and APP processing by M1 mAChR stimulation [26,27] could be explained by the neuroprotection mediated by the Wnt pathway activation, since GSK-3β has been involved in both these processes.

Table 1. AF267B and IBU-PO protect hippocampal neurons from Aβ toxicity activating the Wnt signaling

	Cell survival (%)	GSK-3β activity (%)	Cytoplasmic β-catenin (%)
Control	100	100	100
Aβ	51.7 ± 5.3	162 ± 8.3	34.5 ± 1.2
Aβ-AF267B	94.5 ± 6.3	87.5 ± 3.9	75.9 ± 6.9
Aβ-IBU-PO	97 ± 4	94.3 ± 4.5	98.8 ± 5

On the other hand, mice that over-express GSK-3β have been shown to present active catenin degradation [29], when cultured cortical neurons of post-natal 2 GSK-3β mice were incubated with AF267B an increase in β-catenin levels was observed in a dose dependent manner (Figure 3A). As shown in by inmunofluorescence β-catenin levels are dramatically affected in transgenic mice that over-express GSK-3β. Interestingly, when M1 mAChR are activated, β-catenin is stabilized and shows increased levels in neurites (Figure 3B). This cross-talk between the M1 mAChR and the Wnt signaling that leads to neuroprotection from Aβ-toxicity may be highly relevant in the treatment of AD with M1 agonists. Progress in the development of more specific drugs targeted to cholinergic pathway would offer the possibility of improved therapeutic strategies for AD.

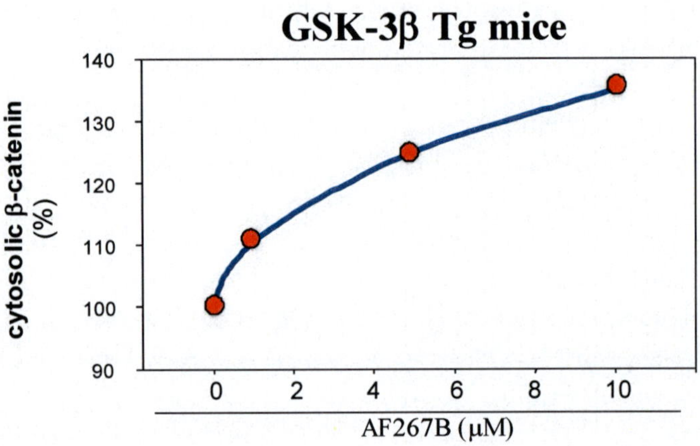

A

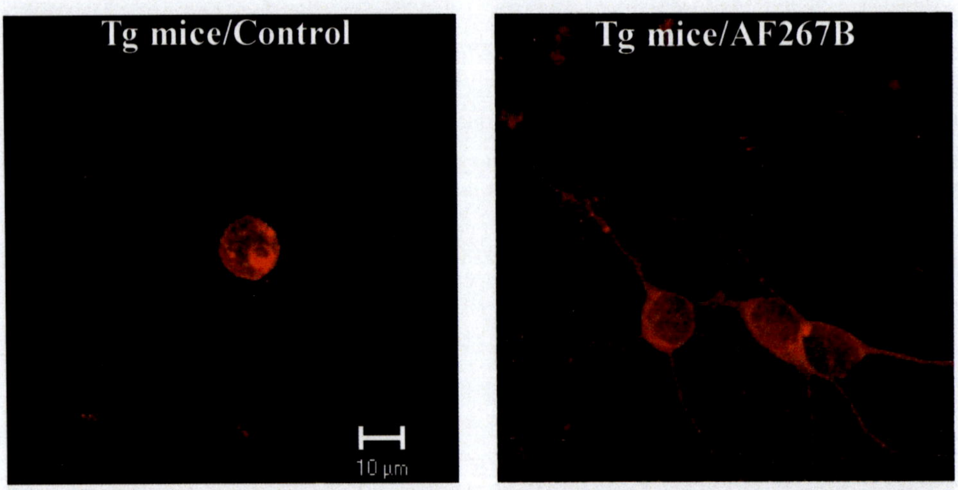

B

Figure 3: m1 mAChR activation protect of β-catenin degradation induced by GSK-3β overexpression. A) Cortical neurons cultures of transgenic mice that overexpress GSK-3β were incubated with AF267B. m1 mAChR activation by AF267B increases the β-catenin level in a dose dependent manner in a Western blott with anti β-catenin antibodies. B) Inmunofluorescence anti β-catenin. In transgenic mice cultures that overexpress GSK-3β showed a dramatically decreases in β-catenin level. In the presence of 10 mM of AF267B β-catenin is stabilized and increases its levels in neurites.

IBU-PO, an Anti-inflammatory and Cholinesterase Inhibitor Bifunctional Compound Protects From Aβ Neurotoxicity Through Wnt Signaling

Functional maintenance of the cholinergic pathway is actually the main therapeutic target in AD; thus, several cholinesterase inhibitors (ChEI) such as donepezil (Aricept), rivastigmine (Exeleon), and galanthamine (Reminyl) have been developed and approved for the treatment of AD patients [30]. The relevance of cholinesterases as a therapeutical target in AD is because it has been showed that butyrylcholinesterase and AChE are present in the senile plaques in brains from AD patients [31], AChE increases Aβ fibrillation [6] and the Aβ-AChE complexes induces the death of neurons *in vitro* more dramatically than Aβ fibrils alone [9,31,32]. In addition, has been showed that cholinesterase inhibitors increase the non-amyloidogenic soluble derivates of APP secretion [33,34].

On the other hand, AD brains show a significant increase of inflammatory markers such as IL-1 [35], TGF-β1 [36] and COX-2 [37]. COX-2 inhibitors have been suggested to be neuroprotective agents in AD [38]. A clinical study demonstrated that a long-term use (>24 months) of non-steroidal anti-inflammatory drugs (NSAIDs) ibuprofen and diclofenac reduce by 5-fold the risk for developing AD [39].

A variety of experimental studies indicate that a subset of NSAIDs dramatically reduce the secretion of $A\beta_{1-42}$ in cells *in vitro*; *in vivo* studies have shown that short-term administration of ibuprofen to transgenic mice overexpressing APP lowered brain levels of $A\beta_{1-42}$; in cultured cells, the decrease in $A\beta_{1-42}$ secretion was accompanied by an increase in the $A\beta_{1-38}$, indicating that NSAIDs subtly alter γ-secretase activity without significantly perturbing other APP processing pathways [40].

A novel NSAID-ChEI bifunctional compounds, ibuprofen-octyl-pyridostigmine (IBU-PO) has been described for its neuroprotective activity against Aβ induced toxicity; since it has been shown to protect neurons from axonal and dendritic dystrophy, and the reduced synaptic vesicles and number of neurites observed in hippocampal neurons exposed to Aβ [41]. In this context, the neuroprotection observed with IBU-PO has been associated with the Wnt signaling pathway since this bifunctional compound inhibits GSK-3β activity induction induced by Aβ and increases β-catenin levels rescuing its reduction elicited by Aβ, finally leading to neuroprotection from Aβ-toxicity. In addition, in transgenic mice that over-express GSK-3β and that show β-catenin reduction, neuronal cultures of this transgenic mice in the presence of IBU-PO present GSK-3β inactivation and an increase in β-catenin levels [41] (Table 1).

On the other hand, it has been shown that this bifunctional compound increases the non-amyloidogenic APP processing and decrease Aβ production. Altogether this data shows that IBU-PO protects hippocampal neurons from Aβ toxicity and neurons from transgenic mice that over-express GSK-3β, acting through some component of the Wnt signaling pathway supporting its role in neuroprotection [42].

These results are consistent with the idea that Aβ induced -neurotoxicity triggers a loss of function of Wnt signaling pathway and indicates that compounds which mimic this signaling cascade may be a putative candidate for therapeutic intervention in AD.

The Wnt Signaling Pathway Crosstalks With PPARγ as a Neuroprotective Mechanism

Peroxisome proliferator activated receptors (PPARs) are members of the nuclear hormone receptor superfamily of ligand-activated transcriptional factors acting as a heterodimer with the 9-cis retinoic acid receptor [43]. PPARγ function has been described in regulation of adipocyte differentiation but recent studies have also shown that some anti-inflammatory drugs which are PPARγ agonists, have neuroprotective actions in different animal models through the prevention of glia and microglia activation [44]. Although PPARγ has been restricted to tissues other than the brain, the observation that PPARγ protein levels change in the temporal cortex of AD patients suggested that it might play a role in AD. At present, it has become clear that anti-inflamatory drugs directly promote neuroprotection in primary neurons exposed to different pro-apoptotic stimuli *in vitro* [45]. Accordingly, thiazolidinedione drugs have been shown to have a potent insulin-sensitizing action that might be mediated through PPARγ-mediated inhibition of GSK-3β and some studies suggest that treatment for insulin resistance may reduce the risk or retard the development of AD [46] In fact, it was recently described that the activation of PPARγ efficiently prevents Aβ-induced neurodegeneration in rat hippocampal neurons [47].

Conclusion

Novel approaches for the understanding the origin as well as the complex molecular cascades involved in AD is of critical importance. Neuroprotective compounds which which have been referred in this manuscript could be used as novel therapeutic strategies in AD patients.

Acknowledgements

This work was supported by a grant FONDAP- Biomedicine N° 13980001, from the Millennium Institute for Fundamental and Applied and Biology (MIFAB) and a pre-doctoral CONICYT Fellowship to MC.

References

[1] Patapoutian, A. and Reichardt, L.F. (2000) Roles of Wnt proteins in neural development and maintenance. *Curr Opin Neurobiol* 10, 392-399.

[2] De Ferrari, G.V. and Inestrosa, N.C. (2000) Wnt signaling function in Alzheimer's disease. *Brain Res Brain Res Rev*. 33: 1-12.

[3] Chen, R.H., Ding, W.V. and McCormick, F. (2000) Wnt signaling to β-catenin involves two interactive components. Glycogen synthase kinase-3β inhibition and activation of protein kinase C. *J Biol Chem* 275: 17894-17899.

[4] Yost, C., Torres, M., Miller, J.R., Huang, E., Kimelman, D. and Moon, R.T. (1996) The axis-inducing activity, stability, and subcellular distribution of β-catenin is regulated in Xenopus embryos by glycogen synthase kinase 3. *Genes Dev* 10: 144314-144354.

[5] Morgan, C., Colombres, M., Nuñez, M.T. and Inestrosa, N.C. (2004) Structure and Function of Amyloid in Alzheimer's Disease. *Progress in Neurobiology*. 74: 323-349.

[6] Inestrosa, N.C., Alvarez, A., Perez, C.A., Moreno, R.D., Vicente, M., Linker, C., et al. (1996). Acetylcholinesterase accelerates assembly of amyloid-β-peptides into Alzheimer's fibrils: possible role of the peripheral site of the enzyme. *Neuron* 16: 881-891.

[7] Alvarez, A., Opazo, C., Alarcón, R., Garrido, J. and Inestrosa, N.C. (1997) Acetylcholinesterase promotes the aggregation of amyloid-β-peptide fragments by forming a complex with the growing fibrils. *J Mol Biol*. 272: 348-361.

[8] Reyes, A.E., Perez, D.R., Alvarez, A., Garrido, J., Gentry, M.K., Doctor, B.P. and Inestrosa, N.C. (1997) A monoclonal antibody against acetylcholinesterase inhibits the formation of amyloid fibrils induced by the enzyme. *Biochem Biophys Res Commun*. 232: 652-655.

[9] Reyes, A.E., Chacon, M.A., Dinamarca, M.C., Cerpa, W., Morgan, C. and Inestrosa N.C. (2004). Acetylcholinesterase-Aβ complexes are more toxic than Aβ fibrils in rat hippocampus: Effect on rat β-Amyloid aggregation, laminin expression, reactive astrocytosis, and neuronal cell loss. *Am J Pathol*. 164: 2163-2174.

[10] De Ferrari, G.V., Chacon, M.A., Barria, M.I., Garrido, J.L., Godoy, J.A., Olivares, G., Reyes, A.E., Alvarez, A., Bronfman, M. and Inestrosa, N.C. (2003) Activation of Wnt signaling rescues neurodegeneration and behavioral impairments induced by β-amyloid fibrils. *Mol Psychiatry* 8: 195-208.

[11] Jones, S.E. and Jomary, C. (2002) Secreted Frizzled-related proteins: searching for relationships and patterns. *Bioessays*. 24:811-820.

[12] Inestrosa, N.C., Urra, S. and Colombres, M. (2004) "Acetylcholinesterase (AChE) – amyloid β peptide complexes in Alzheimer′s Disease. The Wnt signaling pathway" *Curr Alz Res* 1, 249-254.

[13] Lippa, A.S., Pelham, R.W., Beer, B., Critchett, D.J., Dean, R.L. and Bartus, R.T. (1980) Brain cholinergic dysfunction and memory in aged rats. *Neurobiol. Aging* 1: 13-19.

[14] Tran, M.H., Yamada, K. and Nabeshima, T. (2002). Amyloid β-peptide induces cholinergic dysfunction a cognitive deficits: A minireview. *Peptide* 23: 1271-1283.

[15] Giacobini, E. (1996) Cholinesterase inhibitors do more than inhibit cholinesterase. In: *Alzheimer Disease: From Molecular Biology to Therapy*, Edited by Becker, R. and Giacobini, E., pp. 187-204, Birkhauser, Boston.

[16] Colombres, M., Sagal, J. P. and Inestrosa, N.C. (2004) "An Overview of the Current and Novel Drugs for Alzheimer′s Disease with particular references to anti-cholinesterase compounds" *Curr Phar Des* 10, 3121-3130.

[17] Levey, A.I. (1996). Muscarinic acetylcholine receptor expression in memory circuits: implications for treatment of Alzheimer disease. *Proc. Natl. Acad. Sci. USA*. 93: 13541-13546.

[18] Anagnostaras, S.G., Murphy, G.G., Hamilton, S.E., Mitchell, S.L., Rahnama, N.P., Nathanson, N.M., et al., 2003. Selective cognitive dysfunction in acetylcholine M1 muscarinic receptor mutant mice. *Nature Neuroscience* 6: 51-58.

[19] Fisher, A., 2000. Therapeutic strategies in Alzheimer's disease: M1 muscarinic agonists. *Jpn. J. Pharmacol*. 84: 101-112

[20] Haring, R., Fisher, A., Marciano, D., Pittel, Z., Kloog, Y., Zuckerman, A., et al. (1998) Mitogen-activated protein kinase-dependent and protein kinase C-dependent pathways link the m1 muscarinic receptor to β-amyloid precursor protein secretion. *J. Neurochem.* 71: 2094-2103.

[21] Sadot, E., Gurwitz, D., Barg, J., Behar, L., Ginzburg, I. and Fisher, A. (1996). Activation of m1 muscarinic acetylcholine receptor regulates tau phosphorylation in transfected PC12 cells. *J. Neurochem*. 66: 877-880.

[22] Farias, G.G., Godoy, J.A., Hernandez, F., Avila, J., Fisher, A. and Inestrosa, N.C. (2004a). M1 muscarinic receptor activation protects neurons from beta-amyloid toxicity. A role for Wnt signaling pathway. *Neurobiol. Dis*. 17: 337-348.

[23] Racchi, M., Mazzucchelli, M., Pascale, A., Sironi, M. and Govoni, S.(2003). Role of protein kinase Cα in the regulated secretion of the amyloid precursor protein. *Mol. Psychiatry* 8: 209-216.

[24] Xie, J., Guo, Q., Zhu, H., Wooten, M.W. and Mattson, M.P. (2000). Protein kinase C iota protects neural cells against apoptosis induced by amyloid β-peptide. *Brain Res Mol Brain Res*. 82: 107-113.

[25] Garrido, J.L., Godoy, J.A., Alvarez, A., Bronfman, M. and Inestrosa, N.C. (2002). Protein kinase C inhibits amyloid β-peptide neurotoxicity by acting on members of the Wnt pathway. *FASEB J*. 16: 1982-1984.

[26] Hall, A.C., Lucas, F.R. and Salinas, P.C. (2000). Axonal remodeling and synaptic differentiation in the cerebellum is regulated by WNT-7a signaling. *Cell*. 100: 525-535.

[27] Phiel, C.J., Wilson, C.A., Lee, V.M. and Klein, P.S. (2003). GSK-3α regulates production of Alzheimer´s disease amyloid-β peptides. *Nature* 423, 435-439.

[28] Nusse, R. (1997). A versatile transcriptional effector of Wingless signaling. *Cell* 89: 321-323.

[29] Lucas, J.J., Hernandez, F., Gomez-Ramos, P., Moran, M.A., Hen, R. and Avila, J. (2001). Decreased nuclear β-catenin, tau hyperphosphorylation and neurodegeneration in GSK-3β conditional transgenic mice. *EMBO J* 20: 27-39.

[30] Giacobini, E. (1998). Cholinesterase inhibitors for Alzheimer's disease therapy: From tacrine to future applications. *Neurochem. Int*. 32, 413-419.

[31] Alvarez, A., Alarcón, R., Opazo, C., Campos, E.O., Muñoz, F.J., Calderon, F.H., Dajas, F., Gentry, M.K., Doctor, B.P., De Mello, F.G. and Inestrosa, N.C. (1998). Stable complexes involving acetylcholinesterase and amyloid-β-peptide change the biochemical properties of the enzyme and increase the neurotoxicity of Alzheimer's fibrils. *J. Neuroscience* 18: 3213-3223.

[32] Muñoz, F.J. and Inestrosa, N.C. (1999). Neurotoxicity of acetylcholinesterase-amyloid-β-peptide aggregates is dependent on the type of Aβ-peptide and the AChE concentration present in the complexes. *FEBS Lett 450:* 205-209.

[33] Lahiri, D.K., Farlow, M.R., Hintz, N., Utsuki, T. and Greig, N.H. (2000). Cholinesterase inhibitors, β-amyloid precursor protein and amyloid β-peptides in Alzheimer disease. *Acta Neurol. Scand Suppl.* 176: 60-67.

[34] Mori, F., Lai, C.C., Fusi, F. and Giacobini, E. (1995) Cholinesterase inhibitors increase secretion of APPs in rat brain cortex. *NeuroReport* 6, 633-636.

[35] Sheng, J.G., Griffin, W.S., Royston, M.C. and Mrak, R.E. (1998). Distribution of interleukin-1-immunoreactive microglia in cerebral cortical layers: implications for neuritic plaque formation in Alzheimer's disease. Neuropathol. *Appl. Neurobiol* 24: 278-283.

[36] Van der Wal, E.A., Gomez-Pinilla, F. and Cotman, C.W. (1993). Transforming growth factor-β 1 is in plaques in Alzheimer and Down pathologies. *Neuroreport* 4: 69-72.

[37] Kitamura, Y., Shimohama, S., Koike, H., Kakimura, J., Matsuoka, Y., Nomura, Y., Gebicke-Haerter and P.J., Taniguchi, T. (1999) Increased expression of cyclooxygenases and peroxisome proliferator-activated receptor-γ in Alzheimer's disease brains. *Biochem Biophys Res Commun.* 27: 582-586.

[38] Giovannini, M.G., Scali, C., Prosperi, C., Bellucci, A., Pepeu, G. and Casamenti, F. (2003) Experimental brain inflammation and neurodegeneration as model of Alzheimer's disease: protective effects of selective COX-2 inhibitors. *Int J Immunopathol Pharmacol.* 16: 31-40.

[39] In t' Veld, B.A., Ruitenberg, A., Hofman, A., Launer, L.J., van Duijn, C.M., Stijnen, T., et al. (2001). Nonsteroidal anti-inflammatory drugs and the risk of Alzheimer’s disease. *New England J. Med.* 645: 1515-1521.

[40] Gasparini, L., Ongini, E. and Wenk, G. (2004). Non-steroidal anti-inflammatory drugs (NSAIDs) in Alzheimer's disease: old and new mechanisms of action. *J Neurochem.* 91: 521-536.

[41] Farías, G.G., Godoy, J.A., Vázquez, M.C., Adani, R., Meshulam, H., Avila, J., Amitai, G. and Inestrosa, N.C. 2004. The Anti-inflammatory and Cholinesterase Inhibitor Bifunctional Compound IBU-PO Protects from β-Amyloid Neurotoxicity by Acting on Wnt Signaling Components. *Neurobiol. Dis.* 18: 176-183.

[42] Alvarez, A.R., Godoy, J.A., Mullendorff, K., Olivares, G.H., Bronfman, M. and Inestrosa, N.C. 2004. Wnt-3a overcomes β-amyloid toxicity in rat hippocampal neurons. *Exp Cell Res.* 297: 186-196.

[43] Kersten, S., Desvergne, B. and Wahli, W. (2000) Roles of PPARs in health and disease, *Nature* 405:421-424.

[44] Breidert, T., Callebert, J., Heneka, M.T., Landreth, G., Launay, J.M. and Hirsch, E.C. (2002) Protective action of the peroxisome proliferator-activated receptor-γ agonist pioglitazone in a mouse model of Parkinson's disease. *J Neurochem* 82:615-624.

[45] Uryu, S., Harada, J., Hisamoto, M. and Oda, T. (2002) Troglitazone inhibits both post-glutamate neurotoxicity and low-potassium-induced apoptosis in cerebellar granule neurons. *Brain Res* 924:229-236.

[46] Watson, G.S. and Craft, S. (2003) The role of insulin resistance in the pathogenesis of Alzheimer's disease: implications for treatment. *CNS Drugs* 17: 27-45.

[47] Inestrosa, N.C., Godoy, J.A., Quintanilla, R.A., Koenig, C. and Bronfman, M. (2005) Peroxisome proliferator-activated receptor γ (PPAR γ) is expressed in hippocampal neurons and its activation prevents β-amyloid neurodegeneration: Role of Wnt signaling. *Exp Cell Res*. 304: 91-104.

In: Trends In Neurochemistry Research
Editor: Robert M. Coleman, pp. 145-164
ISBN: 1-59454-418-2

Chapter 3

Neuroprotection against Kainic Acid Excitotoxicity by Specific Glutamate Receptor Antagonists in Cultured Organotypic Hippocampal Slices

I. E. Holopainen,* T.-K. Kukko-Lukjanov, and F. R. Lopez-Picon

Department of Pharmacology and Clinical Pharmacology,
University of Turku, Turku, Finland

Abstract

Overactivation of glutamate receptors (excitotoxicity) is considered to be a contributing factor in neuronal death in a number of CNS insults and disorders, including hypoxia, ischemia, trauma, and epileptic seizures. Kainic acid (KA), a competitive agonist for kainate/α-amino-3-hydroxy-5-methyl-4-isoxazole propionic acid (AMPA) type of glutamate receptors is neurotoxic, and it is widely used to study the mechanisms of neuronal death both *in vitro* and *in vivo* models. On the contrary, KA receptor antagonists are suggested to be neuroprotective. We have now studied mechanisms and temporal profile of KA-induced neuronal toxicity, and the neuroprotective effects of the specific N-methyl-D-aspartate (NMDA) and AMPA/KA receptor antagonists in cultured organotypic hippocampal slices (OHC) during the *in vitro* maturation. Fluoro-Jade B (FJB) staining was used as a marker for degenerating neurons, thionin staining to detect neuronal cell loss, and immunoblotting and electron microscopy to clarify the mechanisms of death. In immature slices, neuronal vulnerability to KA treatment was strictly time-dependent, and decreased in order CA3a/b >> CA3c > dentate granule cells,

* Correspondence: I. E. Holopainen, Phone: intl-358-2-333 7549, Fax: intl-358-2-333 7216; e-mail: irma.holopainen@utu.fi

whereas CA1 pyramidal neurons were not affected. The NMDA receptor antagonist, (DL)-2-amino-5-phosphonovaleric acid (APV, 50 μM), and the AMPA/KA receptor antagonist, 6,7-dinitroquinoxaline-2,3-ione (DNQX, 5 μM) were equally protective at the early phase of KA exposure (i.e. at 12 h) in the CA3a/b region, but increased the KA-induced neuronal death at the longer exposure. After the initial treatment (48 h), the 48-h recovery period in normal culture medium did not completely rescue the cells in either immature (11 DIV) or in more mature (18 DIV) slices. However, DNQX significantly attenuated KA neurotoxicity in the CA3 regions in the immature, but not in the more mature slices. In Western blotting studies carried out in the immature slices, the expression levels of the active, 17-kDa form of caspase-3, and the 85-kDa cleaved fragment of poly(ADP ribose)polymerase (PARP) were not altered from the control levels suggesting that neuronal death did not involve the caspase-dependent apoptotic pathways. This was confirmed with electron microscopic analysis of CA3 neurons, which showed necrotic injury after KA treatment in immature and more mature slices. Our results suggest that KA-induced neuronal death in OHCs is a strictly region-specific, mostly irreversible, necrotic process, the early phase of which can partly be ameliorated by glutamate receptor antagonists DNQX and APV in immature slices.

Introduction

Overactivation of glutamate receptors (excitotoxicity) plays an important role in neuronal death in many CNS insults and disorders, including hypoxia, ischemia, trauma, and epileptic seizures (Whetsell, 1996; Mattson et al., 2000; Bengzon et al., 2002). Kainic acid (KA), a competitive agonist for kainate/α-amino-3-hydroxy-5-methyl-4-isoxazole propionic acid (AMPA) type of glutamate receptors is neurotoxic, and it is widely used to study the mechanisms of neuronal death both *in vitro* and *in vivo* models of CNS insults (Bengzon et al., 2002).

The overactivation of glutamate receptors by other agonists, i.e. NMDA, AMPA, and glutamate itself, is also associated with neurotoxicity (Vornov et al., 1991; Bruce et al., 1995; Kristensen et al., 2001), whereas glutamate receptor antagonists are suggested to be neuroprotective (Drian et al., 2001; Zimmer et al., 2000; Zimmermann et al., 2001), although there are also reports at variance with this assumption. For example, the competitive N-methyl-D-aspartate (NMDA) receptor antagonist 2-amino-5-phosponovaleric acid (APV) has dose-dependently impaired the survival of cerebellar Purkinje cell (Yuzaki et al., 1996), and the treatment of 7-day-old rats and cultured cortical neurons with the NMDA receptor antagonist MK-801 has resulted in extensive apoptotic neuronal death in the developing brain (Hansen et al., 2004). The effects of the AMPA/kainate receptor antagonist 6,7-dinitroquinoxaline-2,3-dione (DNQX) has also been controversial. The exposure of dissociated embryonic brain stem neuronal cultures to DNQX has attenuated neuronal survival at the early stage of maturation, but it has had no effect when applied at the later stage (Bardoul et al., 1997). In cultured hippocampal neurons, DNQX has exhibited a dose-dependent, neuron-specific toxicity (Martin et al., 2003), whereas in cortical neurons it has had a dual effect being toxic at the early stages, but neuroprotective at the later stages of maturation (Drian et al., 2001).

In adult rats, KA-induced neuronal death has been suggested to occur through necrosis, apoptosis, or to have features of both mechanisms (Pollard et al.1994; Fujikawa et al., 2000; 2002; Bengzon et al., 2002). Necrosis is characterized by cell swelling, random breakdown of cellular constituents, and finally lysis of the cell, whereas apoptosis leads to cell shrinkage, preservation of membrane integrity, decrease in cellular volume, chromatin condensation, and DNA fragmentation (Bredesen, 2000; Sastry & Dao, 2000). Apoptosis is a tightly regulated process, which accounts for the physiological cell death during the development. In addition to that, the apoptotic pathways are also activated in response to various CNS insults, e.g. trauma, ischemia, and epileptic seizures (Ishimaru et al.,1999; Bittigau et al., 1999; Bredesen, 2000). The apoptotic response depends on the severity of the insult, and on the target cell itself, including its state of maturation (Bredesen, 2000). Several apoptosis-inducing factors have been identified, which may lead to activation of cell membrane death receptors, mitochondrial apoptotic neuronal death pathway, or directly damage DNA (Sastry & Rao, 2000; Zimmermann et al., 2001). The caspase family of enzymes is suggested to play an important role in apoptosis (Earnshaw et al., 1999). The activation of caspase cascade starts with caspase-9, and once activated, it cleaves a 32-kDa inactive caspase-3 into a 17-kDa active caspase-3 (Kroemer & Reed, 2000; Los et al., 2002). Another characteristic event of apoptosis is the proteolytic cleavage of a 116-kDa form of the nuclear DNA repair enzyme poly(ADP ribose)polymerase (PARP) by caspase-3, which results in an inactive, 84-kDa cleaved PARP fragment (Lazebnik et al., 1994; Virag & Szabo, 2002).

The role of apoptosis and necrosis in neuronal death is poorly elucidated in the immature brain, although recent studies suggest that there are fundamental differences between the immature and mature hippocampus in response to various brain insults (Bittigau et al., 1999 ; Joashi et al., 1999; Hu et al., 2000; Khaing et al., 2000; Yakovlev et al., 2001). Moreover, the subunit composition of e.g. glutamate receptors changes during the brain maturation (Monyer et al., 1994; Ritter et al., 2002; Takai et al., 2003), which may effect the pharmacological properties of the receptor proteins resulting in either neuroprotective or neurotoxic effect of drugs depending on the stage of maturation. We have now studied the time course of KA-induced neuronal death, and the hypothesized neuroprotective effect of the specific non-NMDA, and NMDA receptor antagonists, DNQX and APV, respectively, in immature and more mature cultured hippocampal slices. The mechanism of neuronal damage was studied using immunoblotting and electron microscopy.

Materials and Methods

Experimental Procedures

Organotypic Hippocampal Slice Cultures (OHC)

Hippocampal slice cultures were prepared from the hippocampi of 6 to 7 (P6-P7) days old Spraque-Dawley rats using the method of Stoppini et al. (1991) as recently described in detail (Holopainen et al., 2001). Briefly, hippocampi were dissected, 400-μm slices were cut perpendicular to the septotemporal axis of the hippocampus, separated, and placed on top of semipermeable membrane inserts (Millipore Corporation, Bedford, MA) in a six-well plate containing culture medium [50 % of minimum essential medium, 25% Hanks's balanced salt solution (HBSS), 25% heat-inactivated horse serum, 25 mM HEPES, supplemented with 0.5 ml GlutaMaxII (Gibco) and 6.5 mg/ml glucose, pH adjusted to 7.2]. Slices were cultured in stabilized conditions (37 °C, 5% CO_2) for 7-18 days with medium change twice a week. All animal procedures were conducted in accordance with the guidelines set by the European Community Council Directives 86/609/EEC, and had the approval of the Animal Use and Care Committee of the University of Turku.

Pharmacological Treatment of Cultures

Cultures (7 days *in vitro*, DIV) were treated for 6, 12, 24 or 48 h with KA alone (5 μM, Sigma, St Louis, MO) or together with the non-NMDA receptor antagonist 6,7-dinitroquinoxaline-2,3-ione (DNQX, Tocris, Tocris Cookson Ltd., Avonmouth, UK)(final concentration 5 μM), or the NMDA receptor antagonist (DL)-2-amino-5-phosphonovaleric acid (APV, Tocris)(final concentration 50 μM). To study the effect of neuronal maturity on the KA-induced vulnerability, a subset of slices was cultured for 14 days (14 DIV) prior to the treatment. The recovery of damaged neurons after the initial insult of 48 h (KA alone or together with DNQX or APV) was studied in a subset of both immature (7 DIV) and mature (14 DIV) cultures, which were transferred into a new multiwell dish containing fresh culture medium after the treatment, cultured for additional 48 hours, and stained at 11 DIV (regarded as immature), and at 18 DIV (regarded as more mature), respectively.

Fluoro-Jade B Staining

Fluoro-Jade B (FJB) was used to detect degenerating neurons after the treatments in OHCs. This dye stains degenerating neurons and their processes regardless of the mechanism (Schmued et al., 1997; Schmued & Hopkins, 2000). The staining was carried out as recently described in detail (Holopainen et al., 2004). Briefly, inserts with slices were first rinsed with ice-cold 0.1 M phosphate-buffered saline (PBS, pH 7.2), fixed with 4% paraformaldehyde (PFA, Sigma), washed with PBS and water, and transferred to 0.06% potassium permanganate ($KMnO_4$, Sigma) (freshly prepared in H_2O) for 2-5 min. After that, slices were transferred to 0.001% FJB solution for 30 min, washed with water, and carefully mechanically detached from membranes, transferred to gelatin-coated glass slides, dried, immersed in xylene, and coverslipped. The amount of stained cells (i.e. degenerating cells) was analyzed blinded by two investigators (IEH, T-KK-L) with an epifluorescence microscope (Leica DM R, Heerbrugg, Switzerland) in the CA3a/b, CA3c, CA1, and granule cell layers (superior and inferior blades) of the dentate gyrus (DG). The number of stained neurons was scored using the following scheme: 0 = no FJB-stained neurons in the cell layer (regarded as normal), 1 = some (≤ 10) stained neurons; 2 = sparse number of stained neurons; 3 = many stained neurons, 4 = numerous stained neurons.

Thionin Staining

Thionin staining was used to detect neuronal loss after KA treatment in CA3a/b, CA3c, CA1, and granule cell layers. The staining was carried out as recently described in detail

(Holopainen et al., 2001) using PFA-fixed slices attached to the insert membrane. After rinsing with PBS, slices were carefully detached from inserts, transferred to gelatin-coated glass slides, stained in 0.5% thionin solution, dehydrated, cleared in xylene, and mounted in Permount. The stained slices were examined with Leica DM R microscope under bright-field optics. The amount of neurons in CA3a/b, CA3c, CA1, and DG granule cell layers (upper and lower blade) was scored using the following scheme: 0 = no stained neurons in the cell layer (all neurons dead in terms of loss of Nissl staining); 1 = some (1-5) stained neurons; 2 = sparse number of stained neurons; 3 = numerous stained neurons but with slightly disturbed cell layer integrity; 4 = numerous stained neurons with good cell layer integrity (regarded as normal). Digital camera Olympus U-TV1 X (Olympus Optical Co., Ltd., Tokyo, Japan) was used to capture pictures using Olympus BX60 microscope, and images were further processed using Adobe Photoshop (version 6.0) and Corel Draw (version 9.0.).

Confocal Microscopy

The FJB staining in the hippocampus was imaged with a Leica TCS SP confocal microscopy system (Leica, Heidelberg, Germany) equipped with an Argon-Krypton laser (Omnichrome, Melles Griot, Carlsbad, CA, USA). The excitation wavelength for FJB was 388 nm, and the emission wavelength filter settings were 500-550 nm. The images were acquired at 2 μm steps and analyzed with Leica TCS NT/SP Scanware (version 1.6.587) software. The picture of the whole hippocampus was reconstructed in Adobe Photoshop (version 5.5, Adobe Systems, Mountain View, CA) by connecting 6-8 maximum projections of different areas of the hippocampus. Figures were further processed using Corel Draw program (version 9.0).

SDS-PAGE Immunoblotting

Western blot analysis was performed to determine whether caspase-3 and PARP proteins are cleaved in response to KA treatment in 7 DIV OHCs as a consequence of an apoptotic cell death. Immunoblotting was carried out as recently described in detail in OHCs (Holopainen et al., 2001). Briefly, one control group, and the 4-h, 8-h, and 24-h KA-treated (5 μM) groups of slices from two different culture batches (n =93 to 96 slices in both batches for each time group) were collected in ice-cold homogenization buffer containing [50mM TRIS-HCl (pH 7.4), 1% SDS, 2 mM EDTA, 1mM PMSF, and 0.7 mM dithiothreitol], homogenised using Ultra-Turrax T25 homogenizer (Janke and Kunkel, Staufen, Germany), homogenates boiled, centrifuged, supernatants collected, frozen, and stored at –80 °C until used. Protein concentration of the samples was measured using Lowry based Biorad DC Protein assay (Biorad, CA, USA). Fifty micrograms (μg) of protein were applied to each lane for SDS-PAGE and separated by electrophoresis with a 10% acrylamide minigel using Mini-protean III (Biorad), and transferred to a polyvinylidene fluoride Immobilon-P (Millipore) membrane using Mini Trans-blot Cell system (Transblot SD, Biorad). Membranes were incubated overnight with the primary antibodies; polyclonal anti-cleaved caspase-3 (1: 3500) (Cell Signaling Technology, Inc. Beverly, MA), and monoclonal anti-poly(ADP-ribose)polymerase, PARP (1:3500) (Sigma), which detects the uncleaved 116-kDa protein, and the 85-kDa apoptosis-induced cleavage product. This was followed by incubation with the HRP-conjugated secondary antibodies; goat anti-mouse IgG_1 (γ_1 chain specific, 1:3500)

(SouthernBiotech, Birmingham, AL) to detect PARP, and goat anti-rabbit IgG (1:4000)(Sigma) to detect the active caspase-3. The signal was obtained using chemiluminiscence ECL system (Amersham, UK) and Hyperfilm ECL (Amersham). The optical signals were quantified with Image J 1.20s (NIH, USA). The results are given as arbitrary units (a.u.) per mg of protein.

Electron Microscopy

Hippocampal slices were cultured for 7 or 14 days prior to KA treatment (5 μM) for 24 h. After that, slices were briefly washed with PBS and fixed with freshly prepared 5% glutaraldehyde (Electron Microscopy Sciences, Fort Washington, PA, USA). The CA3a/b region was isolated, and the specimen was further prepared according to standard procedures as previously described (Fröjdman et al., 1992). The ultrathin sections (70 nm) were stained with 5% uranyl acetate and 5% lead citrate in Ultrostainer (Leica, Wien, Austria), and examined in a JEM-1200EX (JEOL, Tokyo, Japan) transmission electron microscope.

Statistical Analysis

The overall group differences in the score numbers after the staining with FJB and thionin were assessed by one-way analysis of variance (ANOVA) with the Tukey-Kramer Multiple Comparison Test as a post hoc test. Student's independent two-tailed *t*-test was used to assess the significance of differences between two groups. The nonparametric one-way ANOVA with Kruskal-Wallis Test was used to analyze the group differences in the immunoblots. The level of significance was set at $p<0.05$. Prism 4.0 software (GraphPad Software, San Diego, CA) was used in all statistical analysis.

Results

The Effect of DNQX and APV on KA-induced Neuronal Damage as Verified with Fluoro-Jade B Staining in OHCs

In the immature slices, the vulnerability of hippocampal neurons to KA (5 μM) was strictly region- and time-specific. After KA exposure of 6 h, no significant neuronal damage was detected in any hippocampal subregion, i.e. CA3a/b, CA3c, CA1, and DG granule cells (Figure 5A). KA treatment of 12 h significantly increased the amount of FJB-labeled, i.e. degenerating neurons, in the CA3a/b region. In the presence of DNQX (5 μM) and APV (50 μM), the amount of labelled neurons was dramatically decreased, i.e. these compounds had a neuroprotective effect (Figure 1A). The 24-h KA exposure resulted in equal amount of labelled cells as that of 12 h, but in the presence of DNQX and APV, the amount of labeled cells was significantly further increased (Figure 1B). At 48 h, DNQX and APV did not significantly modify the KA-induced neuronal damage (Figure 1C). In CA3c, the amount of FJB-stained neurons significantly increased at 24 h in the presence of KA +DNQX and KA + APV (Figure 2A-B), but KA treatment alone lead to significantly increased labelling first after the 48-h treatment. At this time point, the addition of DNQX and APV did not significantly change the KA-induced neuronal damage (Figure 2C). In DG granule cells, the 48-h KA treatment resulted in a small, but significant neuronal death (score number 0.9 ± 0.

2, mean ± SEM, n= 10), which was significantly decreased in the presence of DNQX (score number 0.1 ± 0.1, mean ± SEM, n= 16), and APV (score number 0.2 ± 0.2, mean ± SEM, n= 15). In contrast to other hippocampal regions, CA1 had no significant amount of labelled neurons at any time point regardless of the treatment (data not shown, but see Figure 5B).

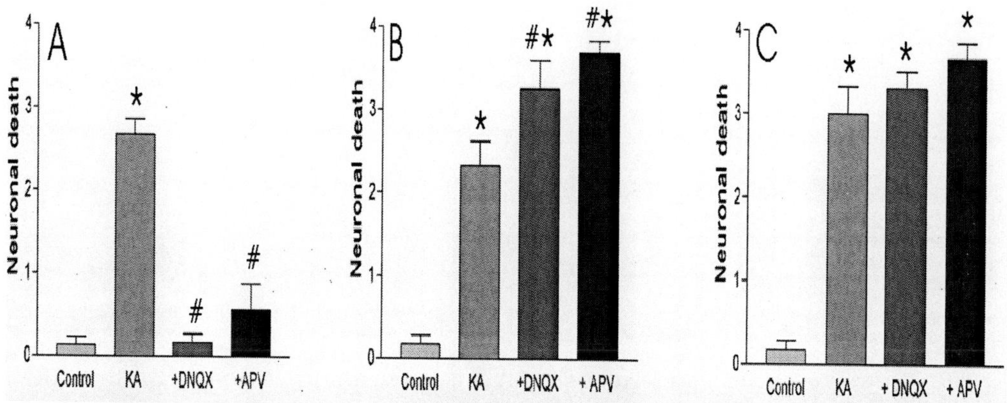

Figure 1. Time course of KA (5 μM)-induced neuronal death in the CA3a/b region, and the effect of DNQX (5 μM) and APV (50 μM). Slices (7 DIV) were treated for 12 h (A), 24 h (B), or 48 h (C) with the drugs prior to FJB staining. The asterics (*) indicate that the value was significantly different from the controls, and # denotes the significance of difference between KA treatment, and KA combined with the antagonist.

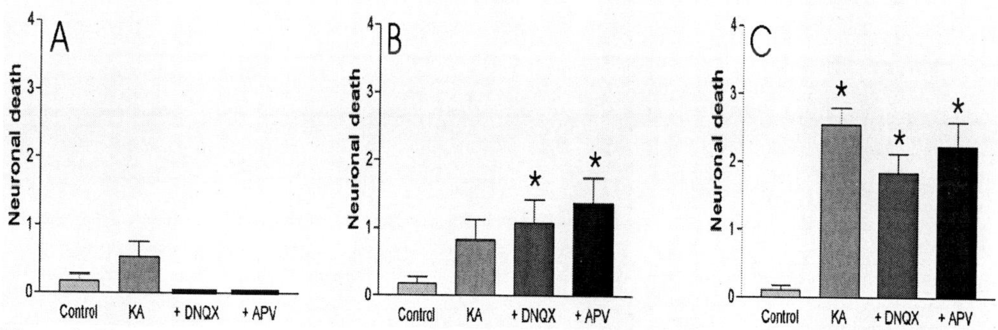

Figure 2. Time course of KA (5 μM)-induced neuronal death in the CA3c region, and the effect of DNQX (5 μM) and APV (50 μM). Slices (7 DIV) were treated for 12 h (A), 24 h (B), or 48 h (C) with the drugs prior to FJB staining. The asterics (*) indicate that the value was significantly different from the controls.

Recovery of Neurons after the Initial KA-induced Damage and the Effect of DNQX and AVP on Recovery

Since the neuroprotective effect of both DNQX and APV was clearly time-dependent, we were interested to know to what extent neurons recovered from the initial 48-h treatment (KA alone, DNQX +KA, APV + KA), and whether the maturity of OHCs effected the survival/damage of neurons. In immature slices (i.e. 11 DIV), the 48-h recovery period in normal culture medium after the 48-h treatment did not completely rescue the neurons in any hippocampal subregion (Figure 3A-C). However, in the presence of DNQX (5 μM), the amount of FJB-labelled neurons, although not totally absent, was significantly reduced in the

CA3a/b and CA3c regions. In 18 DIV cultures, the rescue phase after KA treatment resulted in significant neuronal death in all subfields, the most severely affected regions being CA3a/b and CA3c, followed by CA1 and dentate granule cells (Figure 5 B). In this age group, DNQX (5 μM) and APV (50 μM) did not rescue the degenerating neurons (Figure 4A-D).

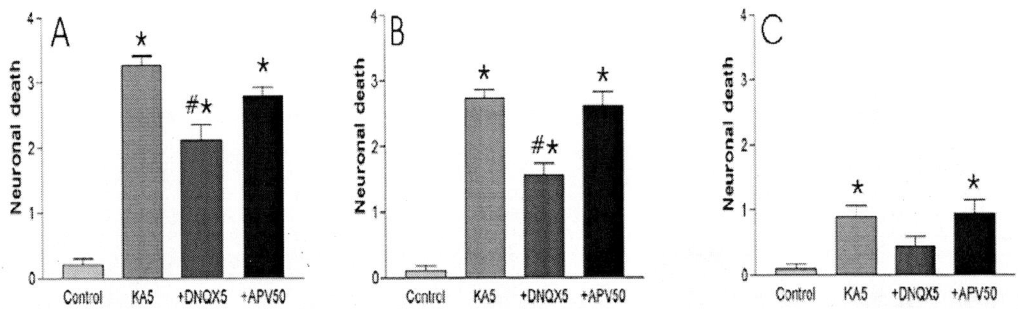

Figure 3. Recovery of hippocampal neurons after KA (5 μM) treatment in CA3a/b (A), CA3c (C), DG (D) region, and the effect of DNQX (5 μM), and APV (50 μM) on the recovery. Slices (7 DIV) were treated with KA (48 h) alone, or together with DNQX or APV for 48 h, and further cultured for 48 h in normal culture medium prior to FJB staining. The asterics (*) indicate that the value was significantly different from the controls, and # denotes the significance of difference between KA-treated, and KA combined with the antagonist.

Neuronal Loss after the Recovery Period as Detected by Thionin Staining

A representative image of a thionin-stained, 11-DIV control slice shows well-preserved neuronal layers in the hippocampus (Figure 5C). To verify the regional and temporal aspects of neuronal loss after the prolonged KA treatment (48 h) alone, or combined with DNQX and APV (48 h), a subset of cultures was stained with thionin after the 48-h rescue phase. In accordance with the results of the FJB staining, the older cultured were more severely damaged by KA treatment (Table 1). The severity of neuronal loss in young cultures (11 DIV) decreased in order CA3a/b = CA3c > dentate granule cells, and in 18-DIV cultures CA3a/b = CA3c > CA1= dentate granule cells. It is noteworthy, that no significant neuronal loss was detected the CA1 region in 11-DIV slices (Table 1, Figure 5D). When studied after the rescue period, the KA-induced neuronal loss was slightly but significantly decreased by in the presence of DNQX and APV in both *in vitro* age groups, most notably in the CA3 regions (Table 1).

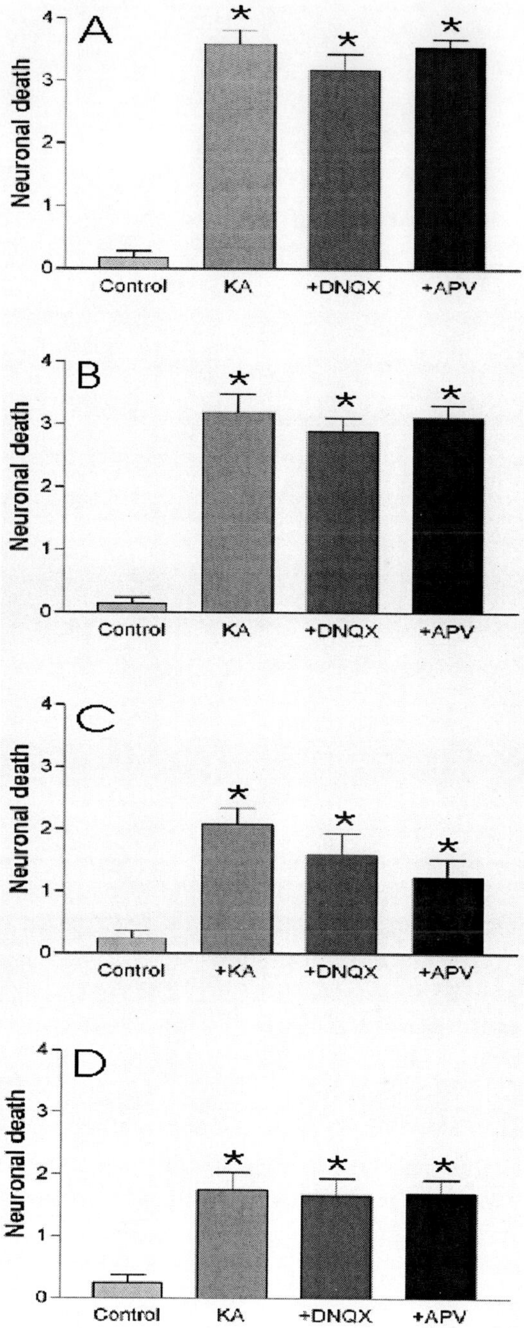

Figure 4. Recovery of hippocampal neurons (18 DIV) after KA (5 μM) treatment in CA3a/b (A), CA3c (C), and DG (D) region, and the effect of DNQX (5 μM) and APV (50 μM). Slices (14 DIV) were treated with KA alone, or together with DNQX or APV for 48 h, and after medium change, further cultured for 48 h in normal culture medium prior to FJB staining. The asterics (*) indicate that the value was significantly different from the controls.

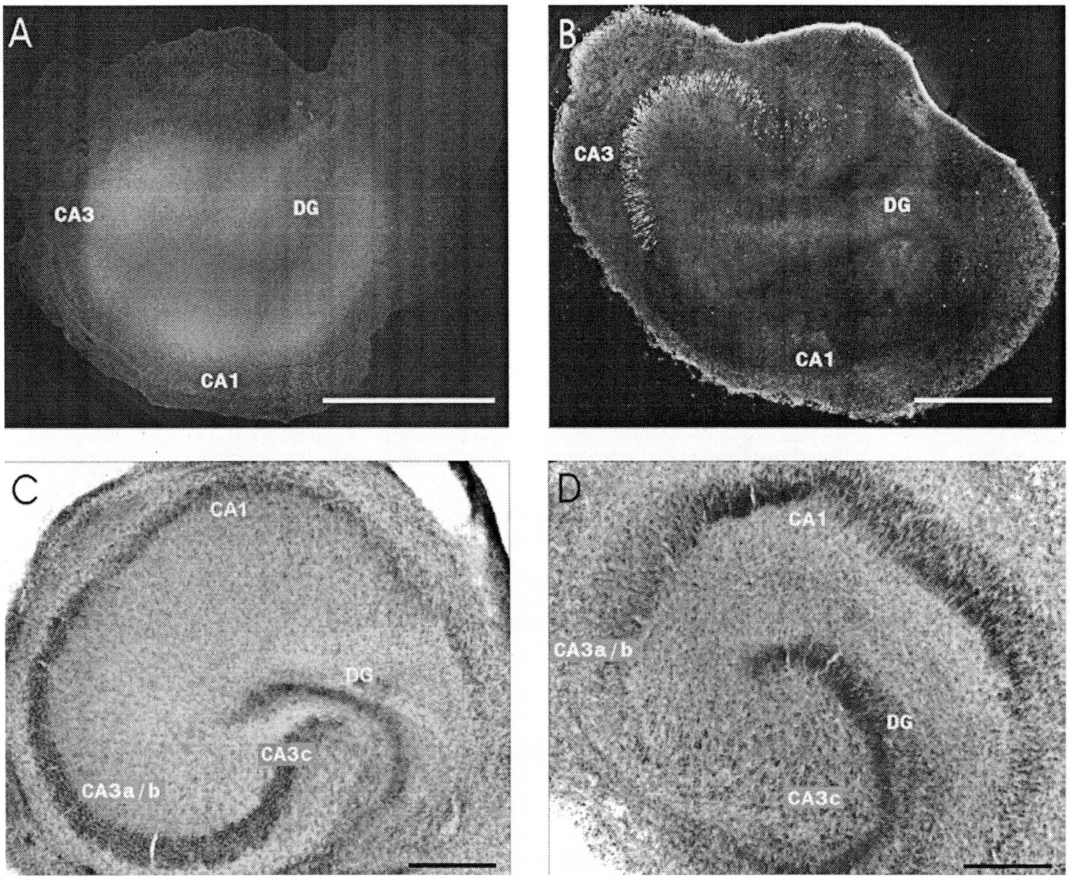

Figure 5. A: A representative confocal image of a FJB- stained 7-DIV-old hippocampal slice after KA (5 μM) treatment of 6 h. B: A 11-DIV-old slice treated with KA (5 μM, 48 h), and after medium change, further cultured for 48 h in normal culture medium (i.e. rescue phase). Note that there are no FJB-labelled neurons after the 6-h treatment, but after the rescue phase, intensely stained neurons are visible in the CA3 regions. C: A thionin-stained slice cultured for 7 days in normal medium. D: A thionin-stained 11-DIV-old slice treated with KA (5 μM, 48 h), and after medium change, further cultured for 48 h (i.e. rescue phase). Note the well-preserved neuronal layers in the control slice, and almost total disappearance of CA3a/b neurons in the treated slice. Scale bars: A and B, 1 mm; C and D, 400 μm. Abbreviation: DG = dentate gyrus.

Table 1. Neuronal loss after the treatments as detected with thionin staining in cultured organotypic hippocampal slices

Hippocampal region	Treatment		
	KA	KA+DNQX	KA+APV
11DIV			
CA3a/b	2.1 ± 0.1*	2.7 ± 0.1* #	3.1 ± 0.1* #
CA3c	2.2 ± 0.2*	3.0 ± 0.1* #	3.1 ± 0.1 * #
CA1	3.3 ± 0.1	3.7 ± 0.1	3.8 ± 0.1
Granule cells	3.0 ± 0.2*	3.6 ± 0.1#	3.7 ± 0.1#
18 DIV			
CA3a/b	1.8 ± 0.1*	2.7 ± 0.1* #	2.4 ± 0.1* #
CA3c	1.4 ± 0.1*	2.4 ± 0.2* #	1.9 ± 0.1*
CA1	2.7 ± 0.1*	3.2 ± 0.1*	2.9 ± 0.1*
Granule cells	2.2 ± 0.2 *	2.5 ± 0.1*	2.6 ± 0.1*

DIV = *days in vitro.*

* denotes that the value was significantly ($p<0.05$) different from the control value

\# denotes the significant difference between KA and KA + DNQX/APV treatment.

Note: Slice cultures were treated with 5 μM kainic acid (KA) for 48 h at 7 and 14 DIV, and thereafter further cultured for 2 days in normal culture medium (recovery phase) prior to staining with thionin. The cell death was scored as described in Experimental Procedures. Results are given as means (± SEM, n= 15-26), and represent slices from 4-5 different culture batches in both groups. The mean score number of control cultures was 4 in each region and at both developmental stages.

Western Blotting of Caspase-3 and PARP Proteins after KA Treatment in Immature OHCs

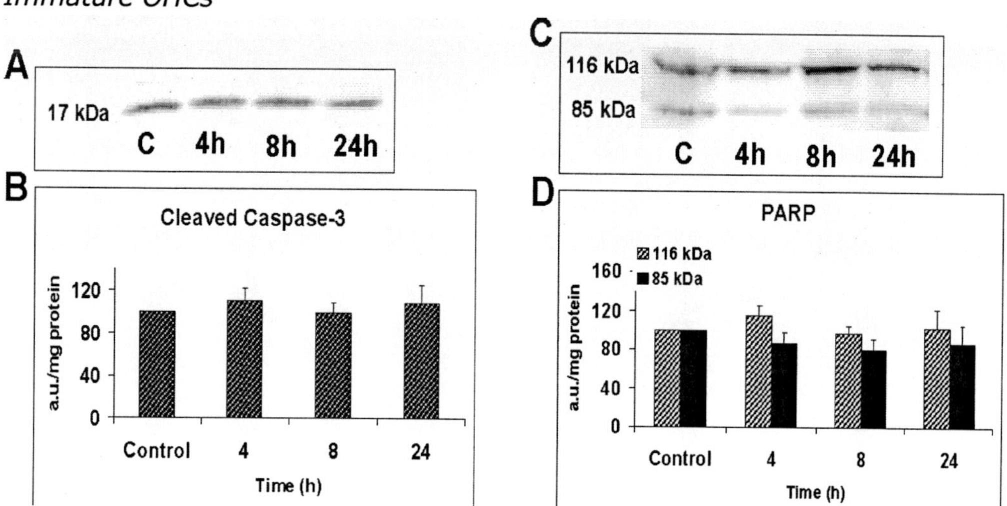

Figure 6. A: A representative immunoblot of the cleaved 17-kDa caspase-3 in control cultures, and after KA treatment (5μM) for 4, 8, and 24 h. B: The signals of the cleaved caspase-3 in control cultures, and in cultures treated with KA (5 μM) (for 4, 8, 24 h). C: A representative immunoblotting of PARP in control cultures, and after KA treatment for 4, 8, and 24 h. D: The signals of the 116-kDa PARP and its 85-kDa cleaved fragment in control cultures, and in cultures treated with KA for 4, 8, 24 h. The results (mean ± SEM, n=8) are given as arbitrary units/mg protein from two different batches of slices repeated 4 times in each batch of slices. Abbreviations: a.u. = arbitrary unit; C = control.

The mechanism of neuronal death was clarified by analyzing the expression of two key factors in apoptosis, caspase-3 and PARP at 4, 8, and 24 h after KA (5μM) treatment. In control cultures, signals of the active, cleaved caspase-3, and the uncleaved and cleaved PARP were detected, but the expression of neither the cleaved caspase-3 (Figure 6A and B) nor the uncleaved or cleaved PARP fragments (Figure 6C and D) were significantly altered ($p>0.05$, one-way ANOVA) after KA treatment.

Ultrastructural Analysis of KA-induced Neuronal Damage

To further analyze the mechanism of neuronal death, the ultrastrucutral analysis of neurons in the CA3a/b region was carried out in 8-DIV- and 15-DIV-old slices after KA treatment (5 μM, 24 h). In control cultures, neurons and their processes were well-preserved as indicated by the intact ultrastructural details of nuclei, cytoplasmic region, intracellular organelles, and plasma membrane in both age groups. Figure 7A shows a representative image of a 8-DIV CA3a/b pyramidal neurons. In KA-treated slices, the CA3a/b neurons had ultrastructurally intact nuclei, and no signs of apoptotic pyknosis or condensation were detected. However, neurons had irregular, broken areas in the cytoplasm, and ruptures of the plasma membrane, features typical for an early phase of necrotic cell death. A representative electron microscopic image of a 8-DIV KA-treated CA3a/b pyramidal neuron is shown in Figure 7B. Similar KA-induced changes were detected in 15-DIV-old slices (data not shown).

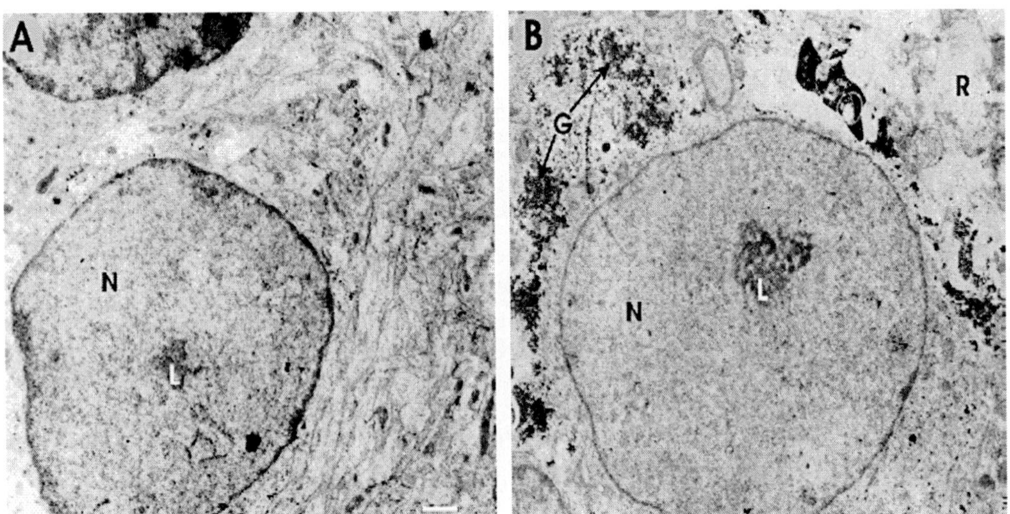

Figure 7. Electron micrographs of a CA3a/b neuron and neuropile in a control (A), and KA-treated (5 μM, 24 h) hippocampal slices cultured for 8 days (B). The nuclear chromatin including the nucleolus (L) is normally structured, and no shrinkage is observed in the nucleus (N) of either control (A) or KA-treated (B) culture. In the control slice (A), the structures in cytoplasm of the CA3a/b neuron are well-preserved, whereas in the KA-treated slice (B), the cytoplasm and the plasma membrane are partially ruptured (R), and accumulations of intracellular glycogen (G) granules are visible. Scale bar is 1 μm in A and B.

Discussion

The Effect of Glutamate Receptor Antagonists on Neuronal Survival in Cultured OHCs Lices

KA treatment of immature OHCs specifically and irreversibly damaged the CA3 pyramidal neurons, but only slightly affected dentate granule cells while leaving CA1 pyramidal neurons intact, in keeping with earlier studies in immature OHCs (Holopainen et al., 2001, 2004). At the early phase of insult (i.e. within 12 h), KA-induced neuronal death was effectively decreased by low doses of DNQX and APV, but the longer exposure (24 h) significantly enhanced the damage in the CA3 regions. This is in accordance with the recent study, which shows that in cultured hippocampal neurons high doses (over 50 μM) of DNQX combined with a long treatment (days) has resulted in toxicity, the main target being neurons (Martin et al., 2003). In the present study, the early neuroprotective effect of a low DNQX dose (5 μM), together with the finding that DNQX increased the survival of immature CA3 pyramidal neurons during the rescue phase are in agreement with the study of Drian et al. (2001). They observed that the maximal neurotoxic effect of DNQX (50 μM or more) was between 2 to 5 days of treatment in cultured cortical neurons. Moreover, the neurotoxic effect gradually decreased during the maturation of neurons, and finally DNQX even had neuroprotective effects in mature neurons. Although this was not very obvious in our study, present results together with earlier studies by other groups suggest that the effect of DNQX depends on the maturity of neurons, exposure time and drug concentration. DNQX can be neuroprotective when given for a short period, but neurotoxic when exposed in high concentrations for longer periods. The neurotoxic effects of long-term (i.e. 46 h) DNQX treatment has been postulated to be due to its binding to AMPA receptors resulting in their long-term internalization, which lead to changes in the AMPA receptor recycling and activation by endogenous glutamate (Martin et al., 2003).

In young rats, the high density of KA1 receptors in the CA3 compared to the CA1 and DG regions (Bahn et al., 1994; Ritter et al., 2002) could explain the selective vulnerability of immature CA3 neurons to KA. Thus the innervation of the CA3 pyramidal cell by mossy fibers, the axons of granule cell, may not be necessary for the KA-induced neurotoxicity. In keeping with this idea, KA-induced CA3a/b neuronal death has also occurred in 3-day old OHCs prepared from 3-day old rats (Gatherer & Sundström, 1998), when the mossy fiber contacts to CA3 pyramidal neurons are not fully mature, and the formation of processes and synaptic contacts are still in progress (Ben-Ari & Cossart, 2000).

Neuronal Recovery from the Initial Insult

The rescue of neurons from the KA-induced insult was poor, and only CA3 neurons showed some recovery after the combined treatment, most notably in immature slices. The poor recovery of neurons is in accordance with the study of Bruce et al. (1995). They showed that after the initial insult (KA 50 μM, 3h), the 48-h recovery phase in normal culture medium resulted in a progressive neuronal loss not only in CA3 but also in CA1, the region well-preserved in our immature slices. The mechanisms whereby the damage is augmented after the withdrawal of KA are poorly elucidated. One reason could be that increased excitatory activity continues within the hippocampal network after KA treatment, and gradually leads to neuronal death as recently suggested for glutamate-induced toxicity in OHCs (Lahtinen et al., 2001). Moreover, our present FJB-staining results suggest that DNQX and APV, when applied together with KA, have a very limited long-term neuroprotective effect. On the other hand, our thionin staining results suggest that DNQX and APV could decreased neuronal death to some extent in both age groups, during which hippocampal neurons can still be rescued.

In more mature hippocampal cultures, KA application (48 h) combined with the 48-h rescue period induced neuronal death in all subfields, the most severely affected regions being CA3a/b and CA3c, but also CA1 neurons were damaged, the region completely preserved in younger cultures. Our results are in accordance with earlier studies in mature OHCs (20-25 DIV), in which KA treatment has damaged CA3 and CA1 pyramidal neurons, and dentate granule cells (Vornov et al., 1991; Bruce et al., 1995; Holopainen et al., 2001; Kristensen et al., 2001). The developmental vulnerability of CA1 neurons has also been observed in acutely dissociated CA1 neurons, in which glutamate-induced death has progressively increased in cultures prepared from newborn to 3-week-old rats (Marks et al., 1996). It was suggested that in immature CA1 neurons, Ca^{2+}-dependent mechanisms, which are proposed to be responsible for neuronal death in more mature neurons, have not yet developed (Marks et al., 1996). In fact, they observed no signs of neuronal death in the absence of an acute, preceding rise in intracellular Ca^{2+}.

Mechanisms of KA-induced neuronal Death in Immature Hippocampal Slices

Our recent study show that in spite of neuronal death within the CA3 regions, neither the expression of the active caspase-3, nor the cleaved and uncleaved PARP proteins were significantly different from those of control slices at any time point (4 up to 24 h) after KA treatment. Moreover, no ultrastructural signs of apoptosis were detected after the 24-h KA treatment either in immature or more mature CA3a/b neurons. Recent studies suggest that apoptotic mechanisms are activated within hours after the initial insult in the immature brain (Joashi et al., 1999; Liu et al., 2001; Yu et al., 2002), thus excluding the possibility that we might have missed the appropriate time for detection. In cultured mature hippocampal slices (18-25 DIV), pyramidal cell death, induction of Bax expression, and increased caspase-3 levels have been detected within 24 h after KA (50 μM) application (Liu et al., 2001) suggesting that in older slices, the mitochondrial apoptotic pathway could be activated in response to the treatment.

It has been suggested that activation of PARP could results in a caspase-independent pathway of programmed cell death (Yu et al., 2002), or to be connected to necrotic death (Ha et al., 1999; Herceg & Wang, 1999; Virag & Szabo, 2002). Our present results, however, suggest that necrotic cell death in immature OHCs does not activate the caspase-3-PARP pathway, at least within the time span (4-24 h) we studied. Although we did not detect any significant alterations in the level of the cleaved (85 kDa) PARP after KA treatment, the rapid PARP cleavage starting immediately after the insult and persisting for 24 h has been detected after hypoxic-ischemic and traumatic insult in the hippocampus of 7-day-old rats followed by morphological features of apoptosis at the later phase after the insult (Joashi et al., 1999). The uncleaved 118-kDa PARP remained also unaltered during the KA treatment, suggesting that there was no increased PARP synthesis, which might indicate activation of

DNA repair processes. In cultured cortical neurons, the intense NMDA treatment (2 mM) has resulted in mostly necrotic neuronal damage together with PARP-1 (one member of the PARP family) overactivation in response to massive DNA damage (Meli et al., 2004). Since PAPR overactivation rapidly consumes cellular energy stores, which are required to proceed with apoptosis (Herceg & Wang, 1999), it has been postulated that the PARP pathway may be more important in necrotic than in caspase-dependent apoptotic neuronal death (Yu et al., 2003, Meli et al., 2004). Although the more comprehensive review of the current huge literature concerning the mechanisms and pathways of neuronal death is not within the scope of this article, recent studies suggest that the intensity and duration of insult, and the stage of neuronal maturation have important roles in determining the mechanisms of neuronal death. It still remains to be clarified what are the death pathways activated by other glutamate receptor agonists in various neuronal types, i.e. pyramidal neurons, dentate granule cells, and inhibitory interneurons in the hippocampus during the maturation.

The activation of both NMDA and non-NMDA receptors increases the intracellular Ca^{2+} (Murphy & Miller, 1988; Holopainen et al., 1989; 1990; 1991), which is believed to be necessary for KA-induced neurotoxicity (Choi, 1988; Whetsell, 1996). At variance, recent studies suggest that also the blockade of ionotropic glutamate receptors may cause caspase-3-mediated neuronal apoptosis due to Ca^{2+} deficiency, which leads to activation of mitochondrial death pathway in immature neurons (Yoon et al., 2003). In particular, blockade of NMDA receptors by its specific antagonist MK-801 has resulted in apoptotic neurodegeneration in the developing brain (Ikonomidou et al., 1999; Turner et al., 2002; Hansen et al., 2004). It was postulated that glutamate acting at NMDA receptors controls neuronal survival, and it is necessary for the normal structural and functional maturation of neurons and synaptic contacts (Ikonomidou et al., 1999; Hansen et al., 2004). In our study, the long-term APV treatment, which might have deteriorated the NMDA receptor-mediated trophic effect combined with the intense KA response, resulted in necrotic death in the CA3a/b neurons as verified with the electron microscopic analysis. These studies suggest that the signaling pathways and mechanisms of neuronal death induced by overactivation/blockade of various glutamate receptors may involve different intracellular cascades, although the intracellular Ca^{2+} seems to be a key player in the vulnerability in both immature and mature neurons.

The Use of Fluoro Jade-B Staining to Detect Degenerating Neurons

The FJB dye specifically stains degenerating neurons regardless of the mechanism a nerve cell dies (Schmued et al., 1997; Schmued & Hopkins, 2000). In general, the amount of FJB-stained neurons correlated well to the magnitude of neuronal loss as measured with thionin staining after the rescue phase. However, the amount of FJB-stained, degenerating neurons may have been underestimated in the present study, since some neurons might have already died, and the dye disappeared after the rescue phase (i.e. 48 h). With this caveat in mind, our results suggest that FJB staining can be used as an additional tool to study the toxicity of drugs in a population of neurons in cultured slices, although it is not the ideal system to study the exact number of degenerating neurons in this model system.

Organotypic Hippocampal Slice Cultures as a Model System for Neurochemical Studies

The main features of the hippocampus, i.e. maturation of synapses, receptors, and intrinsic fiber pathways are well-preserved in OHCs (Caeser & Aertsen, 1991; Frotscher et al., 1995; Holopainen & Lauren, 2003). However, the preparation of slices results in loss of normal afferent input to dentate granule cells, which in long-term cultures favors the gradual growth of aberrant, mainly excitatory connections of dentate granule cell axons, the mossy fibers, into the molecular layers of the dentate gyrus (Frotscher & Gähwiler, 1988; Coltman et al., 1995; Lindroos et al.,2005). In spite of this partial reorganization of the hippocampal circuits in cultured conditions, the slice culture method offers a relative simple and pharmacologically accessible tool to study morphological, biochemical, and functional maturation of the rat hippocampus as well as drug effects on selected hippocampal cell populations *in vitro* in almost unlimited experimental conditions, which are difficult to perform in intact animal experiments.

Conclusion

The KA-induced neuronal death was a necrotic, region-specific, mostly irreversible process, which damaged more widely the mature than immature hippocampus *in vitro*. The neuroprotective effects of the specific NMDA receptor antagonist APV, and the competitive AMPA/KA receptor antagonist DNQX on KA-induced neuronal death were restricted to a short period, and rescued only to modest extent degenerating neurons after the initial insult in immature hippocampal slices, and were of less importance in more mature cultures. However, further investigations are certainly needed to better understand the precise molecular mechanisms of neuronal death, and the role of different glutamate receptors in death processes in the developing, immature brain.

Acknowledgements

We wish to thank Anne Romppanen, M.Lic. for thionin staining. The financial support from the following resources is gratefully acknowledged: Sigrid Juselius Foundation, the Special State Grant for Clinical Research (EVO), and Arvo and Lea Ylppö Foundation to I.E.H.; the Finnish Graduate School of Neuroscience to F.R.L.-P; the Finnish Cultural Foundation, the Research and Science Foundation of Farmos, and the Finnish Society of Sciences and Letters to T.-K. K.-L.

References

Bahn, S., Volk, B. & Wisden, W. (1994) Kainate receptor gene expression in the developing rat brain. *J Neurosci*, 14, 5525-5547.

Bardoul, M., Drian, M.J. & Konig, N. (1997) AMPA/kainate receptors modulate the survival in vitro of embryonic brainstem cells. *Int J Devl Neurosci*, 15, 695-701.

Ben-Ari, Y. & Cossart, R. (2000) Kainate, a double agent that generates seizures: two decades of progress. *Trends Neurosci*, 23, 580-587.

Bengzon, J., Mohapel, P., Ekdahl, C.T. & Lindvall, O. (2002) Neuronal apoptosis after brief and prolonged seizures. *Prog Brain Res*, 135, 111-119.

Bittigau, P., Sifringer, M., Pohl, D., Stadthaus, D., Ishimaru, M., Shimizu, H., Ikeda, M., Lang, D., Speer, A., Olney, J.W. & Ikonomidou, C. (1999) Apoptotic neurodegeneration following trauma is markedly enhanced in the immature brain. *Ann Neurol*, 45, 724-735.

Bredesen, D.E. (2000) Apoptosis: overview and signal transduction pathways. *J Neurotrauma*, 17, 801-810.

Bruce, A.J., Sakhi, S., Schreiber, S.S. & Baudry, M. (1995) Development of kainic acid and N-methyl-D-aspartic acid toxicity in organotypic hippocampal cultures. *Exp Neurol*, 132, 209-219.

Caeser, M. & Aertsen, A. (1991) Morphological organization of rat hippocampal slice cultures. *J Comp Neurol*, 307, 87-106.

Choi, D.W. (1988) Glutamate neurotoxicity and diseases of the nervous system. *Neuron*, 1, 623-634.

Coltman, B.W., Earley, E.M., Shahar, A., Dudek, F.E. & Ide, C.F. (1995) Factors influencing mossy fiber collateral sprouting in organotypic slice cultures of neonatal mouse hippocampus. *J Comp Neurol*, 362, 209-222.

Drian, M.J., Bardoul, M. & Konig, N. (2001) Blockade of AMPA/kainate receptors can either decrease or increase the survival of cultured neocortical cells depending on the stage of maturation. *Neurochem Int*, 38, 509-517.

Earnshaw, W.C., Martins, L.M. & Kaufmann, S.H. (1999) Mammalian caspases: structure, activation, substrates, and functions during apoptosis. *Annu Rev Biochem*, 68, 383-424.

Fröjdman, K., Paranko, J., Virtanen, I. & Pelliniemi, L.J. (1992) Intermediate filaments and epithelial differentiation of male rat embryonic gonad. *Differentiation*, 50, 113-123.

Frotscher, M. & Gähwiler, B.H. (1988) Synaptic organization of intracellularly stained CA3 pyramidal neurons in slice cultures of rat hippocampus. *Neuroscience*, 24, 541-551.

Frotscher, M., Zafirov, S. & Heimrich, B. (1995) Development of identified neuronal types and of specific synaptic connections in slice cultures of rat hippocampus. *Prog Neurobiol*, **45**, 143-164.

Fujikawa, D.G., Ke, X., Trinidad, R.B., Shinmei, S.S. & Wu, A. (2002) Caspase-3 is not activated in seizure-induced neuronal necrosis with internucleosomal DNA cleavage. *J Neurochem*, 83, 229-240.

Fujikawa, D.G., Shinmei, S.S. & Cai, B. (2000) Kainic acid-induced seizures produce necrotic, not apoptotic, neurons with internucleosomal DNA cleavage: implications for programmed cell death mechanisms. *Neuroscience*, 98, 41-53.

Gatherer, M. & Sundström, L.E. (1998) Mossy fibre innervation is not required for the development of kainic acid toxicity in organotypic hippocampal slice cultures. *Neurosci Lett*, 253, 119-122.

Ha, H.C & Snyder, S.H. (1999) Poly(ADP-ribose) polymerase is a mediator of necrotic cell death by ATP depletion. *Proc. Natl. Acad. Sci. U.S.A.* 96, 13978-13982.

Hansen, H.H., Briem, T., Dzietko, M., Sifringer, M., Voss, A., Rzeski, W., Zdzisinska, B., Thor, F., Heumann, R., Stepulak, A., Bittigau, P. & Ikonomidou, C. (2004) Mechanisms leading to disseminated apoptosis following NMDA receptor blockade in the developing rat brain. *Neurobiol Dis*, ***16***, 440-453.

Herceg, Z. & Wang, Z.Q. (1999) Failure of poly(ADP-ribose) polymerase cleavage by caspases leads to induction of necrosis and enhanced apoptosis. *Mol Cell Biol*, 19, 5124-5133.

Holopainen I.E. & Lauren H.B. (2003) Neuronal activity regulates GABA-A receptor subunit expression in organotypic hippocmapal slice cultures. *Neuroscience,* 118, 967-974.

Holopainen, I., Enkvist, M.O. & Åkerman, K.E. (1989) Glutamate receptor agonists increase intracellular Ca^{2+} independently of voltage-gated Ca^{2+} channels in rat cerebellar granule cells. *Neurosci Lett*, 98, 57-62.

Holopainen, I., Louve, M., Enkvsit, M.O.K. & Åkerman, K.E.O. (1990) Coupling of glutamatergic receptors to changes in intracellular Ca^{2+} in rat cerebellar granule cells in primary culture. *J Neurosci Res*, 25, 187-193.

Holopainen, I., Louve, M. & Åkerman, K.E.O. (1991) Interactions of glutamate receptor agonists coupled to changes in intracellular Ca^{2+} in rat cerebellar granule cells in primary culture. *J Neurochem,* 57, 1729-1734.

Holopainen, I.E., Järvelä, J., Lopez-Picon, F.R., Pelliniemi, L.J. & Kukko-Lukjanov, T.K. (2004) Mechanisms of kainate-induced region-specific neuronal death in immature organotypic hippocampal slice cultures. *Neurochem Int*, **45**, 1-10.

Holopainen, I.E., Lauren, H.B., Romppanen, A. & Lopez-Picon, F.R. (2001) Changes in neurofilament protein-immunoreactivity after kainic acid treatment of organotypic hippocampal slice cultures. *J Neurosci Res*, 66, 620-629.

Hu, B.R., Liu, C.L., Ouyang, Y., Blomgren, K. & Siesjö, B.K. (2000) Involvement of caspase-3 in cell death after hypoxia-ischemia declines during brain maturation. *J Cereb Blood Flow Metab*, 20, 1294-1300.

Ikonomidou, C., Bosch, F., Miksa, M., Bittigau, P., Vockler, J., Dikranian, K., Tenkova, T.I., Stefovska, V., Turski, L. & Olney, J.W. (1999) Blockade of NMDA receptors and apoptotic neurodegeneration in the developing brain. *Science*, 283, 70-74.

Ishimaru, M.J., Ikonomidou, C., Tenkova, T.I., Der, T.C., Dikranian, K., Sesma, M.A. & Olney, W. 1999. *J Comp Neurol,* 408, 461-476.

Joashi, U.C., Greenwood, K., Taylor, D.L., Kozma, M., Mazarakis, N.D., Edwards, A.D. & Mehmet, H. (1999) Poly(ADP ribose) polymerase cleavage precedes neuronal death in the hippocampus and cerebellum following injury to the developing rat forebrain. *Eur J Neurosci*, 11, 91-100.

Khaing, Z.Z., Weickert, C.S., Weinberger, D.R. & Lipska, B.K. (2000) Differential DNA damage in response to the neonatal and adult excitotoxic hippocampal lesion in rats. *Eur J Neurosci*, 12, 4424-4433.

Kristensen, B.W., Noraberg, J. & Zimmer, J. (2001) Comparison of excitotoxic profiles of ATPA, AMPA, KA and NMDA in organotypic hippocampal slice cultures. *Brain Res*, 917, 21-44.

Kroemer, G. & Reed, J.C. (2000) Mitochondrial control of cell death. *Nat Med*, **6**, 513-519.

Lahtinen, H., Autere, A.M., Paalasmaa, P., Lauri, S.E. & Kaila, K. (2001) Post-insult activity is a major cause of delayed neuronal death in organotypic hippocampal slices exposed to glutamate. *Neuroscience*, 105, 131-137.

Lazebnik, Y.A., Kaufmann, S.H., Desnoyers, S., Poirier, G.G. & Earnshaw, W.C. (1994) Cleavage of poly(ADP-ribose) polymerase by a proteinase with properties like ICE. *Nature*, 371, 346-347.

Lindroos, M.M., Soini, S.L., Kukko-Lukjanov, T.-K., Korpi, E.R., Lovinger, D. & Holopainen, I.E. (2005) Maturation of cultured hippocampal slices results in increased excitability in granule cells. *Int J Devl Neurosci,* 23, 65-73.

Liu, W., Liu, R., Chun, J.T., Bi, R., Hoe, W., Schreiber, S.S. & Baudry, M. (2001) Kainate excitotoxicity in organotypic hippocampal slice cultures: evidence for multiple apoptotic pathways. *Brain Res*, 916, 239-248.

Los, M., Mozoluk, M., Ferrari, D., Stepczynska, A., Stroh, C., Renz, A., Herceg, Z., Wang, Z.Q. & Schulze-Osthoff, K. (2002) Activation and caspase-mediated inhibition of PARP: a molecular switch between fibroblast necrosis and apoptosis in death receptor signaling. *Mol Biol Cell*, 13, 978-988.

Marks, J.D., Friedman, J.E. & Haddad, G.G. (1996) Vulnerability of CA1 neurons to glutamate is developmentally regulated. *Brain Res Dev Brain Res*, 97, 194-206.

Martin, A., Recasens, M. & Guiramand, J. (2003) DNQX-induced toxicity in cultured rat hippocampal neurons: an apparent AMPA receptor-independent effect? *Neurochem Int*, 42, 251-260.

Mattson, M.P., LaFerla, F.M., Chan, S.L., Leissring, M.A., Shepel, P.N. & Geiger, J.D. (2000) Calcium signaling in the ER: its role in neuronal plasticity and neurodegenerative disorders. *Trends Neurosci*, 23, 222-229.

Meli, E., Pangallo, M., Picca, R., Baronti, R., Moroni, F. & Pellegrini-Giampietro, D.E. (2004) Differential role of poly(ADP-ribose) polymerase-1in apoptotic and necrotic neuronal death induced by mild or intense NMDA exposure in vitro. *Mol Cell Neurosci*, 25, 172-180.

Monyer, H., Burnashev, N., Laurie, D.J., Sakmann, B. & Seeburg, P.H. (1994) Developmental and regional expression in the rat brain and functional properties of four NMDA receptors. *Neuron,* 12, 529-540.

Murphy, S.N. & Miller, R.J. (1988) A glutamate receptor regulates Ca^{2+} mobilization in hippocampal neurons. *Proc Natl Acad Sci U.S.A*. 85, 8737-8741.

Pollard, H., Charriaut-Marlangue, C., Cantagrel, S., Represa, A., Robain, O., Moreau, J. & Ben-Ari, Y. (1994) Kainate-induced apoptotic cell death in hippocampal neurons. *Neuroscience*, 63, 7-18.

Ritter, L.M., Vazquez, D.M. & Meador-Woodruff, J.H. (2002) Ontogeny of ionotropic glutamate receptor subunit expression in the rat hippocampus. *Brain Res Dev Brain Res*, 139, 227-236.

Sastry, P.S. & Rao, K.S. (2000) Apoptosis and the nervous system. *J Neurochem*, 74, 1-20.

Schmued, L.C., Albertson, C. & Slikker, W., Jr. (1997) Fluoro-Jade: a novel fluorochrome for the sensitive and reliable histochemical localization of neuronal degeneration. *Brain Res*, 751, 37-46.

Schmued, L.C. & Hopkins, K.J. (2000) Fluoro-Jade B: a high affinity fluorescent marker for the localization of neuronal degeneration. *Brain Res*, 874, 123-130.

Stoppini, L., Buchs, P.A. & Muller, D. (1991) A simple method for organotypic cultures of nervous tissue. *J Neurosci Methods*, 37, 173-182.

Takai, H., Katayama, K.-I., Uetsuka, K, Nakayama, H. & Doi, K. (2003) Distribution of *N*-methyl-D-aspartate receptors (NMDARs) in the developing rat brain. *Exp Mol Pathol,* 75, 89-94.

Turner, C.P., Pulciani, D. & Rivkees, S.A. (2002) Reduction in intracellular calcium levels induces injury in developing neurons. *Exp Neurol*, 178, 21-32.

Whetsell, W.O. (1996) Current concepts of excitotoxicity. *J Neuropathol Exp Neurol*, 55, 1-13.

Virag, L. & Szabo, C. (2002) The therapeutic potential of poly(ADP-Ribose) polymerase inhibitors. *Pharmacol Rev*, 54, 375-429.

Vornov, J.J., Tasker, R.C. & Coyle, J.T. (1991) Direct observation of the agonist-specific regional vulnerability to glutamate, NMDA, and kainate neurotoxicity in organotypic hippocampal cultures. *Exp Neurol*, 114, 11-22.

Yakovlev, A.G., Ota, K., Wang, G., Movsesyan, V., Bao, W.L., Yoshihara, K. & Faden, A.I. (2001) Differential expression of apoptotic protease-activating factor-1 and caspase-3 genes and susceptibility to apoptosis during brain development and after traumatic brain injury. *J Neurosci*, 21, 7439-7446.

Yoon, W.J., Won, S.J., Ryu, B.R. & Gwag, B.J. (2003) Blockade of ionotropic glutamate receptors produces neuronal apoptosis through the Bax-cytochrome C-caspase pathway: the causative role of Ca2+ deficiency. *J Neurochem*, 85, 525-533.

Yu, S.-W., Wang, H., Poitras, M.F., Coombs, C., Bowers, W.J., Federoff, H.J., Poirier, G.G., Dawson, T.M. & Dawson, V.L. (2002) Mediation of poly(ADP-ribose) polymerase-1-dependent cell death by apoptosis-inducing factor. *Science*, 297, 259-263.

Yu, S.-W., Wang, H., Dawson, T.M. & Dawson, V.L. (2003) Poly(ADP-ribose) polymerase-1 and apoptosis inducing factor in neurotoxicity. *Neurobiol Dis,* 14, 303-317.

Yuzaki, M., Forrest, D., Verselis, L.M., Sun, S.C., Curran, T. & Connor, J.A. (1996) Functional NMDA receptors are transiently active and support the survival of Purkinje cells in culture. *J Neurosci*, 16, 4651-4661.

Zimmer, J., Kristensen, B.W., Jakobsen, B. & Noraberg, J. (2000) Excitatory amino acid neurotoxicity and modulation of glutamate receptor expression in organotypic brain slice cultures. *Amino Acids*, 19, 7-21.

Zimmermann, K.C., Bonzon, C. & Green, D.R. (2001) The machinery of programmed cell death. *Pharmacol Ther*, 92, 57-70.

In: Trends In Neurochemistry Research
Editor: Robert M. Coleman, pp. 165-177
ISBN: 1-59454-418-2

Chapter 4

Bipolar Disorder: An Update

Barkur S. Shastry [*]
Department of Biological Sciences, Oakland University, Rochester, MI

Abstract

Bipolar disorder (BPD) is one of the most severe forms of mental illness and is characterized by swinging moods. It affects both sexes equally in all age groups and its worldwide prevalence is approximately one percent. The clinical course of illness can vary from a mild depression to a severe form of mania. The condition has a high rate of recurrence and if untreated it has an approximately 15% risk of death by suicide. It is the third leading cause of death among people aged 15-24 and is a burden on society and families. The pathophysiology of the disorder is poorly understood. However, a variety of imaging studies suggests the involvement of structural abnormalities in the amygdala, basal ganglia and prefrontal cortex. There are two main biological models that have been proposed for depression. These are called the serotonin and norepinephrine hypotheses. Multiple lines of evidence support both of them. It is a life-long disease and runs in families but has a complex mode of inheritance. Family, twin and adoption studies suggest genetic factors but the candidate susceptibility genes which when mutated can account for a substantial portion of BPD patients, have not yet been conclusively identified. There has been an increasing number of new generation antidepressant drugs developed to treat BPD. However, lithium salt is only the drug that is most efficient in long - term preventive treatment and it also has an anti-suicidal effect. The condition can be well managed by physicians and psychiatrists along with family and patient education. Identification of risk genes in the future may provide a better understanding of the nature of pathogenesis that may lead to a better therapeutic target.

Key words: gene, chromosome, association study, pathophysiology, lithium

[*] Correspondence to: B. S. Shastry. Tel: 248-370-3577; FAX: 248-370-4225; E-mail: shastry@oakland.edu

Introduction

Bipolar disorder (BPD) is one of the most common, severe and devastating psychiatric disorders. It is a clinically and socially extremely costly condition [1] and is associated with decreased quality of life for patients and their families. The disorder affects approximately 1.3 to 1.5% of the US population [2 – 3] and its worldwide prevalence is about 1 percent. It has been predicted that by the year 2020, depression will be the second greatest disease - burden worldwide [4]. It is a life-long disease and affects both sexes almost equally [5]. The disorder has a high rate of recurrence (90%) and is one of the leading causes of disability worldwide [6]. The condition is also known as manic depressive illness and is characterized by changing moods that could be due to abnormal neuronal communication.

The clinical course of the illness can vary from a mild depression and a brief hypomania to a severe form of mania or depression. There are two main types of this disorder: unipolar (major depression) and bipolar (manic depression). Bipolar is further divided into bipolar I which is characterized by one or more manic or mixed episodes and bipolar II which involves recurrent episodes of major depression and hypomania. The condition usually occurs in both men and women between 18 and 24 years of age but it can affect all age groups. The diagnostic criteria for major depressive and manic episodes have been described [7]. In addition to depression, it has many other symptoms such as thoughts of suicide or death, trouble concentrating, too much or too little sleep, alcohol and drug abuse, involvement in high - risk behavior and loss of judgement. Many patients with early onset BPD experience psychotic features such as delusions or hallucinations that can be fatal from suicide [8]. It is the third leading cause of death in young individuals. In children, hyperactivity is the most common behavioral symptom and the condition appears to be more severe than in adult onset. In some patients it coexists with attention - deficit hyperactivity disorder [9-11]. In addition, an increased cardiac morbidity is shown to be associated with depression [12].

Neuropathophysiology

Because of the complexity of brain function, the etiology and pathophysiology of the disorder is not well understood. However, a variety of studies such as neuroimaging, biochemical and neuropathological analyses [13–16] suggest that structural abnormalities in the amygdala, hippocampus, basal ganglia and several regions of the prefrontal cortex [17-20] may be responsible for severe depression. This has been supported by the observation that patients who were successfully treated with drugs showed changes in the prefrontal cortex. There are two main neuronal circuitry systems that have been proposed for depression. One of them is the basal ganglia – thalamocortical circuitry and the second one is the hypothalamic – pituitary – adrenocortical (HPA) system. The observation that patients with severe depression showed an increased HPA axis activity that can be normalized by antidepressant pharmacotherapy, supports the involvement of the above circuitry. These patients have an elevated level of corticotrophin releasing factor (CRF) in the cerebrospinal fluid. This connection is further supported by the development of numerous animal models that exhibit similar pathological characteristics to that of patients. For instance, mice overproducing CRF exhibited endocrine abnormalities of the HPA axis. These animal models are useful for further understanding of the interaction between HPA axis and depression. However, at present none of the animal model can reproduce mood cyclicity that is the characteristic feature of BPD [21]. Furthermore, magnetic resonance imaging (MRI) studies have shown that in children with BPD, distinct brain regions in the prefrontal cortex and other brain areas are affected [22]. In older adolescents and adults, a reduction in N-acetylaspartate was also reported compared to healthy control samples [23] that again suggests neuronal dysfunction in BPD. However, there is no compelling evidence to relate the gross anatomic pathology with BPD.

Biochemical abnormalities include those within the serotonergic and dopaminergic system, deficit of the membrane Na, K – ATPase and disturbance of thyroid hormone metabolism [24-26]. The observation that selective serotonin re-uptake inhibitor (SSRI) can be used to effectively treat the condition as well as evidence that dietary depletion of tryptophan results in depression in patients previously treated with SSRI strongly suggests that alteration in serotonergic function may contribute to the symptoms of depression [27, 28]. Similarly, patients can also be successfully treated with norepinephrine re-uptake inhibitors (NRI). When these patients were administered with α - methylparatyrosine (AMPT), which is a competitive inhibitor of tyrosine hydroxylase, (an enzyme involved in catecholamine synthesis) it resulted in the recurrence of depressive symptom [29]. These results once again strongly reinforce the idea that the noradrenergic system may have a role in modulating the depressive disorder. Additionally, in a small scale study, it has been reported that patients with type I disorder have more monoamine releasing cells in their brain than that of control samples [30]. These observations suggest that different neurochemical systems may mediate depression and all of these systems may affect a common brain circuitry [31]. Similarly, major depressive disorder is found to be associated with a reduced volume of the left subgenual prefrontal cortex and hippocampus [32, 33] suggesting that subtle alterations in the central nervous system may be responsible for the disorder. Unfortunately however, these anatomical alterations could not be used as markers for the diagnosis of BPD.

Genetic Factors

BPD has a complex etiology. It is a disease of neural circuitry and neural communication. Family, twin and adoption studies suggest that it is a heritable disorder [34-36]. Family studies have provided (a) familial of the condition (increased risk in first degree relatives such as parents, siblings and offspring), (b) a complex mode of inheritance and (c) a range of disease phenotypes associated with defective gene (s). In some cases, the sex of the transmitting parent [37, 38] may influence the transmission of BPD. The results of twin studies are consistent and indicate that among monozygotic twins (MZ) there is a higher concordance rate (79%) for the disorder compared to dizygotic (DZ) twins (19%). This observation strongly supports an etiological role for genetic factors although shared uterine environmental factors could not be disputed. Interestingly, in the case of unipolar disorder, the concordant rate is much lower (54% for MZ and 24% for DZ twins) which may indicate that unipolar and BPD may not be one and the same condition. Additionally, the less than 100% concordance rate between monozygotic twin may indicate that non-genetic factors such as stochastic developmental and environmental factors (e. g. stressful life events) contribute to the development of the disorder [39]. Similarly, adoption studies also support the genetic basis of the disorder although the results are variable.

To understand the genetic contribution to the pathophysiology of BPD, a variety of studies such as family-based linkage (which looks for a correlation between gene and phenotype within families) and population-based association analyses (which looks for a correlation between gene and phenotype in a population), sib-pair and large pedigree studies have been carried out over the past two decades [40-44]. These analyses have identified several suggestive loci. For instance, the whole genome scan has identified multiple loci (1p14, 4p16, 6q14, 6q24, 10p12, 10q26, 11p15, 12q24, 16p13, 17q, 18p11, 18q22, 20p12, 21q21) among which 6q22, 17q and 12q24.3 showed significant linkage [45-47]. Furthermore, interaction between loci on chromosomes 6q and 6p was shown to increase susceptibility to BPD [48]. Although, these studies have suggested that several genes contribute to BPD, no specific gene related to the condition has been isolated. One of the difficulties in identifying the BPD susceptibility gene is the heterogeneous patient population with a bipolar spectrum of disorders. In addition, locus heterogeneity may pose another challenging problem.

Additionally, family-based and case-control association analyses suggest that variants in genes such as serotonin receptor HTR3A [49], serotonin transporter [50], G-protein alpha subunit galphaz [51], neural cell adhesion molecule 1 [52], serotonin 4 receptor [53], phosphatidylinositide – 3 kinase [54], phospholipase C - gamma 1 [55], corticotrophin - releasing hormone (CRH) and proenkaphalin [56] and mitochondrial DNA [57] may confer susceptibility to mood change. Moreover, functional promoter polymorphism may also be involved in the pathogenesis [58]. However, it should be noted that some of these studies are either controversial or not replicated by others [59] which could be due to the small sample size employed in the analyses. Similarly, several studies excluded direct involvement of genes such as serotonin receptor 1D – beta [60], 5-hydroxy tryptamine 2A receptor [61], ADARB1 [62] and dopamine receptor [63]. To date, none of the candidate susceptibility genes can account for a substantial number of BPD patients.

Treatment

BPD has two major problems: suicide and recurrence. Therefore, any treatment for BPD must focus on these two aspects. There are two broad classes of pharmacological agents: antidepressants and mood stabilizers (e. g. lithium) which have been used to treat this devastating mental illness. It is managed by using a multiphase treatment strategy that includes acute, continuation and maintenance phases [64]. In the acute phase, the condition may be improved or the patient may fail to respond or may not be able to tolerate the adverse effects of treatment. In such cases treatment will be adjusted or discontinued. The continuation phase is designed to avoid relapse and it ends when the patient has recovered. The purpose of the maintenance phase is to prevent the recurrence of the future episodes.

Over the past several years, accumulated evidence from several controlled studies (double – blind trials) indicate that lithium salts and lamotrigine are the most effective drugs for maintenance treatment of BPD [65-68]. Initial treatment of BPD includes lithium or valproic acid or SSRI such as bupropion, sertraline, citalopram and escitalopram [69]. One advantage of lithium salt is that it also possesses an anti-suicidal effect [70-72] compared to another anticonvulsant drug divalproex, which is at least, equally good as lithium. However, in order to achieve the maximum response of lithium, it is necessary to identify patients with classical BPD. In that respect, the anticonvulsant carbamazepine is quite effective in the long-term treatment of patients with bipolar spectrum disorder. When lithium and carbamazepine are not effective, valproic acid, which has received approval in the United States of America, has been used to treat patients with mixed state or acute mania. The advantage of valproic acid is that it does not interact with other drugs, hence it can be used in combination with other medications. Lithium and carbamazepine have also been used to treat acute mania. However, the disadvantage of lithium is that when it is co-administered with rofecoxib it may produce a life-threatening lithium intoxication [73]. Rapid cycling BPD often does not respond to mood stabilizers. However, in the case of non-cycling patients clozapine appears to be more effective [74]. There are also an increasing number of new anticonvulsant drugs such as topiramate and gabapentin but they have not been approved for treatment. More data are needed from controlled studies regarding their efficacy and safety.

The mechanism of action and anti-suicidal effect of lithium is not known. However, this simple ion may exert its effect on the signal transduction through the second messenger system by controlling either their synthesis or release or functions [75, 76]. For instance, it has been shown that long-term lithium treatment increases serotonin release and decreases serotonin receptor in the hippocampus [77]. It is possible that this serotonin agonistic action of lithium may account for its anti-suicidal effect. In addition, it may also have neurotrophic and neuroprotective effects [78]. Interestingly, there is also evidence that a sudden interruption rather than a gradual discontinuation of long – term lithium treatment may have a rebound effect and increase the risk of suicide. Whatever the mechanism of its action, lithium is only the drug that is most efficient for the maintenance treatment of BPD.

In addition to drug treatment, it is necessary that family members must learn the warning signs of suicide because, the condition most often (15%) involves suicide [79] in young patients in an early phase of illness. In order to handle this problem, family members and patients must be educated. This is very useful in the maintenance phase of treatment [80]. Moreover, it is also recommended that alcohol use must be discouraged because it may interfere with medication. In short, BPD can be well managed by physicians and psychiatrists.

Summary and Conclusion

BPD is one of the most common debilitating mood disorders. It can affect equally both sexes of all age groups. There are no diagnostic tests for BPD. Because of its high rate of recurrence and suicide in young patients, it is a major public health burden. The clinical course of illness can vary from mild depression to a severe form of mania. Although several regions of the brain have been implicated in the pathogenesis, it is not known how the brain changes from a mild condition to a severe from of depression. BPD is associated with central nervous system function [81] and it has been proposed that a subtle alteration in signal transduction may play a role in the pathophysiology of BPD.

BPD is a phenotypically and genetically complex disorder. Population and family-based association studies have identified numerous chromosomal loci as well as several candidate genes that may confer susceptibility to the disorder. However, none of the genes identified so far can account for the major portion of BPD patients. In addition to genetic factors, environmental factors can play a part in depressive episodes. This notion is consistent with the non-Mendalian nature, clinical differences in females and males, age group for susceptibility and fluctuation of BPD course. It is now thought that gene-gene and gene-environmental (divorce, job loss, poor health, loss of a family member, child abuse, drugs, hormones, diet) interaction may contribute to the disorder [82, 83]. This gene-environmental interaction may be operating at the level of DNA methylation. For instance, valproate, a drug that has been used to treat acute mania, produces hypomethylated DNA. However, more studies are needed to define the role of epigenetic factors (DNA modification) in BPD. Identification of the BPD gene in the future may provide genetic markers for diagnosis, an understanding of pathophysiological mechanism of BPD and a new therapeutic target [84].

Initial treatment of BPD includes lithium or valproic acid or SSRI or bupropion (antidepressant). However, recently antidepressant drugs of SSRI group have not been recommended to treat major depressive disorders in children and adolescents [85]. This is because, review of clinical data revealed that they might increase the risk of suicidal thoughts. Bipolar depression is a more challenging condition than unipolar depression and is associated with an increased suicidal risk. It is recommended that along with lithium, inclusion of another antidepressant may be beneficial for the treatment of bipolar depression. The dopamine agonist pramipexloe was shown to have significant antidepressant effects in patients with bipolar II depression [86]. This suggests that dopaminergic system may also a role in the pathophysiology of BPD. It should be noted that there is no direct evidence that suggests that antidepressants prevent suicide. In addition to drug therapy, there are other types of therapies such as vagus nerve stimulation (VNS), neurosurgical procedures and transcranial magnetic stimulation (TMS) that are available for major depressive disorder, but all of them either involve surgery or high-risk side effects or a proper procedure need to be established. In the future, new classes of drugs with rapid effects must be developed. Since antidepressant drugs have been proposed to act at the transcriptional level changing the pattern of gene expression [87], ultimately it may be possible to design such drugs that may help to alter the defective synaptic function. In addition, it is necessary that family members and patients must be educated to handle the warning signs of suicide and management of depression [80]. Moreover, structured schedules such as regular awakening and sleep-time as well as psychotherapy are found to be effective in reducing the symptoms of acute depression [88]. Psychotherapy is considered to be an effective addition to drug therapy. BPD is a treatable disease with the help of physicians and psychiatrists but it is not curable. In the future, advances in basic neuroscience research may provide a better understanding of the underlying neurobiological states of mood disorders that will enable physicians treat the condition most effectively.

References

[1] Peele, P. B., Xu, Y. & Kupfer, D. J. (2003). Insurance expenditure on bipolar disorder: clinical and parity implications. *Am J Psychiatry* 160: 1286-1290.

[2] Narrow, W. E., Rae, D. S. & Robins, L. N. (2002). Revised prevalence estimates of mental disorders in the United States: using a clinical significance criterion to reconcile 2 surveys' estimates. *Arch Gen Psychiatry* 59: 115-123.

[3] Muller-Oerlinghausen, B., Berghofer, A. & Bauer, M. (2002). Bipolar disorder. *Lancet* 359: 241-247.

[4] Murray, C. J. & Lopez, A. D. (1996). Evidence based health policy – lessons from the global burden of disease study. *Science* 274: 740-743.

[5] Faedda, G. L., Baldessarini, R. J., Suppes, T., Tondo, L., Becker, I. & Lipschitz, D. (1995). Pediatric onset bipolar disorder: a neglected clinical and public health problem. *Harv Rev Psychiatry* 3: 171-195.

[6] Gitlin, M. J., Swendsen, J., Heller, T. L. & Hammen, C. (1995). Relapse and impairment in bipolar disorder. *Am J Psychiatry* 152: 1635-1640.

[7] Griswold, K. S. & Pessar, L. F. (2000). Management of bipolar disorder. *Am Fam Physician* 62: 1343-1353.

[8] Angst, F., Stassen, H. H., Clayton, P. J. & Angst, J. (2002). Mortality of patients with mood disorders: follow-up over 34-38 years. *J affect Disord* 68: 167-181.

[9] Lewinsohn, P. M., Klein, D. N. & Seeley, J. R. (1995). Bipolar disorder in a community sample of older adolescents: prevalence, phenomenology, comorbidity and course. *J Am Acad Child Adolese Psychiatry*. 34: 454-463.

[10] Shastry, B. S. (2004). Molecular genetics of attention – deficit hyperactivity disorder (ADHD): an update. *Neurochem Int* 44: 469-474.

[11] Stahlberg, O., Soderstrom, H., Rastam, M. & Gillberg, C. (2004). Bipolar disorder, schizophrenia, and other psychotic disorders in adults with childhood onset AD/HD and/or autism spectrum disorders. *J Neural transm* 111: 891-902.

[12] Frasure-Smith, N., Lesperance, F. & Talajic, M. (1993). Depression following myocardial infraction: impact on six months survival. *JAMA* 270: 1819-1825.

[13] Soares, J. C. & Mann, J. J. (1997). The functional neuroanatomy of mood disorders. *J Psychiatr Res* 31: 393-432.

[14] Drevets, W. C. (2000). Neuroimaging studies of mood disorders. *Biol Psychiatry* 48: 813-829.

[15] Strakowski, S. M., Adler, C. M. & DelBello, M. P. (2002). Volumetric MRI studies of mood disorders: do they distinguish unipolar and bipolar disorder? *Bipolar Disord* 4: 80-88.

[16] Vawter, M. P., Freed, W. J. & Kleinman, J. E. (2000). Neuropathology of bipolar disorder. *Biol Psychiatry* 48: 486-504.

[17] Strakowski, S. M., DelBello, M. P., Adler, C., Cecil, K. M. & Sax, K. W. (2000). Neuroimaging in bipolar disorder. *Bipolar Disord* 2: 148-164.

[18] Bearden, C. E., Hofmann, K. M. & Cannon, T. D. (2001). The neuropsychology and neuroanatomy of bipolar affective disorder: a critical review. *Bipolar Disord* 3: 106-150.

[19] Drevets, W. C., Price, J. L., Simpson, J. R. Jr, Todd, R. D., Reich, T., Vannier, M. & Raichle, M. E. (1997). Subgenual prefrontal cortex abnormalities in mood disorders. *Nature* 386: 824-827.

[20] Fava, M. & Kendler, K. S. (2000). Major depressive disorder. *Neuron* 28: 335-341.

[21] Machado-Vieira, R., Kapczinski, F. & Soares, J. C. (2002). Perspectives for the development of animal models of bipolar disorder. *Prog Neuropsychopharmacol Biol Psychiatry* 28: 209-224.

[22] Vastag, B. (2003). Imaging studies reveal brain changes in children with bipolar disorder. *JAMA* 289: 2057.

[23] Cecil, K. M., DelBello, M. P., Morey, R. & Strakowski, S. M. (2002). Frontal lobe difference in bipolar disorder as determined by proton MR spectroscopy. *Bipolar Disord* 4: 357-365.

[24] Drevets, W. C., Frank, E., Price, J. C., Kupfer, D. J., Holt, D., Greer, P. J., Huang, Y., Gautier, C. & Mathis, C. (1999). PET imaging of serotonin 1A receptor binding in depression. *Biol Psychiatry* 46: 1375-1387.

[25] Looney, S. W. & el-Mallakh, R. S. (1997). Meta-analysis of erythrocyte Na, K – ATPase activity in bipolar illness. *Depress Anxiety* 5: 53-65.

[26] Bauer, M. & Whybrow, P. C. (2001). Thyroid hormone, neural tissue and mood modulation. *World J Biol Psychiatry* 2: 57-67.

[27] Spillmann, M. K., Van der Daes, A. J., Rankin, M. A., Vuolo, R. D., Alpert, J. E., Nierenberg, A. A., Rosenbaum, J. F., Hayden, D., Schoenfeld, D. & Fava, M. (2001). Tryptophan depletion in SSRI-recovered depressed outpatients. *Psychopharmacology* 155: 123-127.

[28] Delgado. P. L., Miller, H. L., Salomon, R. M., Licinio, J., Krystal, J. H., Moreno, F. A., Heninger, G. R. & Charney, D. S. (1999). Tryptophan-depletion challenge in depressed patients treated with desipraamine or fluoxetine: implications for the role of serotonin in the mechanism of antidepressant action. *Biol Psychiatry* 46:212-220.

[29] Miller, H. L., Delgado, P. L., Salomon, R. M., Berman, R., Krystal, J. H., Heninger, G. R. & Charney, D. S. (1996). Chemical and biochemical effect of catecholamine depletion on antidepressant induced remission of depression. *Arch Gen Psychiatry* 53: 117-128.

[30] Zubieta, J. K., Huguelet, P., Ohl, L. E., Koeppe, R. A., Kilbourn, M. K., Carr, J. H., Giordani, B. J. & Frey, K. A. (2000). High vesicular monoamine transporter binding in asymptomatic bipolar I disorder: sex differences and cognitive correlates. *Am J Psychiatry* 157: 1619-1628.

[31] Bremner, J. D., Vythilingam, M., Ng, C. K., Vermetten, E., Nazeer, A., Oren, D. A., Berman, R. M. & Charney, D. S. (2003). Regional brain metabolic correlates of alpha-methylparatyrosine induced depressive symptoms: implications for the neural circuitry of depression. *JAMA* 289: 3125-3134.

[32] Gaddman, D. (2002). DBH and functional taxonomy of major depressive disorder. *Biol Psychiatry* 51: 347-348.

[33] Vythilingam, M., Heim, C., Newport, J., Miller, A. H., Anderson, E., Bronen, R., Brummer, M., Staib, L., Vermetten, E., Charney, D. S., Nemeroff, C. B. & Bremner, J. D. (2002). Childhood trauma associated with smaller hippocampal volume in women with major depression. *Am J Psychiatry* 159: 2072-2080.

[34] Craddock, N. & Jones, I. (1999). Genetics of bipolar disorder. *J Med Genet* 36: 585-594.

[35] Mitchell, P., MacKinnon, A. & Waters, B. (1993). The genetics of bipolar disorder. Aust N Z *J Psychiatry* 27: 560-580.

[36] Taylor, L., Faraone, S. V. & Tsuang, M. T. (2002). Family, twin and adoption studies of bipolar disease. *Curr Psychiatry Rep* 4: 130-133.

[37] McMahon, F. J., Stine, O. C., Meyers, D. A., Simpson, S. G. & DePaulo, J. R. (1995). Patterns of maternal transmission in bipolar affective disorder. *Am J Hum Genet* 56: 1277-1286.

[38] Stine, O. C., Xu, J., Koskela, R., McMahon, F. J., Gschwend, M., Friddle, C., Clark, C. D., McInnis, M. G., Simpson, S. G., & Breschel, T. S. et al (1995). Evidence for linkage of bipolar disorder to chromosome 18 with a parent-of-origin effect. *Am J Hum Genet* 57: 1384-1394.

[39] Duffy, A., Grof, P., Robertson, C. & Alda, M. (2000). The implications of genetic studies of major mood disorders for clinical practice. *Clin Psychiatry* 61: 630-637.

[40] Gershon, E. S., Hamovit, J., Guroff, J. J., Dibble, E., Leckman, J. F., Sceery, W., Targum, S. D., Nurnberger, J. I. Jr, Goldin, L. R. & Bunney, W. E. Jr (1982). A family study of schizoaffective, bipoolar I, bipolar II, unipolar and normal control probands. *Arch Gen Psychiatry* 39: 1157-1167.

[41] Ewald, H., Flint, T., Kruse, T. A. & Mors, O. (2002). A genome-wide scan shows significant linkage between polar disorder and chromosome 12q24.3 and suggestive linkage to chromosomes 1p14-21, 4p16, 6q14-22, 10q26 and 16p13.3. *Mol Psychiatry* 7: 734-744.

[42] Dick, D. M., Foroud, T., Flury, L., Bowman, E. S., Miller, M. J., Rau, N. L., Moe, P. R., Samavedy, N., el-Mallakh, R., Manji, H., Glitz, D. A., Meyer, E. T., Smiley, C., Hahn, R., Widmark, C., McKinney, R., Sutton, L., Ballas, C., Grice, D., Berrettini, W., Byerley, W., Coryell, W., DePaule, R., MacKinnon, D. F., Gershon, E. S., Kelsoe, J. R., McMahon, F. J., McInnis, M., Murphy, D. L., Reich, T., Scheftner, W. & Nurnberger, J. I. Jr. (2003). Genome-wide scan analysis of bipolar disorder: a new sample of 250 pedigrees from the National Institute of Mental Health genetic initiative. *Am J Hum Genet* 73: 107-114 (erratum 73: 979).

[43] Segurado, R., Detera-Wadleigh, S. D., Levinson, D. F., Lewis, C. M., Gill, M., Nurnberger, J. I. & Craddock, N. et al. (2003). Genome scan meta-analysis of schizophrenia and bipolar disorder: part III – bipolar disorder. *Am J Hum Genet* 73: 49-62.

[44] Pato, C. N., Pato, M. T., Kirby, A., Petryshen, T. L., Medeiros, H., Carvalho, C., Macedo, A., Dourado, A., Coelho, I., Valente, J., Soares, M. J., Ferreira, C. P., Lei, M., Verner, A., Hudson, T. J., Morley, C. P., Kennedy, J. L., Azevedo, M. H., Daly, M. J. & Sklar, P. (2004). Genome-wide scan in Portuguese island families implicates

multiple loci in bipolar disorder: fine mapping adds support on chromosome 6 and 11. *Am J Med Genet* 127B: 30-34.

[45] Middleton, F. A., Pato, M. T., Gentile, K. L., Morley, C. P., Zhao, X., Eisener, A. F., Brown, A., Petryshen, T. L., Kirby, A. N., Medeiros, H., Carvalho, C., Macedo, A., Dourado, A., Coelho, I., Valente, J., Soares, M. J., Ferreira, C. P., Lei, M., Azevedo, M. H., Kennedy, J. L., Daly, M. J., Sklar, P. & Pato, C. N. (2004). Genome-wide linkage analysis of bipolar disorder by use of a high – density- single –nucleotide polymorphism (SNP) genotyping assay: a comparison with microsatellite marker assays and finding of a significant linkage to chromosome 6q22. *Am J Hum Genet* 74: 886-897.

[46] McInnis, M. G., Dick, D. M., Willour, V. L., Avramopoulos, D., MacKinnon, D. F., Simpson, S. G., Patash, J. B., Edenberg, H. J., Bowman, E. S., McMahon, F. J., Smiley, C., Chellis, J. L., Huo, Y., Diggs, T., Meyer, E. T., Miller, M., Matteini, A. T., Rau, N. L., DePaulo, J. R., Gershon, E. S., Badner, J. A., Rice, J. P., Goate, A. M., Detera-Wadleigh, S. D., Numberger, J. I., Reich, T., Zandi, P. P. & Foroud, T. M. (2003). Genome-wide scan and conditional analysis in bipolar disorder: evidence for genomic interaction in the National Institute of Mental Health genetic initiative bipolar pedigrees. *Biol Psychiatry* 54: 1265-1273.

[47] Willour, V. L., Zandi, P. P., Huo, Y., Diggs, T. L., Chellis, J. L., MacKinnon, D. F., Simpson, S. G., McMahon, F. J., Potash, J. B., Gershon, E. S., Reich, T., Foroud, T., Nurnberger, J. I. Jr., DePaulo, J. R. Jr. & McInnis, M. G. (2003). Genome scan of the 56 bipolar pedigrees from the NIMH genetic initiative replication samples: Chromosomes 4, 7, 9, 18, 19, 20 and 21. *Am J Med Genet* 121B: 21-27.

[48] Schulze, T. G., Buervenich, S., Badner, J. A., Steele, C. J., Detera-Wadleigh, S. D., Dick, D., Foroud, T., Cox, N. J., MacKinnon, D. F., Potash, J. B., Berrettini, W. H., Byerley, W., Coryell, W., DePaulo, J. R. Jr., Gershon, E. S., Kelsoe, J. R., McInnis, M. G., Murphy, D. L., Reich, T., Scheftner, W., Nurnberger, J. I. Jr. & McMahon J. J. (2004). Loci on chromosomes 6q and 6p interact to increase susceptibility to bipolar affective disorder in the national institute of mental health genetics initiative pedigrees. *Biol Psychiatry* 56: 18-23.

[49] Niesler, B., Flohr, T., Nothen, M. M., Fischer, C., Rietschel, M., Franzek, E., Albus, M., Propping, P. & Rappold, G. A. (2001). Association between the 5'-UTR variant C178T of the serotonin receptor gene HTR 3A and bipolar affective disorder. *Pharmacogenetics* 11: 471-475.

[50] Kirov, G., Rees, M., Jones, I., MacCandless, F., Owen, M. J. & Craddock, N. (1999). Bipolar disorder and serotonin transporter gene: a family based association study. *Psychol Med* 29: 1249-1254.

[51] Saito, T., Papolos, D. F., Chernak, D., Rapaport, M. H., Kelsoe, J. R. & Lachman, H. M. (1999). Analysis of GNAZ gene polymorphism in bipolar affective disorder. *Am J Med Genet* 88: 324-328.

[52] Arai, M., Itokawa, M., Yamada, K., Toyota, T., Arai, M., Haga, S., Ujike, H., Sora, I., Ikeda, K. & Yoshikawa, T. (2004). Association of neural cell adhesion molecule 1 gene polymorphisms with bipolar affective disorder individuals. *Biol Psychiatry* 55: 804-810.

[53] Ohtsuki, T., Ishiguro, H., Detera-Wadleigh, S. D., Toyota, T., Shimizu, H., Yamada, K., Yoshitsugu, K., Hattori, E., Yoshikawa, T. & Arinami, T. (2002). Association between serotonin 4 receptor gene polymorphisms and bipolar disorder in Japanese case-control samples and the NIMH genetic initiative bipolar pedigrees. *Mol Psychiatry* 7: 954-961.

[54] Stopkova, P., Saito, T., Papolos, D. F., Vevera, J., Paclt, I., Zukov, I., Bersson, Y. B., Margolis, B. A., Strous, R. D. & Lachman, H. M. (2004). Identification of PIK3C3 promoter variant associated with bipolar disorder and schizophrenia. *Biol Psychiatry* 55: 981-988.

[55] Turecki, G., Grof, P., Cavazzoni, P., Duffy, A., Grof, E., Ahrens, B., Berghofer, A., Muller-Oerlinghausen, B., Dvorakova, M., Libigerova, E., Vojtechovsky, M., Zvolsky, P., Joober, R., Nilsson, A., Prochazka, H., Licht, R. W., Rasmussen, N. A., Schou, M., Vestergaard, P., Holzinger, A., Scumann, C., Thau, K., Rouleau, G. A., & Alda, M. (1998). Evidence for a role of phospholipase C-gamma 1 in the pathogenesis of bipolar disorder. *Mol Psychiatry* 3:. 534-538.

[56] Alda, M., Turecki, G., Gorf, P., Cavazzoni, P., Duffy, A., Grof, E., Ahrens, B., Berghofer, A., Muller-Oerlinghausen, B., Dvorakova, M., Libigerova, E., Vojtechovsky, M., Zvolsky, P., Joober, R., Nilsson, A., Prochazka, H., Licht, R. W., Rasmussen, N. A., Schou, M., Vestergaard, P., Holzinger, A., Scumann, C., Thau, K. & Rouleau, G. A. (2000). Association and linkage studies of CRH and PENK genes in bipolar disorder: a collaborative IGSLI study. *Am J Med Genet* 96: 178-181.

[57] Kato, T., Kunugi, H., Nanko, S. & Kato, N. (2001). Mitochondrial DNA polymorphisms in bipolar disorders. *J Affect Disord* 62: 151-164.

[58] Hoogendoorn, B., Coleman, S. L., Guy, C. A., Smith, S. K., O'Donovan, M. C. & Buckland, P. R. (2004). Functional analysis of polymorphisms in the promoter regions of genes on 22q11. *Hum Mutat* 24: 35-42.

[59] Mendes de Oliveira, J. R., Otto, P. A., Vallada, H., Lauriano, V., Elkis, H., Lafer, B., Vasquez, L., Gentil, V., Passos-Bueno, M. R. & Zatz, M. (1998). Analysis of a novel functional polymorphism within the promoter region of the serotonin transporter gene (5-HTT) in Brazilian patients affected by bipolar disorder and schizophrenia. *Am J Med Genet* 81: 225-227.

[60] Mundo, E., Zai, G., Lee, L., Parikh, S. V. & Kennedy, J. L. (2001). The 5HT1D beta-receptor gene in bipolar disorder: a family based association study. *Neuropsychopharmacology* 25: 608-613.

[61] Arranz, M. J., Erdmann, J., Kirov, G., Rietschel, M., Sodhi, M., Albus, M., Ball, D., Maier, W., Davies, N., Franzek, E., Abusaad, I., Weigelt, B., Murray, R., Shimron-Abarbanell, D., Kerwin, R., Propping, P., Sham, P., Nothen, M. M. & Collier, D. A. (1997). 5HT2A receptor and bipolar affective disorder: association studies in affected patients. *Neurosci Lett* 224: 95-98.

[62] Amore, M., Strippoli, P., Laterza, C., Tagariello, P., Vitale, L., Casadei, R., Frabetti, F., Canaider, S., Lenzi, L., D'Addabbo, P., Carinei, P., Torroni, A., Ferrari, G. & Zannotti, M. (2004). Sequence analysis of ADARB1 gene in patients with familial bipolar disorder. *J Affect Disord* 81: 79-85.

[63] Li, T., Liu, X., Sham, P. C., Aitchison, K. J., Cai, G., Arranz, M. J., Deng, H., Liu, J., Kirov, G., Murray, R. M. & Collier, D. A. (1999). Association analysis between dopamine receptor gene and bipolar affective disorder. *Psychiatry Res* 86: 193-201.

[64] Sachs, G. S. (2001). A 25 year old woman with bipolar disorder. JAMA 285: 454-462.

[65] Schou, M. (1997). Forty years of lithium treatment. *Arch Gen Psychiatry* 54: 9-13.

[66] Maj, M. (2000). The impact of lithium prophylaxis on the course of bipolar disorder: a review of the research evidence. *Bipolar Disord* 2: 93-101.

[67] Prien, R. F., Caffey, E. M. Jr. & Klett, J. (1973). Prophylactic efficiency of lithium carbonate in manic depressive illness. *Arch Gen Psychiatry* 28: 337-341.

[68] Bowden, C., Calabrese, J. R., Sachs, G., Yatham, L. N., Asghar, S. A., Hompland, M., Montgomery, P., Earl, N., Smoot, T. M. & DeVeaugh-Geiss, J. (2003). A placebo-controlled 18 month trail of lamotrigine and lithium maintenance treatment in recent manic or hypomanic patients with bipolar I disorder. *Arch Gen Psychiatry* 60: 392-400.

[69] Hartmann, P. M. (1996). Strategies for managing depression complicated by bipolar disorder, suicidal ideation or psychotic features. *J Am Board Fam Pract* 9: 261-269.

[70] Baldessarini, R. J., Tondo, L. & Hennen, J. (2001). Treating the suicidal patients with bipolar disorder: reducing suicidal risk with lithium. *Ann NY Acad Sci* 932: 24-38.

[71] Goodwin, F. K., Fireman, B., Simon, G. E., Hunkeler, E. M., Lee, J. & Revicki, D. (2003). Suicide risk in bipolar disorder during treatment with lithium and divalproex. *JAMA* 290: 1467-1473.

[72] Baldessarini, R. J. & Tondo, L. (2003). Suicide risk and treatments for patients with bipolar disorder. *JAMA* 290: 1157-1159.

[73] Ratz, B. A. E., Egger, S. S., Crespo, S., Probst, W. L. & Krahenbuhl, S. (2004). Lithium intoxication as aresult of an interaction with rofecoxi. *Ann Pharmacother PMID* 15187206.

[74] Suppes, T., Erkan, O. M. & Carmody, T. (2004). Response to clozapine of rapid cycling versus non-cycling patients with a history of mania. *Bipolar Disord* 6: 329-332.

[75] Stoll, A. L. & Severus, E. (1996). Mood stabilizers: shared mechanism of action at postsynaptic signal transduction and kindling processes. Harv Rev Psychiatry 4: 77-89.

[76] Shastry, B. S. (1997). On the function of lithium: the mood stabilizer. *Bioessay* 19: 199-200.

[77] Treiser, S. L., Cascio, C. S., O'Donohue, T. L., Thoa, N. B., Jacobowitz, D. M. & Kellar, K. J. (1981). Lithium increases serotonin release and decreases serotonin receptor in the hippocampus. *Science* 213: 1529-1531.

[78] Manji, H. K., Moore, G. J. & Chen, G. (2000). Clinical and preclinical evidence for the neurotrophic effect of mood stabilizers: implication for the pathophysiology and treatment of manic depressive illness. *Biol Psychiatry* 48: 740-754.

[79] Strakowski, S. M., McElory, S. L., Keck, P. E. & West, S. A. (1996). Suicidality among patients with mixed and manic bipolar disorder. *Am J Psychiatry* 153: 674-676.

[80] Gilbody, S., Whitty, P., Grimshaw, J. & Thomas, R. (2003). Educational and organizational interventions to improve the management of depression in primary care: a systematic review. *JAMA* 289: 3145-3151.

[81] Merali, Z., Du, L., Hrdina, P., Palkovits, M., Faludi, G., Poulter, M. O. & Anisman, H. (2004). Dysregulation in the suicide brain: mRNA expression of corticotropin – releasing hormones and $GABA_A$ receptor subunits in frontal cortical brain region. *J Neurosci* 24: 1478-1485.

[82] Abdolmaleky, H. M., Smith, C. L., Faraone, S. V., Shafa, R., Stone, W., Glatt, S. J. & Tsuang, M. T. (2003). Methylomics in psychiatry: modulation of gene-environment interactions may be through DNA methylation. *Am J Med Genet* 127B: 51-59.

[83] Petronis, A. (2003). Epigenetics and bipolar disorder: new opportunities and challenges. *Am J Med Genet* 123C: 65-75.

[84] Insel, T. R. & Charney, D. S. (2003). Research on major depression: strategies and priorities. *JAMA* 289: 3167-3168.

[85] Ramchandani, P. (2004). Treatment of major depressive disorder in children and adolescents. *BMJ* 328: 3-4.

[86] Zarate, C. A. Jr., Payne, J. L., Singh, J., Quiroz, J. A., Luckenbaugh, D. A., Denicol, K. D., Charney, D. S. & Manji, H. K. (2004). Pramipexole for bipolar II depression: a placebo-controlled proof of concept study. *Biol Psychiatry* 56: 54-60.

[87] Duman, R. S., Heninger, G. R. & Nestler, E. J. (1997). A molecular and cellular theory of depression. *Arch Gen Psychiatry* 54: 597-606.

[88] Bolton, P., Bass, J., Neugebauer, R., Verdeli, H., Clougherty, K. F., Wickramaratne, P., Speelman, L., Ndogoni, L. & Weissman, M. (2003). Group interpersonal psychotherapy for depression in rural Uganda. *JAMA* 289: 3117-3124.

In: Trends In Neurochemistry Research
Editor: Robert M. Coleman, pp. 179-188

ISBN: 1-59454-418-2

Chapter 5

Serotonin 2a (5-HT$_{2A}$) Receptors in Human Brains Postmortem from Psychiatric Patients

Donatella Marazziti,* Bernardo Dell'Osso, Mario Catena, Stefano Barone, Irene Masala and Gino Giannaccini
Dipartimento di Psichiatria, Neurobiologia, Farmacologia e Biotecnologie,
University of Pisa, Italy

Abstract

The results of our study reported the presence of a high-affinity and saturable binding of [^{3}H]-ketanserin to frontal and parietal brain membranes obtained from normal control subjects and patients affected by unipolar depression, schizophrenia and bipolar disorder. The human brain samples (60 frontal cortex and 51 parietal cortex), were donated by the Stanley Foundation Brain Collection. A higher density in the frontal than in the parietal cortex was found in normal controls, depressed and schizophrenic patients, while bipolar patients did not show any difference. When the data were analysed according to the 2 hemispheres, [^{3}H]-ketanserin binding sites did not show any difference in the two frontal cortices, while they were less represented in the right parietal cortex of normal and bipolar patients and more dense in schizophrenic patients. In conclusion, our findings have showed the presence of heterogenous alterations of [^{3}]H-ketanserin binding sites in normal controls and different psychiatric disorders. These results may be of help in a further elucidation of the specific role that 5-HT$_{2A}$ receptors may play in these disorders.

* Author for correspondence: Dr. Donatella Marazziti, "Dipartimento di Psichiatria, Neurobiologia, Farmacologia e Biotecnologie", University of Pisa, via Roma, 67, 56100 Pisa, Italy. Tel. +39 050/835412; Fax: +39 050/21581; e-mail address: dmarazzi@psico.med.unipi.it

Introduction

Serotonin (5-HT) receptors are classified on the basis of pharmacological and functional effects into seven families: 5-HT_1, 5-HT_2, 5-HT_3, 5-HT_4, 5-HT_5, 5-HT_6 and 5-HT_7 (Lovenberg et al., 1993). It is not surprising that serotonin, involved in the modulation of different neuronal activities, has been found to act through multiple receptors with different pattern of distribution in the brain (Martin and Humphrey, 1994). The interest of the researchers has been directed towards 5-HT_{1a} (Srinivas et al., 2001; Yoshino et al., 2002) and 5-HT_2 receptors, including the 5-HT_{2A}, 5-HT_{2B} and 5-HT_{2C} subtypes, that are considered involved in the pathophysiology of different neuropsychiatric disorders (Arango et al., 1990; Hrdina et al., 1993; Blumensohn et al., 1995). In fact, 5-HT_2 receptors have been proposed as mediator of the effects of hallucinogens, such as diethylamine lysergic acid (LSD) (Glennon, 1990), and consequently it has been hypothesized their possible role in the development of psychotic symptoms (Heninger, 1991; Gerlach, 1991; Roth et al., 1992). The recent development of atypical antipsychotics, behaving as 5-HT_2 antagonist, has supported this hypothesis (Kapur and Remington, 2001). Post-mortem studies have showed that the 5-HT_{2A} receptor density is altered in brain tissues obtained from both schizophrenic patients and suicide victims, irrespectively from diagnoses (Joyce et al., 1993; Hashimoto et al., 1993; Hrdina et al., 1993; Gurevich et al., 1997; Rosel et al., 1998). Furthermore, PET studies have demonstrated a decreased density of 5-HT_{2A} receptors in brains of suicide attempters (Audenaert et al., 2001), while findings obtained in schizophrenic patients have been more controversial (Joyce et al., 1993; Hashimoto et al., 1993; Gurevich and Joyce 1997; Dean, 2001). However, 5-HT_{2A} receptors seemed to be also involved in the development of anxiety and depression since ritanserin, a 5-HT_{2A} antagonist, has been found to have anxiolytic and antidepressant properties in humans (Zifa and Fillion, 1994) and different antidepressants, such as mirtazapine, have been shown to interact at the level of these receptors (Attar-Levy et al., 1999; Dekeyne and Millan, 2003). Furthermore, in patients with major depression a decrease in cortical 5-HT_{2A} has been reported (Yatham et al., 2000) and in patients not responding to SSRI a reduced expression of these receptors in comparison to responders has been observed (Zanardi et al., 2001). Increased level of 5-HT_{2A} receptors has been observed in platelets obtained from drug-free bipolar or schizoaffective patients and the receptors density was even more increased in suicidal patients (Pandey et al., 2003). Recently some authors have shown a possible involvement of the 5-HT_{2A} not only in psychopathological conditions, but also in personality traits, such as the harm avoidance: the higher the binding capacity of 5-HT_{2A} receptors, the lower the tendency to escape danger (Moresco et al., 2002). To address the specific distribution in the brain of 5-HT_{2A} receptors radioligand binding tecniques, histochemistry, mRNA expression and PET studies have been performed in the brain of several mammals (Pazos et al., 1987; Mengod et al., 1990; Burnet et al., 1995; Pasqualetti et al., 1996; Marazziti et al., 1997; Hall et al., 2000): neocortex seemed to be the richest areas in these receptors, while hippocampus, basal ganglia and thalamus showed a lower density. Ketanserin, a N-alkylpiperidine behaving as an antagonist, is considered the radioligand of choice for binding studies (Leysen et al., 1982) although it binds also to 5-HT_{2C} receptors (Zifa and Fillion, 1994). In order to elucidate the role of 5-HT_{2A} in the pathophysiology of neuropsychiatric disorder, the aim of the present study has been to compare the specific binding of [^{3}H]-ketanserine in human frontal and parietal cortex obtained postmortem from brains of bipolar, depressed, schizophrenic patients and normal controls.

Materials and Methods

Materials

[^{3}H]-ketanserin (specific activity: 77.1 Ci/mmol) was purchased from New England Nuclear (NEN, Milan, Italy); mianserine and the Trizma base were obtained from Sigma Chemical Co. (Milan, Italy), and all other reagents used were of analytical grade.

Human Brain Tissue

The postmortem human brain samples (60 frontal cortex and 51 parietal cortex of both emishperes), were donated by the Stanley Foundation Brain Collection, courtesy of Drs. M.B. Knable, E.F. Torrey, M.J. Webster and R.H. Yolken. They were collected during autoptic session of 60 subjects (36 male and 24 female), their age ranging between 25 and 68 years (mean ± S.D.: 45.4 ± 11.2). The post-mortem delay (time in hours between death and tissue dissection/freezing, PMD) was between 7 and 62 hours (mean ± S.D.: 29.4 ± 13.4). Tissues appeared well conserved, as previously shown by the measurement of mRNA and proteins levels (Torrey et al., 2000).

Causes of death were suicide (n = 20), cardiopulmonary failure (n = 32), accident (n = 5) and other (n = 3) (Torrey et al., 2000): the suicides were equally distributed in the three groups of patients.

These subjects were divided in 4 groups, each consisting of 15 subjects (9 male and 6 female) according to different psychiatric disorders or healthy conditions.

Patients, whose diagnoses had been made on the basis of DSM IV criteria (APA, 1994) , were so grouped: bipolar patients (mean age ± S.D.: 42.3 ± 11.7; mean PMD ± S.D.: 32.5 ± 16.1 hours); unipolar depressed patients (mean age ± S.D.: 46.5 ± 9.3; mean PMD ± S.D.: 27.5 ± 10.7 hours); schizophrenic patients (mean age ± S.D.: 44.5 ± 13.1; mean PMD ± S.D.: 33.7 ± 14.6 hours).

Treatment was based on flufenazine for bipolar and schizophrenic patients, while depressed patients were undergoing psychoanalytic therapy only.

Normal controls had no previous personal or familial history of evidence of psychiatric disorders or treatment with psychotropic drugs before demise, their age ranged between 29 and 68 years (mean ± S.D.: 48.1 ± 10.7) and PMD ranged between 8 and 42 hours (mean ± S.D.: 23.7 ± 9.9).

Age of death, PMD, age at onset and duration (time passed between the onset of a disorder and death of the subject) of the disorders, and administered treatments (equivalent of flufenazine) resulted to be homogeneous in the different groups.

Membrane Preparation

Human brain membranes were prepared according to the method of Marazziti et al. (1997). Tissue blocks were weighed, homogenised by means of an Ultraturrax T25 homogenizer at the half-maximum speed for about 30 seconds in 10 volumes of ice-cold 0.32 M Sucrose, containing protease inhibitors (200 μg/ml benzamidine, 160 μg/ml bacitracin and 20 μg/ml soybean trypsin inhibitor) and then centrifuged at 1,000 × g for 5 min at 4°C. The resulting pellet was discarded, whereas the supernatant was centrifuged at 49,000 g for 15 min at 4°C. The pellet was re-suspended in 10 volumes (w/v) of ice-cold Tris HCl 50 mM, pH 7.4 with protease inhibitors and subsequently incubated, with stirring, at 37°C for 15 min to destroy endogenous 5-HT. The homogenate was then aliquoted and centrifuged as above. The final pellets were stored at –80°C until assay, performed within 2 weeks.

Binding Assay

On day of assay, membranes were defrosted, resuspended in 40 volumes of 50 mM Tris-HCl buffer (pH 7.4) without protease inhibitors (T1), yelding a final protein concentration of about 0.39 mg/ml. Proteins were measured according to the method of Peterson (1977).

Membranes were incubated in a final volume of 1.0 ml T1 buffer: each test tube contained about 0.350 mg protein and [^{3}H]-ketanserin ranging from 0.15 to 2.5 nM. Non specific binding was accounted for by incubating similar samples with 500 nM mianserine. After 30 min at 37°C, the incubation was halted by the addition of 5 ml of ice-cold T1 buffer. Samples were rapidly vacuum-filtered through Whatman GF/B glass filters, 2.5 cm in diameter, and then washed twice with 5 ml of ice-cold T1 buffer. Filters were then placed in vials with 4 ml of Ready Protein scintillation cocktail (Beckman) and radioactivity was measured by means of a beta-counter (Packard 1600 Tricarb).

Data Analyses and Statistics

The equilibrium-saturation binding parameters, i.e., the maximum binding capacity (B_{max}, fmol/mg protein) and the dissociation constant (K_d, nM), were analysed with the iterative curve-fitting computer programmes KINETICS, EBDA and LIGAND by means of an IBM compatible personal computer (McPherson, 1985).

B_{max} values were normally distributed according to the Gaussian distribution, while K_d values were standardised by means of the transformation in pK_d ($-LogK_d$). Therefore, statistical analyses were performed by means of the multigroup variance analysis (ANOVA) followed by the post-hoc Scheffé's procedure. Differences between the two hemispheres or between sex were evaluated through Student's t-test. Correlations between B_{max} or pK_d values and the different parameters (age, PMD, age at onset, duration, cerebral weight and cerebral pH) were performed according to the Spearman's method, while considering $p < 0.05$ as significant values. All these tests were carried out by means of personal computer programmes (The GraphPad ISI Software, Philadelphia, PA, USA).

Results

Saturation curves and Scatchard plots obtained in the frontal (panel a) and parietal (panel b) cortex of the 4 groups are reported in fig. 1: all curves were monophasic.

Significant correlations between the different parameters were observed. Age correlated significantly and negatively with frontal B_{max} of schizophrenic patients (r= -0.56) and positively with parietal B_{max} of depressed patients (r= 0.81). Age at onset correlated negatively with frontal B_{max} in depressed patients (r= -0.47).

As far as gender-related differences were concerned, male schizophrenic patients had a significantly higher B_{max} in the frontal than in the parietal cortex (136±31 and 45±5, respectively, p= 0.04) (table 1).

Table 1. [^{3}H]-ketanserin binding parameters in frontal and parietal cortices of male and female patients and controls

	Frontal Cortex				*Parietal Cortex*			
	Male		*Female*		*Male*		*Female*	
	B_{max}	pK_d	B_{max}	pK_d	B_{max}	pK_d	B_{max}	pK_d
Normal Controls	92 ± 14	8.78 ± 0.06	96 ± 15	8.91 ± 0.06	34 ± 14	8.94 ± 0.07	53 ± 23	8.74 ± 0.13
Bipolar Patients	94 ± 19	8.84 ± 0.04	69 ± 16	8.76 ± 0.1	77 ± 26	8.88 ± 0.09	94 ± 23	8.6 ± 0.26
Depressed Patients	92 ± 20	8.81 ± 0.07	131 ± 32	8.84 ± 0.07	57 ± 26	8.83 ± 0.03	59 ± 12	8.75 ± 0.18
Schizophrenic Patients	136 ± 31*	8.64 ± 0.13	87 ± 21	8.78 ± 0.09	45 ± 5*	8.71 ± 0.16	64 ± 13	8.73 ± 0.17

*Significant p = 0.05.

The B_{max} and pK_d values (mean ± S.E.M.) did not show any intergroup difference in the frontal or in the parietal cortex. However, when the two areas were compared inside each group, it turned out that normal controls, depressed and schizophrenic patients had a significant higher B_{max} in the frontal than in the parietal cortex (93±10 and 45±14, p = 0.02; 106±17 and 58±13, p= 0.03; 112±22 and 55±5, p = 0.01, respectively) (table 2).

Table 2. [^{3}H]-ketanserin binding parameters in frontal and parietal cortices of patients and controls

	Frontal Cortex		*Parietal Cortex*	
	B_{max}	pK_d	B_{max}	pK_d
Normal Controls	93 ± 10*	8.83 ± 0.04	45 ± 14*	8.82 ± 0.09
Bipolar Patients	84 ± 13	8.81 ± 0.05	83 ± 18	8.79 ± 0.11
Depressed Patients	106 ± 17§	8.83 ± 0.05	58 ± 13§	8.79 ± 0.08
Schizophrenic Patients	112 ± 22#	8.69 ± 0.09	55 ± 5#	8.71 ± 0.16

*, §, # Significant: p = 0.05.

Data were also analysed according to the two hemispheres: while no difference in the right versus the left frontal cortex was observed, the B_{max} resulted to be significantly lower in the right than in the left parietal cortex of normal controls and bipolar patients (21±0.5 and 57±19 vs 59±16 and 113±43, p= 0.05 and p= 0.02, respectively), and higher in schizophrenic patients (67±3 vs 42±2, p= 0.004). Left parietal B_{max} of bipolar patients was significantly higher than that of normal controls, depressed patients and schizophrenic patients (113±43 vs 59±16, 54±16 and 42±2, respectively, F-test= 18, p= 0.0001) (table 3).

Table 3. [^{3}H]-ketanserin binding parameters in the two hemispheres of patients and controls

	Frontal Cortex				*Parietal Cortex*			
	Right Hemisphere		*Left Hemisphere*		*Right Hemisphere*		*Left Hemisphere*	
	B_{max}	pK_d	B_{max}	pK_d	B_{max}	pK_d	B_{max}	pK_d
Normal Controls	94 ± 17	8.8 ± 0.07	93 ± 12	8.85 ± 0.06	21 ± 0.5$^{\wedge}$	8.75 ± 0.05	59 ± 16$^{§\wedge}$	8.8 ± 0.1
Bipolar Patients	94 ± 20	8.8 ± 0.05	73 ± 18	8.82 ± 0.08	57 ± 19°	8.84 ± 0.12	113 ± 43$^{§°}$	8.73 ± 0.21
Depressed Patients	100 ± 32	8.9 ± 0.06	111 ± 20	8.75 ± 0.06	52 ± 2	8.98 ± 0.02	54 ± 16§	8.76 ± 0.09
Schizophrenic Patients	108 ± 18	8.6 ± 0.14	115 ± 39	8.75 ± 0.12	67 ± 3	8.9 ± 0.03	42 ± 2$^{§¥}$	8.56 ± 0.25

$^{\wedge, °, §}$ Significant: p = 0.05

Discussion

The results of our study reported the presence of a high-affinity and saturable binding of [^{3}H]-ketanserin to frontal and parietal brain membranes obtained from normal control subjects and patients affected by unipolar depression, schizophrenia and bipolar disorder. All saturation curves were monophasic, thus confirming the existence of one binding site only for [^{3}H]-ketanserin, as already reported (Marazziti et al., 1997).

While in a previous study no effect of age or postmortem delay on the binding parameters was observed (Marazziti et al., 1997), the present study showed some significant correlations: age was negatively correlated with the density of [^{3}H]-ketanserin binding site in the frontal cortex of schizophrenic patients, and positively with the density in parietal cortex of depressed patients. Depressed patients also showed a negative correlation between the density of receptors in the frontal cortex and age at onset: the earlier the age at onset, the lower the number of receptors. It may be that a younger age at onset of depression is a factor affecting negatively 5-HT_{2A} receptors in the frontal cortex.

Therefore the presence of a psychiatric disorder could make the individual more vulnerable to the effect of age with a sort of specificity and regionality. In fact schizophrenia seems to provoke a receptor decrease in the frontal cortex of older patients, while depression tends to have a contrary effect in parietal cortex. A decreased 5-HT_{2A} receptor density in chronic schizophrenic patients has been already documented in other studies and it may be related to the longer course and duration of neuroleptic treatment (Hashimoto et al., 1993; Gurevitch et al., 1997).

Another parameter that in our study seems to influence 5-HT_{2A} receptors in schizophrenic patients is represented by gender: in fact, male patients had an higher density of [^{3}H]-ketanserin binding in the frontal than in the parietal cortex: such a gender effect might provide a possible explanation of the controversial data obtained in this psychiatric disorder (Joyce et al., 1993; Hashimoto et al., 1993; Gurevich and Joyce 1997; Dean, 2001).

When the binding parameters were compared in the two areas, the heterogeneous effects of different psychiatric disorders on [^{3}H]-ketanserin binding was even more evident. Normal controls, depressed and schizophrenic patients had a higher density in the frontal than in the parietal cortex, while bipolar patients did not show any difference.

Data about 5-HT_{2A} receptor density in the brain of depressed patients are controversial: in fact increased (Yates et al., 1990; Hrdina et al., 1993), decreased (Attar-Levy et al., 1999; Yatham et al., 2000), or no significant changes (Crow et al., 1984) have been reported: however, in most cases, detailed diagnoses were quite few.

Our finding in schizophrenic patients are consistent with a large body of evidences (Joyce et al., 1993; Hashimoto et al. 1993; Gurevich et al., 1997); on the contrary, the observation in bipolar patients is original to our knowledge.

When the data were analysed according to the 2 hemispheres, [^{3}H]-ketanserin binding sites did not show any difference in the two frontal cortices, while they were less represented in the right parietal cortex of normal and bipolar patients and more represented in schizophrenic patients. Depressed patients did not present this asymmetry: this result is consistent with other studies (Arato et al., 1991) showing that depressive illness may revert the normal brain asymmetry of other serotonergic parameters such as the 5-HT transporter.

The main bias of our study was that the size of each subgroup was small and the significant correlations observed must be interpreted with cautions and only further studies in larger samples, that are very difficult to perform, may confirm or reject our findings. It is, however, worth nothing the relative homogeneity of the brain areas and of the subjects in each diagnostic subgroups, as well as the strict inclusion criteria in the brain collection by the Stanley Foundation.

In conclusion, our study has demonstrated the presence of heterogenous alterations of [^{3}]H-ketanserin binding sites in normal controls and different psychiatric disorders that may be of help in a further elucidation of the specific role that 5-HT_{2A} receptors may play in these disorders.

References

Arango, V., Ernsberger, P., Marzuk, P.M., Chen, J.S., Tierney, H., Stanley, M., Reis, D.J., Mann, J.J., 1990. Autoradiographic demonstration of increased serotonin 5-HT_2 and beta-adrenergic receptor binding sites in the brain of suicide victims. *Archives of General Psychiatry* 47, 1038-1047.

Arato, M., Frecska, E., Tekes, K., MacCrimmon, D.J., 1991. Serotonergic interhemispheric asymmetry: gender difference in the orbital cortex. *Acta Psychiatrica Scandinavica* 84(1), 110-111.

Attar-Levy, D., Martinot, J.L., Blin, J., Dao-Castellana, M.H., Crouzel, C., Mazoyer, B., Poirier, M.F., Bourdel, M.C., Aymard, N., Syrota, A., Feline, A., 1999. The cortical serotonin 2 receptors studied with positron-emission tomography and [^{18}F]-spiperone during depressive illness and antidepressant treatment with clomipramine. *Biological Psychiatry*, 45, 180–186.

Audenaert, K., Van Laere, K., Dumont, F., Slegers, G., Mertens, J., van Heeringen, C., Dierckx, R.A., 2001. Decreased frontal serotonin 5-HT $_{2A}$ receptor binding index in deliberate self-harm patients. *European Journal of Nuclear Medicine* 28(2), 175-182.

Blumensohn, R., Ratzoni, G., Weizman, A,., Israeli, M., Greuner, N., Apter, A., Tyano, S., Biegon, A., 1995. Reduction in serotonin 5-HT$_2$ receptor binding on platelets of delinquent adolescents. *Psychopharmacology* 118, 354-356.

Burnet, P.W., Eastwood, S.L., Lacey, K., Harrison, P.J., 1995. The distribution of 5-HT$_{1A}$ and 5-HT$_{2A}$ receptor mRNA in human brain. *Brain Research* 676(1), 157-168.

Crow, T.J., Cross, A.J., Cooper, S.J., Deakin, J.F., Ferrier, I.N., Johnson, J.A., Joseph, M.H., Owen, F., Poulter, M., Lofthouse, R., 1984. Neurotransmitter receptors and monoamine metabolites in the brains of patients with Alzheimer-type dementia and depression, and suicides. *Neuropharmacology* 12B, 1561-1569.

Dean, B., 1991. A predicted cortical serotonergic/cholinergic/GABAergic interface as a site of pathology in schizophrenia. *Clinical Experimental Pharmacology and Physiology*, 28(1-2), 74-78.

Dekeyne, A., Millan, M.J., 2003. Discriminative stimulus properties of antidepressant agents: review. *Behav Pharmacol* 14, 391-407.

Gerlach, J., 1991. New antipsychotics: classification, efficacy and adverse effects. *Schizophrenia Bulletin* 155(Suppl. 5), 33-36.

Glennon, R.A., 1990. Do classical hallucinogens act as 5-HT$_2$ agonist or antagonist? *Neuropsychopharmacology* 118, 354-356.

Gurevich, E.V., Joyce, J.N., 1997. Alterations in the cortical serotonergic system in schizophrenia: a postmortem study. *Biological Psychiatry* 42(7), 529-545.

Hall, H., Farde, L., Halldin, C., Lundkvist, C., Sedvall, G., 2000. Autoradiographic localization of 5-HT(2A) receptors in the human brain using [^{3}H]-M100907 and [^{11}C]-M100907. *Synapse* 38(4), 421-431.

Hashimoto, T., Kitamura, N., Kajimoto, Y., Shirai, Y., Shirakawa, O., Mita, T., Nishino, N., Tanaka, C., 1993. Differential changes in 5-HT$_{1A}$ and 5-HT$_{2A}$ receptor binding in patients with chronic schizophrenia. *Psychopharmacology* 112, S35-S39.

Heninger, G.R., 1991. The role of serotonin in clinical disorders. In Bloom F.E. and Meltzer H.Y. (eds.), *Psychopharmacology: The Fourth Generation of Progress*, Raven Press, New York, pp. 471-482.

Hrdina, P.D., Demeter, E., Vu, T.B., Sotonyi, P., Palkovits, M., 1993. 5-HT uptake sites and 5-HT$_2$ receptors in brain of antidepressant-free suicide victims/depressives: increase in 5-HT$_2$ sites in cortex and amygdala. *Brain Research* 614(1-2), 37-44.

Joyce, N.J., Shane, A., Lexow, N., Winokur, A., Casanova, M., Kleinman, J.E., 1993. Serotonin uptake sites and serotonin receptors are altered in the limbic system of schizophrenics. *Neuropsychopharmacology* (Berl.) 124(1-2), 95-106.

Kapur, S., Remington, G., 2001. Atypical antipsychotics: new directions and new challenges in the treatment of schizophrenia. *Ann Rev Med* 52, 503-17.

Leysen, J.E., Niemegeers, C.J., van Neuten, J.M., Laduron, P.M., 1982. [^{3}H]-ketanserin (R41468), a selective ligand for 5-HT$_2$ receptor binding sites. *Molecular Pharmacology* 21, 301-314.

Lovenberg, T.W., Erlander, M.G., Baron, B.M., Sutcliffe, J.G., 1993. Cloning of new 5-HT receptors. *Int. Clinical Psychopharmacology* 8(S2), 19-23.

Marazziti, D., Rossi, A., Palego, L., Giannaccini, G., Naccarato, A., Lucacchini, A., Cassano, G.B., 1997. [^{3}H]-ketanserin binding in human brain postmortem. *Neurochemistry* Research 22(6), 753-757.

Martin, G.R., Humphrey, P.P.A., 1994. Receptors for 5-hydroxytryptamine: current perspectives on classification and nomenclature. *Neuropharmacology* 33, 261-273.

McPherson, G.A., 1985. KINETIC, EBDA, LIGAND, LOWRY: *A collection of ratio ligand binding analysis programs*. Biosoft, Cambridge, UK.

Mengod, G., Pompeiano, M., Martinez-Mir, M.I., Palacios, J.M., 1990. Localization of the mRNA for the 5-HT$_2$ receptor by in situ hybridization histochemistry. Correlation with the distribution of receptor sites. *Brain Research* 524, 139-143.

Moresco, F.M., Dieci, M., Vita, A., Messa, C., Gobbo, C., Galli, L., Rizzo, G., Panzacchi, A., De Peri, L., Invernizzi, G., Fazio, F., 2002. In vivo serotonin 5HT$_{2A}$ receptor binding and personality traits in healthy subjects: a positron emission tomography study. *Neuroimaging* 17, 1470-78.

Pandey, G.N., Pandey S.C., Ren, X., Dwivedi, Y., Janicak, P.G., 2003. Serotonin receptors in platelets of bipolar and schizoaffective patients: effect of lithium treatment. *Psychopharmacology* 170, 115-23.

Pasqualetti, M., Nardi, I., Ladinsky, H., Marazziti, D., Cassano, G.B., 1996. Comparative anatomical distribution of serotonin 1A, 1D alpha and 2A receptor mRNAs in human brain postmortem. *Brain Research Molecular Brain Research* 39(1-2), 223-33.

Pazos, A., Probst, A., Palacios, J.M., 1987. Serotonin receptors in human brain. I. Autoradiographic mapping of serotonin-2 receptors. *Neuroscience* 21, 123-139.

Peterson, G.L., 1977. A simplification of the protein assay method which is more generally applicable. *Analitical Biochemistry* 83, 356-366.

Rosel, P., Arranz, B., Vallejo, J., Oros, M., Crespo, J.M., Menchon, J.M., Navarro, M.A., 1998. Variations in [^{3}H]-imipramine and 5-HT$_{2A}$ but not [^{3}H]-paroxetine binding sites in suicide brains. *Psychiatry Research* 82(3), 161-170.

Roth, B.L., Ciaranello, R.D., Meltzer, H.Y., 1992. Binding of typical and atypical antipsychotic agents with transiently expressed 5-HT$_{1C}$ receptors. *Journal of Pharmacological & Experiment Therapy* 260, 1361-1365.

Torrey, E.F., Webster, M., Knable, M., Johnston, N., Yolken, R.H., 2000. The Stanley Foundation brain collection and Neuropathology Consortium. *Schizophrenia Research* 44, 151–155.

Yates, M., Leake, A., Candy, J.M., Fairbairn, A.F., McKeith, I.G., Ferrier, I.N., 1990. 5HT$_2$ receptor changes in major depression. Biological Psychiatry 27(5), 489-496.

Yatham, L.N., Liddle, P.F., Shiah, I.S., Scarrow, G., Lam, R.W., Adam, M.J., Zis, A.P., Ruth, T.J., 2000. Brain serotonin 2 receptorsin major depression: a positron emission tomography study. *Arch Gen Psychiatry* 57, 850-58.

Zanardi, R., Artigas, F., Moresco, R., Colombo, C., Messa, C., Gobbo, C., Smeraldi, E., Fazio, F. 2001. Increased 5-hydroxytryptamine-2 receptor binding in the frontal cortex of depressed patients responding to paroxetine treatment: a positron emission tomography scan study. *J Clin Psychopharmacol* 21, 53-58.

Zifa, E., Fillion, G., 1994. 5-Hydroxytryptamine receptors. *Pharmacological Review* 44(3), 402-457.

In: Trends In Neurochemistry Research
Editor: Robert M. Coleman, pp. 189-222

ISBN: 1-59454-418-2

Chapter 6

Plasticity of the Spinal Neural Circuitry in Nitric Oxide Synthase Immunoreactive Pathways After Spinal Cord Hemisection

Nadežda Lukáčová, Jaroslav Pavel, Karolína Kuchárová and Jozef Maršala
Institute of Neurobiology, Slovak Academy of Sciences, Košice, Slovak Republic

Abstract

This chapter focuses on the current issue of nitric oxide in the pathogenesis of traumatic spinal cord injury with special emphasis on alterations of catalytic nitric oxide synthase activity after hemisection performed at the thoracic level. Based on experiments in animal models, our new data suggest that hemisection may cause extensive changes to NOS expression in effected neurons and in areas located far beyond the epicenter of the injury. A considerable decrease in NOS activity at the injured site supports the view that a rapid formation and a release of NO in affected spinal cord regions may accelerate neurodestructive events and lead to a secondary inactivation of NOS activity, possibly by NO binding to the hem iron of NOS and/or by the phosphorylation of NOS by proteinkinases. After hemisection, when half of the spinal cord is cut, the spared portion of the spinal cord might partly assume the lost functions (motor, sensory and autonomic), and, manage at least in part the restructuring of the supraspinal and intersegmental connections. Clinical studies have suggested that patients suffering from spinal cord lesions very often regain a limited ability to walk. However, permanent disturbances in standing and stepping, as well as some disorder in the intralimb and interlimb coupling are observed. A crucial role in such disturbances may consist in the recovery of long ascending and descending propriospinal pathways connecting the cervical and lumbar intumescence. Recent tract-tracing studies have documented that in some large axonal bundles, NOS is a part of the axoplasmic transport and, that NO may be released at a great distance from the cell bodies. The disconnection of the ascending premotor propriospinal NOS-IR pathway interconnecting neurons in the lumbosacral enlargement with the ventral motor nucleus at the cervicothoracic level causes a strong depletion of NOS-IR in the ventral part of the ventral horn in C7-Th1 segments after hemisection performed at the thoracic level. Our results show that NO represents a neuromodulator that may participate in the processing of informations between the cervicothoracic and lumbosacral intumescence. The functional significance of these findings and the probable mechanisms are critically discussed.

Introduction

The role of nitric oxide (NO) was originally identified by Murad et al., [1978], who demonstrated the ability of NO to relax smooth cells of the blood vessels. Two years later Furchgott & Zawadski, [1980] confirmed this effect in experiments *in vitro*, and demonstrated that agonist-induced vascular relaxation required the presence of an intact endothelium. The molecule mediating vascular relaxation was termed by them as endothelium-derived relaxing factor (EDRF). Subsequent studies identified the EDRF as NO [Ignarro et al., 1988], produced from the amino acid L-arginine in vascular endothelial cells [Palmer et al., 1988a,b] and as a messenger molecule involved in a variety of cellular functions, including synaptic plasticity in the central nervous system [Garthwaite et al., 1988]. NO is known as a diffusible, short-lived (less then 5s) and highly reactive free radical molecule, that is very difficult to measure *in vivo* conditions [Archer, 1993]. Thus, the studies on the role of NO under physiological circumstances and in the pathophysiology of the spinal cord injury are based on the activity and/or the expression of the nitric oxide synthase (NOS) [Bredt et al., 1991] and its histochemical marker, nicotinamide adenine dinucleotide phosphate diaphorase (NADPHd) [Scherer-Singler et al., 1983].

In this chapter we review recent information on the rostrocaudal distribution of NOS in the spinal cord and the role of NO in acute and chronic spinal cord injuries performed at different spinal cord levels. New research carried out in our laboratory shows that hemisections performed at the cervical and thoracic levels may influence the NOS pools in remote parts of the spinal cord injury.

Nitric Oxide as a Neurotransmitter/Neuromodulator

For a long time it was believed that interneuronal communication was realized exclusively *via* synaptic contacts. However, since the discovery of NO as a neuronal transmitter [Snyder & Bredt, 1991; Dawson et al., 1992], an increasing interest has focused on its ability to modulate synaptic functions. NO is an unique neurotransmitter/modulator that differs from classical neurotransmitters in that it passes through the cell membranes without specific release or uptake mechanisms and acts directly on surrounding neural tissue in extended spatial limits of approximately 300-400μm [Wood & Garthwaite, 1994; Vincent, 1994]. After its synthesis in soma of neurons, NO induces changes in signal-related functions by several ways. In particular, the activation of the soluble guanylyl cyclase (sGC), the formation of cyclic guanosine 3`,5`-monophosphate (cGMP) and the action of cGMP-dependent protein kinases have been identified as the main signal transduction pathway of NO in the nervous system [Bred & Snyder, 1992; Wang & Robinson, 1997; Smolenski et al., 1998; see Figure 1].

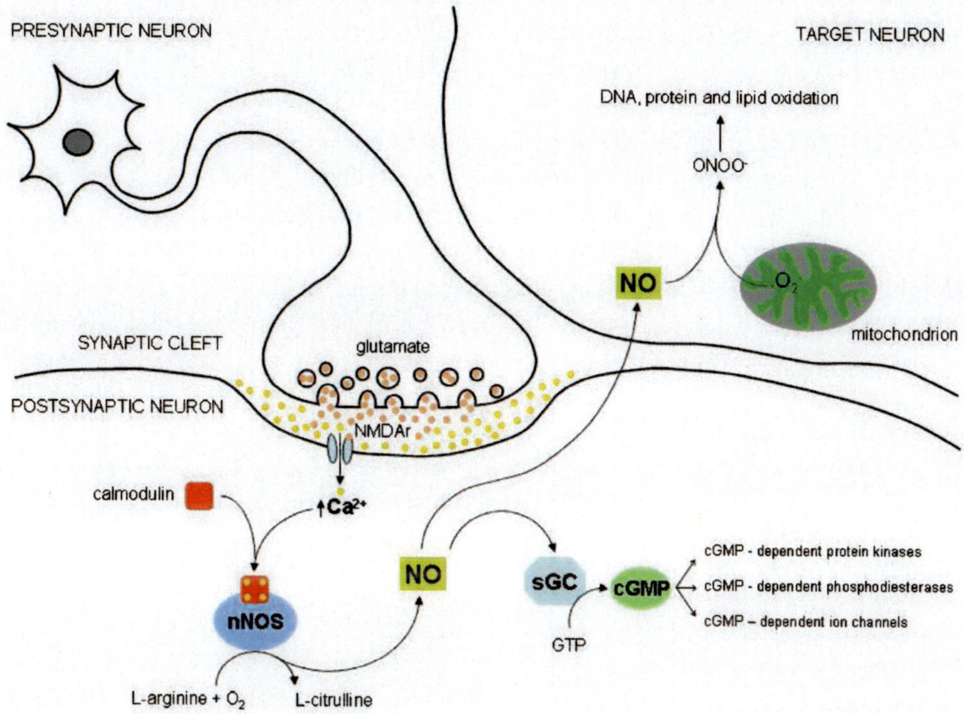

Figure 1. Schematic drawing of the NO/cyclic GMP signal transduction pathway. The binding of excitatory neurotransmitter – glutamate, released from the presynaptic neuron into the synaptic cleft, induces structural changes in the N-methyl-D-aspartate receptor (NMDAr), permitting Ca^{2+} influx into postsynaptic neuron. Increased intracellular concentration of Ca^{2+} ions facilitates its binding to calmodulin, that together with other co-factors activates the neuronal nitric oxide synthase (nNOS). This enzyme catalyzes oxidation-reduction reaction of the L-arginine to L-citrulline and nitric oxide (NO). Primary cellular response to NO production is the activation of soluble isoform of guanylate cyclase (sGC), which is followed by the production of cyclic guanosine monophosphate (cGMP). The increase of cGMP level can influence the activity of three different targets, i. e. cGMP-dependent protein kinases, cGMP-dependent phosphodiesterases, and cGMP-dependent ion channels. In addition, produced NO passes through the cytoplasmatic membrane and acts directly on surrounding neural cells. Together with superoxide anion radical ($.O_2^-$) in the target neuron, NO can create a highly toxic peroxynitrite anion ($ONOO^-$), which causes a damage of important structural and functional components of the cell, including DNA, proteins and lipids.

Isoforms of NOS

NOS was found in many cell types of mammalian organism, such as neurons, glial cells, endothelial cells and macrophages [Moncada et al., 1991; Dawson et al., 1991]. According to the occurrence and the ability to form NO, three isoforms of NOS were identified, neuronal (nNOS, type I), inducible (iNOS, type II) and endothelial (eNOS, type III). Neuronal and endothelial isoforms have very similar properties; both of them are expressed constitutively and their activities are regulated by Ca^{2+} and calmodulin. The major differences being in that, the neuronal NOS is present in the cytosol while the endothelial isoform is predominantly membrane-associated protein [Förstermann et al., 1991; Hecker et al., 1991, 1994]. Neuronal NOS is able to generate superoxide and hydrogen peroxide in the absence of a substrate [Pou et al., 1992]. This property of enzyme has major implications for neurotoxicity in some pathological conditions. Inducible isoform of NOS was initially described in macrophages [Moncada et al., 1991]. It is a high-output enzyme [Förstermann et al., 1998], whose expression is largely or completely calcium-independent and occurs in response to cytokines or other appropriate stimuli in many other cell types [Geller et al., 1993]. All three isoforms of NOS produce NO oxidizing a guanidino nitrogen of L-arginine to L-citrulline utilising molecular oxygen, NADPH and tetrahydrobiopterin as co-substrates [Mayer et al., 1991].

NO/cGMP Signal Transduction System

NO and cGMP have emerged as important signal transduction mediators of inter- and intracellular signals. Although the precise role of cGMP has not been defined as yet, there is evidence suggesting that cGMP produced by the activation of a soluble guanylyl cyclase (sGC) may influence the activity of three different targets, i.e. protein kinases, phosphodiesterases and ion channels. Several recent studies confirmed the existence of a NO/cGMP signal transduction pathway in the brain [Denninger & Marletta, 1999; Meller et al., 1992; Morris et al., 1994; Schmidt & Pehl, 1996] and the spinal cord [Meller et al., 1992; Pavel et al., 2001; Lukáčová et al., 2002b]. Although NOS- and cGMP immunoreactive neurons are often found to be distributed in close proximity to each other [Schmidt et al., 1992; Allaerts et al., 1998], a quantitative assessment of catalytic NOS activity [Bredt et al., 1991] *versus* cGMP or sGC [Steiner et al., 1972] has revealed considerable regional and segmental differences in the distribution of both compounds [Lukáčová et al., 1999; Pavel et al., 2000]. The NO/cGMP signal transduction pathway [Murad, 1994] is known to control glutamate and cholinergic neurotransmission in CNS [George et al., 1970] and appears to mediate many of the physiological actions of NO in those parts of the CNS having a well-defined synaptic circuitry [Garthwaite & Boulton, 1995; Hobbs, 1997; Maršala et al., 1998; Maršala et al., 1999]. However, limited data are available for the white matter composed of glial cells, axons and entirely devoid of synapses [Garthwaite et al., 1999a,b; Lukáčová et al., 2002a].

NOS in the Spinal Cord under Physiological Conditions

First, information concerning the physiological role of NO in the spinal cord comes from the beginning of the nineties [Faraci, 1990]. At this time an overwhelming majority of relevant studies described the location, morphologic appearance and ramification patterns of nicotinamide adenine dinucleotide phosphate diaphorase (NADPHd) exhibiting neurons. In 1991, a homotopic distribution of NADPHd positivity and NOS immunoreactivity was identified at various, exactly specified sites of the brain [Dawson et al., 1991; Bredt et al., 1991] and, one year later, both stainings confirmed such neurons in the spinal cord laminae [Brüning, 1992; Anderson, 1992; Valtschanoff et al., 1992; Dun et al., 1992]. Since that time the NADPHd histochemistry was taken as a histochemical marker for neuron-containing nNOS. However, there are numerous loci in the spinal cord where a fewer number of nNOS immunoreactive than NADPHd-exhibiting neurons are present [Vizzard et al., 1994; Lukáčová et al., 1999]. These findings pointed out the species-related differences and moreover, in many cases provided controversal results.

A detailed analysis of the neuronal pools in the spinal cord lead to the discoveries of the occurrences of NADPHd positive and NOS-IR neurons along the rostrocaudal axis in the spinal cord of rat, mouse, cat, rabbit, dog, squirrel monkey and man [Valtschanoff et al., 1992; Vincent, 1994; Egberongbe et al., 1994; Dun et al., 1993; Maršala et al., 1998, Maršala et al., 1999; Lukáčová et al., 1999], and in combination with a radioassay detection of catalytic NOS activity [Lukáčová & Pavel, 2000; Lukáčová et al., 2002a] and a densitometric study of the NADPHd staining [Maršala et al., 2001a] exactly to localize the segmental, laminar and funicular [Maršala et al., 2003b] distribution of NOS spinal cord pools. A localization of the constitutive NOS in the spinal cord under physiological conditions provides the basis for a determination of the participation of NO on the modulation of the spinal microcircuits after various experimental interventions.

Segmental Distribution of Catalytic NOS Activity in the Spinal Cord

Since no significant differences in constitutive NOS activity were found in L1/L2 and L3/L4 segments of the rat [Yezierski et al., 1996], it was concluded, that a different segmental distribution of NOS activity does not exist. However, recently published data from our laboratory have shown a heterogeneous distribution of NOS-IR/NADPHd positivity and NOS activity in the segments along the rostrocaudal axis of the spinal cord [Lukáčová et al., 1999; Lukáčová & Pavel., 2000; Lukáčová et al., 2002a; see Table 1]. The comparison of the constitutive NOS activity at the cervical, thoracic, lumbar and sacral level of the rabbit and dog spinal cord clearly demonstrates that both the constitutive NOS activity and the number of NOS-IR somata are considerably lower at the lower cervical and upper thoracic levels than in the sacral segments. Since it is known that a long-term systemic inhibition of NOS activity in the rat causes a high incidence of spinal cord infarcts predominantly at the cervical or cervicothoracic levels, followed by front leg motor dysfunctions [Blot et al., 1994], it seems possible that spinal cord segments, having a low constitutive NOS activity, may be the primary site, where the effects of NOS inhibition are becoming apparent.

Table 1. Rostrocaudal and laminar distribution of NOS pools in the spinal cord of the rabbit

Laminae		Segments					
		C6	Th2	L1	L5	S2	S3
I-VI	NADPHd	48.3 ± 1.3	21.9 ± 0.8	79.3 ± 2.1	109.3 ± 3.1	114.1 ± 3.4	110.9 ± 1.8
	NOS-IR	28.1 ± 0.4	10.4 ± 0.3	25.7 ± 0.4	29.4 ± 0.5	55.0 ±1.2	46.3 ± 0.8
	NOS	80.1 ± 5.0	72.9 ± 2.5	115.3 ± 6.6	206.7 ± 6.6	207.5 ± 5.0	187.2 ± 6.2
VII	NADPHd	2.5 ± 0.2	3.1 ± 0.1	3.1 ± 0.3	4.2 ± 0.2	2.6 ± 0.3	2.2 ± 0.2
	NOS-IR	1.3 ± 0.1	3.5 ± 0.2	2.9 ± 0.1	2.6 ± 0.1	2.2 ± 0.1	1.8 ± 0.1
	NOS	29.1 ± 2.7	61.1 ± 2.8	38.8 ± 3.7	62.3 ± 1.8	120.7 ± 6.2	113± 13
VIII-IX	NADPHd	3.6 ± 0.1	2.3 ± 0.4	4.9 ± 0.3	4.5 ± 0.4	5.1 ± 0.3	3.9 ± 0.5
	NOS-IR	2.0 ± 0.2	1.6 ± 0.1	2.1 ± 0.2	2.8 ± 0.2	2.4 ± 0.2	2.2 ± 0.1
	NOS	57.7 ± 2.7	33.3 ± 2.9	42.7 ± 3.2	66.7 ± 2.4	58.9 ± 2.8	88.1 ± 5.1
X	NADPHd	5.2 ± 0.4	3.5 ± 0.5	7.3 ± 0.5	9.6 ± 0.2	12.2 ± 1.2	13.8 ± 1.1
	NOS-IR	2.9 ± 0.2	2.8 ± 0.1	4.6 ± 0.4	5.6 ± 0.2	6.8 ± 0.3	7.4 ± 0.2
	NOS	46.9 ± 3.2	33.9 ± 2.7	45.9 ± 4.5	73.4 ± 1.9	114.0 ± 9.3	126.2 ± 9.4

NADPHd- number of NADPH diaphorase positive neurons; NOS-IR- number of nitric oxide synthase immunoreactive neurons; NOS- the activity of constitutive nitric oxide synthase.

Laminar Distribution of NOS Pools in the Gray Matter of the Spinal Cord

In the spinal cord, NADPHd-staining and nNOS immunoreactivity are localized to discrete populations of neurons in the superficial (laminae I-III) and deep layers of the dorsal horn (laminae IV-VI), in the pericentral region (lamina X), in the dorsal commissure all along the spinal cord and in the intermediolateral cell column (IML) of the thoracolumbar segments in different species studied [Valtschanoff et al., 1992; Dun et al., 1992; Maršala et al., 1998; Maršala et al., 1999; see Table 1]. The occurrence of NADPHd-exhibiting and/or NOS immunoreactive neurons in the ventral horn was the subject of controversial descriptions. While only a few intensely-stained bipolar neurons could be found in the ventral horn, then along its medial border and in the thin sprouts of gray matter into the ventral column in rat [Valtschanoff et al., 1992], few isolated NOS-IR neurons were scattered in the ventral horn of rat, mouse, cat and squirrel monkey [Dun et al., 1993]. Rare, presumed NOS-positive motoneurons were noted in the ventral horn of the rat, particularly in the medial portion of lamina VIII and IX [Saito et al., 1994]. A radioassay detection of the constitutive NOS activity in the regions of the gray matter [Lukáčová et al., 1999; Lukáčová & Pavel, 2000; Lukáčová et al., 2002a; see Table 1] confirms the morphological findings and reflects the occurrence of the three different patterns of histochemically analyzed and detected NADPHd-staining and/or NOS-immunoreactivity in cell bodies, the fiber-like expression of the occurrence of nNOS in axons and dendrites and, a highly important punctate, non-somatic NADPHd-staining or NOS immunoreactivity seen with a varying intensity in the neuropil of various gray matter regions all along the rostrocaudal axis of the spinal cord.

Localization of Nitrergic Nerve Fibers in the Spinal Cord

Several observations raise the possibility that NO may be released at great distance from the neuronal cell body [Lee et al., 1993; Miyazaki et al., 1996]. It seems possible, that an anterograde axoplasmic transport of NOS along with other synaptically acting enzymes and neurotransmitters may serve as a mechanism, allowing to promote and realize synaptic activity in remote neurons. Sporadic notes describing the white matter NADPHd positivity and NOS fiber immunoreactivity appeared in different animal species [Valtschanoff et al., 1992; Dun et al., 1993; Maršala et al., 2003b]. Some of them include the description of NADPHd positive fibers, perhaps axons in all planes of sectioning, traceable in longitudinal planes along the whole length of the section. The NADPHd-positive fibers were found to be dense in the ventral part of the dorsal funiculus and sparser in the lateral and ventral funiculi [Valtschanoff et al., 1992; Maršala et al., 2003b]. Similarly, NOS immunoreactive fibers forming strand-like tracts could be traced in the dorsal, ventral and lateral white matter [Maršala et al., 2002; Maršala et al., 2003a,b; Maršala et al., 2004]. They were most numerous in the thoracic and lower lumbar/sacral segments, where they seemed to terminate in the adjacent gray matter regions [Dun et al., 1993]. It should be noted that these descriptions are based purely on histochemical and immunocytochemical analyses.

NOS Activity in the White Matter Versus the Gray Matter

The results from our laboratory clearly define the constitutive NOS activity in the white matter of both the cervicothoracic and lumbosacral enlargements of rabbit and dog spinal cords [Lukáčová & Pavel, 2000; Lukáčová et al., 2002a; Maršala et al., 2003b; Maršala et al., 2004]. In the white matter of the cervicothoracic segments of the rabbit taken as a whole, the enzyme activity closely resembles that detected in the gray one [Lukáčová & Pavel, 2000]. No significant differences in the constitutive NOS activity were noted in the gray matter regions (dorsal horn, intermediate zone and ventral horn) and the white matter columns (dorsal, lateral and ventral) within the upper cervical segments (C1-C4), except for a significant increase in the ventral horn of the C4 segment. In the C6 segment of the dog, the enzyme's activity displayed significant differences in the intermediate zone, ventral and lateral columns [Lukáčová et al., 2002a]. Surprisingly, an extremely high NOS activity was noted in the lowest cervical segment, especially in the dorsal horn and dorsal column. The constitutive NOS activity in the gray matter *versus* white matter of L5-S3 segments clearly prevails and strongly indicates a high quantitative distribution of NO synthesizing NADPHd-exhibiting and/or NOS immunoreactive neurons detected in the lower lumbar and sacral segments [Maršala et al., 1999; Lukáčová et al., 1999; Lukáčová et al., 2002a]. Comparing the enzyme's activity in the upper and lower lumbar segments of the spinal cord, the constitutive NOS activity prevailed in L4-L7 segments in all of the above mentioned white matter columns.

The Case for Nitrergic Projections in the Spinal Cord after Spinal Hemisections

NOS-IR pathways in the spinal cord have been understood only recently through the deficits resulting from its damage. In this connection detailed information has become available regarding the participation of long ascending and/or descending NOS-immunoreactive (NOS-IR) pathways [Xu et al., 1998; Xu et al., 2000; Maršala et al., 2001b; Maršala et al., 2002; Maršala et al., 2003b; Maršala et al., 2004; Lukáčová et al., 2003] in the spinal cord injury. It was shown that hemisection at Th10-Th11 caused a strong upregulation of bNOS-IR neurons, found mostly ipsilaterally in lamina VIII, the medial part of lamina VII and in the pericentral region (lamina X) of the lumbosacral segments and, a strong depletion of bNOS-IR in the ventral motor nucleus, i.e. in the terminal field of an ascending premotor propriospinal NOS-IR pathway at cervicothoracic level [Maršala et al., 2004]. Similarly, after a lower thoracic hemisection, nNOS expression increased dramatically from the basal level in the red nucleus and, *de novo* synthesis of nNOS was noted in the nucleus dorsalis, that normally lack nNOS [Wu et al., 1994; Xu et al., 1998; Xu et al., 2000]. These data support the evidence that intrinsic properties of neurons may contribute to a different cellular response, thus inducing the NOS expression in the neurons located far away from the epicenter of the spinal cord injury. The case for two NOS-IR pathways identified in the spinal cord after cervical and thoracic hemisections is given in the next sub-headings.

Bulbospinal Respiratory NOS-IR Pathway in the Dog

Motoneurons of the phrenic nucleus being located in the ventral gray matter in the extent of C3-C5 segments [Goshgarian & Rafols, 1981] are NOS immunonegative [Maršala et al., 2002]. They receive excitatory and inhibitory inputs from the brainstem bulbospinal neurons located within the dorsal (DRG) and rostral ventral (rVRG) respiratory groups of the medulla [Feldman, 1986], and provide axons to the ipsilateral phrenic muscles [Goshgarian & Rafols, 1984]. Previous investigations from our laboratory have shown that a hemisection performed at C2-C3 segment level followed by nine days survival greatly decreased the density of punctate NOS-immunoreactivity in the neuropil of the phrenic nucleus ipsilaterally to the site of the hemisection [Maršala et al., 2002]. The NOS-IR of the small and middle-sized NOS-containing neurons in the ventral horn of C3-C5 segments and the NOS-IR of interneurons that do not innervate the phrenic motoneurons but are located in the ventral horn in association with the phrenic nucleus were not influenced by C2-C3 hemisection. The retrograde axotomy-induced changes, consisting in a strong upregulation of NOS containing neurons were found in the dorsal and rostral ventral respiratory group of the medulla on the side contralateral to the hemisection. These experiments show that the crossed premotor bulbospinal respiratory pathway contains a fairly high number of nitric oxide synthase immunopositive fibers terminating in the phrenic nucleus. The injections of the fluorescent tracer Fluorogold (FG) into the phrenic nucleus, combined with immunoprocessing for NOS revealed many FG fluorescent and NOS-IR axons in the lateral and ventral funiculi at C1-C2 level ipsilateral to the site of injection. A large number of double-labeled neurons were found in both respiratory groups at the contralateral side and, contrary to this, a considerably reduced number of such neurons could be detected in the ipsilateral DRG and rVRG (Table 2). A quantitative assessment of identified double-labeled (FG+NOS-IR) neurons in the DRG and rVRG revealed that approximately 74% of crossed premotor bulbospinal respiratory NOS immunoreactive axons originate in the rostral ventral respiratory group and 26% is given by NOS-containing neurons of the dorsal respiratory group. These results provide the evidence that premotor bulbospinal respiratory pathway connecting the bulbar respiratory centers with the motor neurons of the phrenic nucleus in the dog is NOS immunoreactive.

Table 2. Quantitative assessment of neurons labeled for Fluorogold (FG) only, NOS-immunoreactive (NOS-IR) only and double-labeled for Fluorogold and nitric oxide synthase immunopositivity (FG+NOS-IR) in the dorsal (DRG) and rostral ventral (rVRG) respiratory groups of medulla after injections of FG into the phrenic nucleus at C3 level and processing the sections for NOS-IR*

Experimental	DRG		rVRG	
group	contralateral	ipsilateral	contralateral	ipsilateral
FG only	3.5±0.8	2.9±0.7	10.2±1.5	2.0±0.7
FG+NOS-IR	7.6±0.7	2.4±0.5	20.3±2.5	2.9±0.8
NOS-IR only	3.6±0.7	3.9±0.7	4.0±1.0	3.3±1.1

*Ten transverse sections (24 μm thick) cut through the lower brainstem of each animal and positioned 150-450 μm rostrally to the obex were used for quantitative assessment.

Premotor NOS-IR Pathway Underlies the Connection between Cervical and Lumbar Enlargement in the Dog

The long ascending and descending propriospinal pathways connecting the cervical and lumbar intumescence in a bidirectional way, play a crucial role in the interlimb coupling both in quadrupeds and bipeds [Miller & van der Burg, 1973; Miller & van der Meché, 1976; Altman & Bayer, 2001]. A previous study from our laboratory examined the occurrence and origin of the punctate NOS immunoreactivity in the neuropil of the ventral motor nucleus (VMN) in C7-Th1 segments of the control spinal cord of the dog, supposing that NOS-IR belongs to the terminal field of an ascending propriospinal NOS immunoreactive pathway [Maršala et al., 2004]. A hemisection performed at Th10-Th11 segment level caused a strong depletion of the punctate NOS-IR in the neuropil of the VMN ipsilaterally with the hemisection, thus revealing that an uncrossed ascending propriospinal pathway, containing a fairly high number of NOS-immunoreactive fibers terminates in the VMN. Concurrently, a strong upregulation of NOS expression was detected in some NOS-containing neurons located in the lumbosacral segments. The bulk of the retrogradely-reacting NOS containing neurons were found to be located in L4-S2 segments. With regard to the laminar distribution, the number of retrogradely-responding NOS-IR neurons localized in laminae VII+X and laminae VIII+IX of L4-S2 segments considerably exceeded the number occurring in the superficial (laminae I-II) and deep (laminae III-VI) dorsal horn layers (Table 3). Segmental and laminar distribution of double-labeled (FG+bNOS-IR) neurons after injections of FG into the VMN in C7-Th1 segments revealed a noticeable number of the double-labeled neurons in the midlumbar (L4-L5), lower lumbar (L6-L7) and sacral (S1-S2) segments (Table 4). Double-labeled (FG+bNOS-IR) neurons clearly prevailed in the ventral horn (laminae VIII-IX), pericentral region (lamina X) and medial part of the intermediate zone (lamina VII) of L4-L7 and S1-S2 segments.

Table 3. The number of normally appearing brain nitric oxide synthase immunoreactive (bNOS-IR) neurons in laminae I-II, laminae III-VI, lamina VII and X and laminae VIII-IX in the lumbosacral enlargement (L4-S2 segments) of the dog and the number of retrogradely reacting brain nitric oxide synthase immunoreactive neurons (RRbNOS-IR) 10 days after hemisection at Th10-Th11*

Segments	Experimental groups		Laminae			
			I-II	III-VI	VII +X	VIII- IX
L4-L5	bNOS-IR	(C)	11.8±0.26	14.38±0.49	14.01±0.54	41.36±0.93
	RRbNOS-IR	(H)	1.45±0.13	2.85±0.09	4.28±0.13	8.38±0.58
L6-L7	bNOS-IR	(C)	21.81±0.67	12.37±0.72	11.15±0.40	21.03±0.53
	RRbNOS-IR	(H)	2.01±0.06	3.5±0.44	4.26±0.14	10.85±0.51
S1-S2	bNOS-IR	(C)	18.24±0.38	12.75±0.35	11.34±0.39	20.46±0.56
	RRbNOS-IR	(H)	2.53±0.10	3.21±0.07	3.46±0.06	6.53±0.58

*Laminar quantitative distribution of bNOS-IR and RRbNOS-IR somata counted from 20 serial sections of three animals (n=3) subjected to hemisection. Each second section was used to assess the number of normally appearing and retrogradely reacting nitric oxide synthase-immunoreactive neurons. Transverse sections were prepared from L4-S2 segments. C-control; H-hemisection. Data are given as means ± SEM.

Table 4. Fluorogold-labeled and brain nitric oxide synthase immunoreactive (FG+NOS-IR) neurons after Fluorogold injections into the ventral motor nucleus in C7-Th1 segments and processing the sections cut from L4-S2 segments for bNOS immunocytochemistry*

Segments	Laminae			
	I-II	III-VI	VII + X	VIII-IX
L4- L5	0.35±0.03	0.82±0.09	1.45±0.03	3.98±0.22
L6- L7	0.56±0.08	1.68±0.03	1.65±0.03	2.95±0.14
S1- S2	1.38±0.09	1.87±0.06	2.31±0.14	2.85±0.10

*For segmental and laminar distribution of Fluorogold-labeled and bNOS-IR somata 20 transverse serial sections from all L4-S2 segments were used consecutively, one for Fluorogold-labeled neurons and the following for bNOS-IR. Data are given as means ± SEM.

These findings demonstrate that NOS-IR neurons in the lumbar and sacral segments of the dog issue NOS-IR axons that ascend contralaterally and ipsilaterally through the ventral part of the lateral funiculus and lateral and medial parts of the ventral funiculus of the thoracic segments and project as far as the VMN in C7-Th1 segments.

Intra and Interlimb Coordination After a Thoracic Injury

It is evident that the intra- and interlimb coordination of the movement in quadrupeds and bipeds is mediated with highly predictable patterns from supraspinal centers when the spinal cord is functionally symmetrical, or by spinal networks without supraspinal input [Rossignol & Barbeau, 1993; Brustein & Rossignol, 1998; de Leon et al., 2004]. A major question concerning the locomotor recovery after spinal cord injury is to what extent the hind limb coordination depends on the supraspinal symmetry. In stroke patients absent of the spinal cord injury, the assymmetry in stepping is common, suggesting that a unilateral impairment can disrupt the intrinsic lumbosacral spinal cord symmetry. It was found recently, that after injury the spinal circuitry exhibits a great degree of plasticity. The experiments performed in rats and cats have shown the ability to perform a relatively complex motor task following midthoracic hemisection, depending predominantly on the difficulty of the task. Assymmetrical descending input to a normal spinal cord can cause the symmetrical stepping [de Leon et al., 2004]. In connection with the intra- and interlimb coordination, a phenomenon clearly demonstrable after spinal injury, several studies appeared in recent years indicating that there is a differential control of leg flexor and extensor muscles under various functional conditions. More importantly, there is convincing evidence, that in adult cats after a complete low-thoracic spinal transection, the animals can step bipedally over a range of speeds while full weights bearing may indicate postural adjustments during stepping and standing and apprehension of a specific task e.g. stepping and standing [de Leon et al., 1998a; de Leon et al., 1998b; de Leon et al., 1999; Edgerton & Roy, 2002]. The performance of both specific tasks is complicated by the facts, indicating that there is centrally determined dominance in the control of leg flexor activity, but that proprioceptive afferentation determines extensor activation [Dietz, 2002]. Reminding the fact that under physiologic conditions NOS is present all along the afferent limb of the stretch H-reflex [Maršala et al., 2005] and there are powerful presynaptic inhibitory effects of flexor group Ia afferents on extensor group Ia afferents, but these effects are weak from extensor to flexors [Iles & Roberts, 1987], it seems justified to assume that nNOS and NO may play a significant role in the stepping cycle both in quadrupeds and bipeds.

Spinal Cord Hemisection at Thoracic Level affects NO Pools in Remote Parts of the Injury in the Rabbit

Since no biochemical studies were made to specify the changes in NOS pools in remote areas of the spinal cord hemisection, known to reflect the deep differences in the distribution of neurotransmitters and second messenger receptor sites, we aimed to determine the Ca^{2+}-dependent NOS activity, nNOS protein and cGMP level in the cervicothoracic and lumbosacral segments after Th9-Th10 hemisection in the rabbit, and in combination with nNOS-immunocytochemistry to gain the data needed for a critical appraisal of changes affecting the NOS neurotransmission.

An almost equal decrease of Ca^{2+}-dependent NOS activity was detected in contra- and ipsilateral dorsal part of the lumbosacral segments at 7th day of hemisection (Figure 2). This finding was accompanied by statistically significant changes in the number of nNOS-IR neurons (Table 5). While a quantitative assessment of nNOS immunostaining disclosed a strong decrease in the number of small neurons in superficial laminae (laminae I-III), the number of large nNOS-IR neurons located in the deep dorsal horn layers (laminae IV-VI) was not significantly affected. A considerable decrease of enzyme activity was also noted in the ventral part on contralateral side of Th9-Th10 hemisection (Figure 2), a finding strongly contrasting with a statistically significant increase of Ca^{2+}-dependent NOS activity assessed on ipsilateral side. Such an increase may account for a high number of retrogradely-responding NOS-IR neurons localized in laminae VIII+IX of L3-L6 segments [Maršala et al., 2003b]. Surprisingly, non-significant changes were detected in neuronal NOS protein level (Figure 3) proposing the participation of endothelial NOS activity in Th9-Th10 hemisection.

Table 5. The number of neuronal nitric oxide synthase-immunoreactive (nNOS-IR) somata in laminae I-III, IV-VI, VII, X and VIII-IX in the lumbosacral segments of the rabbit´s spinal cord 7days after hemisection at Th10-Th11.

Laminae	Group side	Segments		
		L5-L6	L7-S1	S2
LI-LIII	Control	11.13±1.82	17.85±1.1	16.17±1.27
	Ipsilateral	7.98±0.32**	13±1.11**	8.14±1.31**
	Contralateral	11.88±2.72	14.8±1.27*	7.7±0.51***
LIV-LVI	Control	2.88±0.22	3.06±1.18	3.5±0.75

Ipsilateral
3.43±0.25
2.1±0.74
1.71±0.91

Contralateral
4.1±1.97
1.8±0.51
1.71±0.5

LVII
Control
3.3±1.47
3.25±0.97
3.17±1.06

Ipsilateral
2.8±1.55
2.7±0.42
1.06±0.32

Contralateral
3.38±0.53
1.75±0.51
1.42±0.59

LX
Control
1.97±0.65
1.6±0.46
2.05±0.58

Ipsilateral
1.65±0.23
1.55±0.35
1.86±0.72

Contralateral

1.81±0.11

1.35±0.72

2.14±0.2

LVIII-IX

Control

2.2±0.32

1.99±0.67

1.67±0.75

Ipsilateral

3.38±2.3

1.3±0.63

1.14±0.2

Contralateral

1.85±0.49

0.85±0.44

1.24±0.68

nNOS-IR somata were counted from 10 transverse serial sections from tree animals. *<0.05, **<0.01, ***<0.001 with respect to controls; Control-intact control; Ipsilateral- ipsilateral side to hemisection; Contralateral- contralateral side to hemisection.

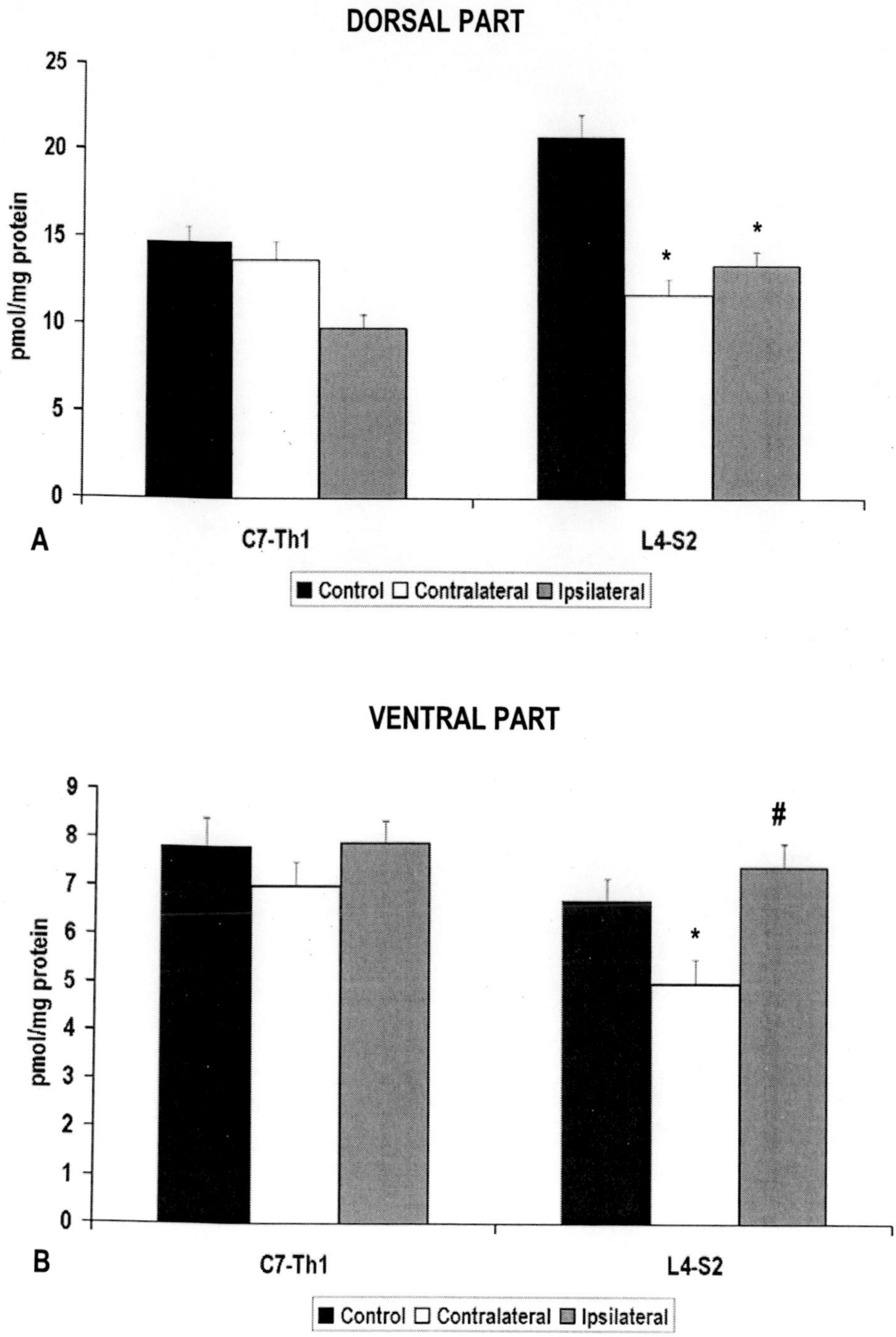

Figure 2. The catalytic NOS activity in the dorsal (A) and ventral (B) part of cervicothoracic (C7-Th1) and lumbosacral (L4-S2) segments of the spinal cord of rabbit. The enzyme activity was detected by conversion of ^{14}[C]arginine to ^{14}[C]citrulline [Bredt et al., 1991] in the spinal cord segments of control and on contralateral and ipsilateral side to hemisection performed at Th9-Th10 level. The experimental animals survived for 7 days. The results are expressed as pmol/mg protein. Data are given as means ± SEM (n=6); *p<0.05 with respect to control; $^{\#}$p<0.05 with respect to contralateral side.

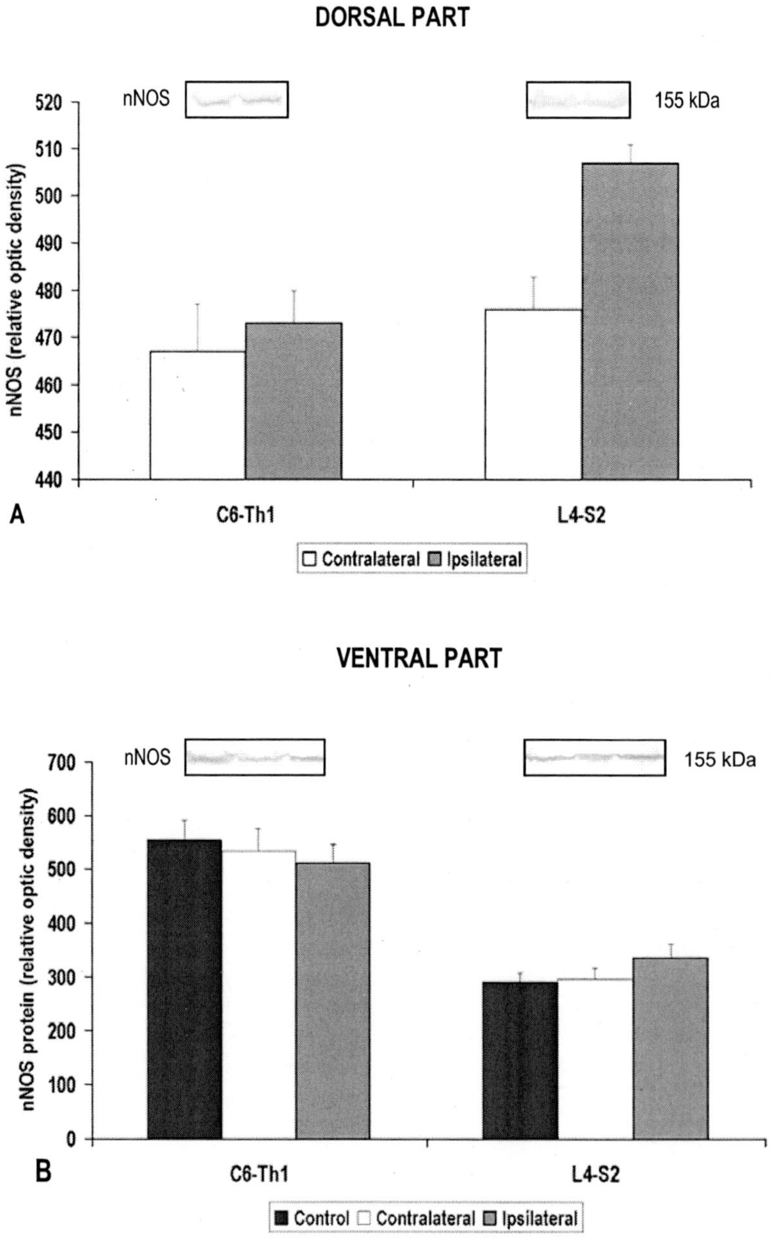

Figure 3. Western blot for the detection of neuronal NOS protein level in the dorsal (A) and ventral (B) part of cervicothoracic (C7-Th1) and lumbosacral (L4-S2) segments of the spinal cord of the rabbit after hemisection at Th9-Th10 level followed by 7 days of survival. Neuronal NOS protein level was detected on contra- and ipsilateral side of hemisection. Sixty micrograms of protein was subjected to 10% polyacrylamide SDS-PAGE and analyzed for nNOS by immunoblotting using sheep nNOS antibody diluted 1:1500 in T-TBS containing 5% milk. nNOS antibody complex was identified with the anti-sheep IgG horseradish peroxidase conjugate and visualized using 0.05% 3,3,4,4-diaminobenzidine hydrochloride. Results represent a immunoblot from four separate experiments. The results are plotted as means ± SEM and are expressed as relative optic density.

A hemisection induced for 7 days, produced a considerable upregulation of nNOS-IR in lamina X of the lower lumbar (Figure 4) and sacral segments both on the contralateral and the ipsilateral side of spinal cord injury. Considering the shape and extent of the densitograms taken from somata in lamina X, nNOS-IR seems to be low in control neurons (Figure 5A) and appeared to be more intensely stained for nNOS-IR in neurons on the contralateral and the ipsilateral side (Figure 5B,C). However, the number of nNOS-IR neurons in lamina X was not significantly changed (Table 5).

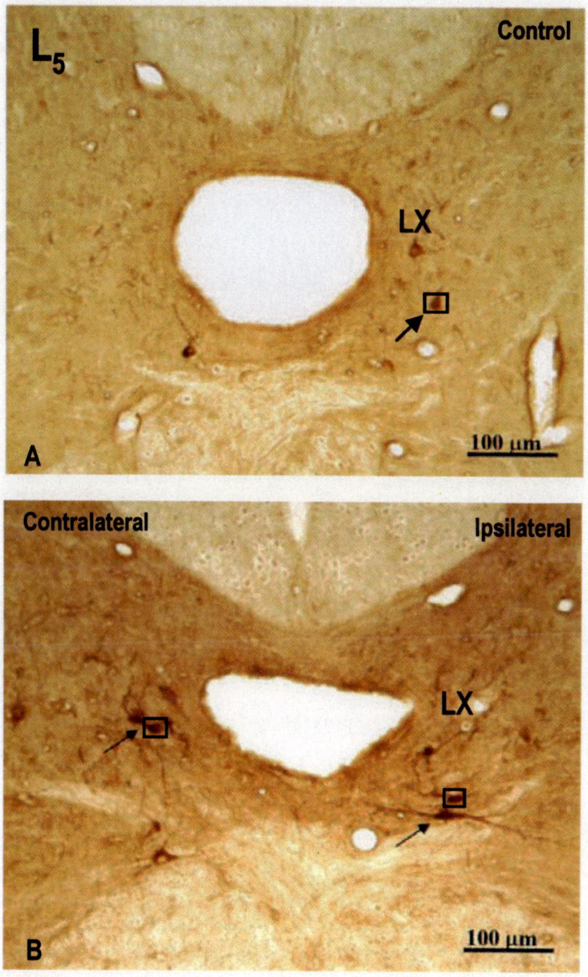

Figure 4. Microphotographs showing the distribution of nNOS immunostaining in lamina X of L5 segment of the spinal cord. A: nNOS immunoreactivity in control; arrow points to immunolabelled somata in lamina X (LX). B: a transverse section cut through the L5 segment of the rabbit subjected to Th10-Th11 hemisection and 7 days survival. The up-regulation of nNOS (arrows) in the projection neurons was observed on both the contralateral and ipsilateral side of lamina X (LX).

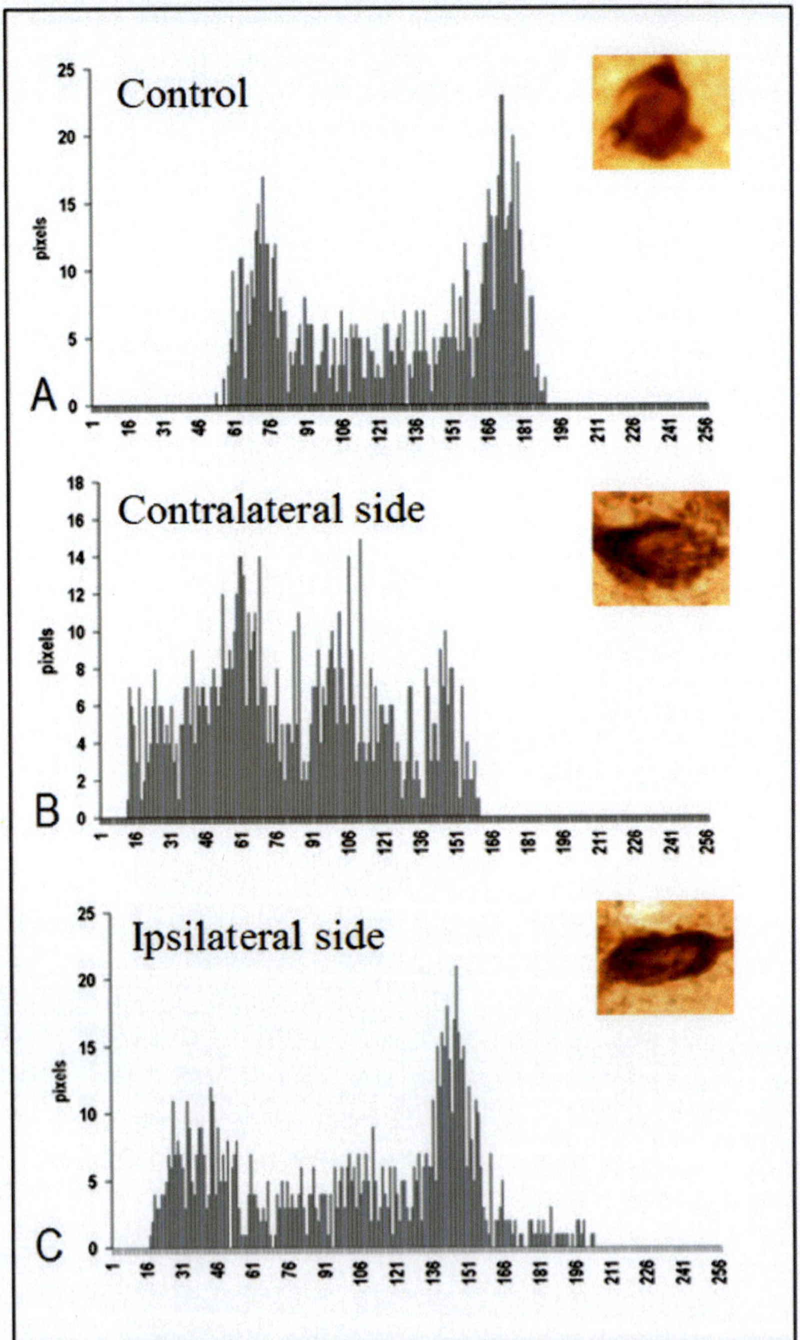

Figure 5. A high-power microphotographs of the boxed areas seen in Figure 4 showing microphotographs and densitograms of nNOS immunostained somata in lamina X of L5 segment after Th10-Th11 hemisection followed by 7 days. (A) control; (B) contralateral side and (C) ipsilateral side to hemisection. Densitometric analysis of nNOS immunostaining was performed on a PC using the UTHSCSA *Image Tool* program, which enabled a reliable detection of subtle differences in the optical density of nNOS immunostaining. The average optical density of the pixels contained within the boundaries of each outlined nNOS immunoreactive somata using a 256-unit gray scale (0=black, 256=white).

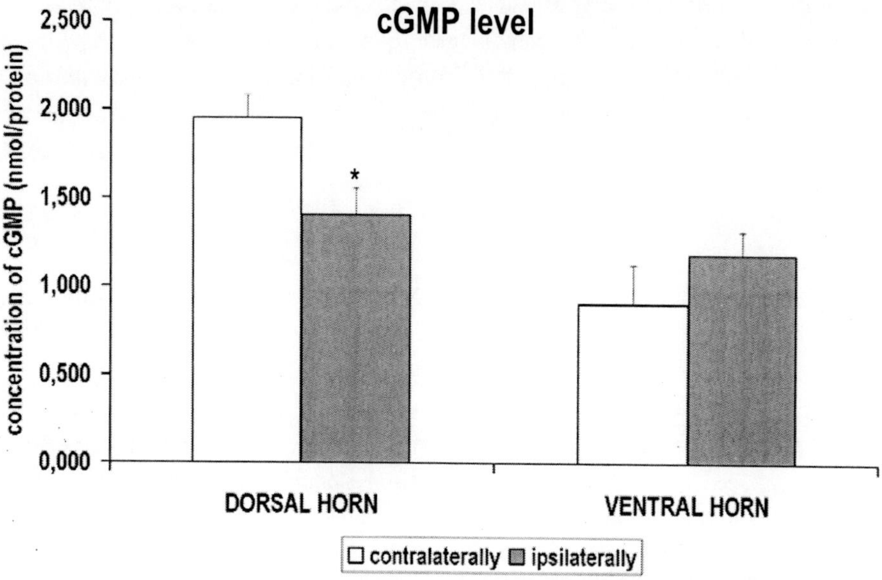

Figure 6. The cGMP level in the dorsal and ventral horn of the rabbit's spinal cord after hemisection at Th9-Th10 level followed by 7 days survival. A radioimmunoassay kit (Immunotech, Ca#1118) was used to assay cGMP, based on the competition between the succinylated cGMP of the sample and a ^{125}I-labeled tracer for binding to the polyclonal antibody [Steiner et al., 1972]. Data are the mean values of 3 experiments ± S.E.M. The results were expressed as nmol cGMP/mg protein and statistically evaluated by ANOVA as well as by Tukey-Kramer test; *$p<0.05$ with respect to controls; contralaterally-contralateral side to hemisection; ipsilaterally-ipsilateral side to hemisection.

A close correlation was seen between the constitutive NOS activity, measured by radioassay detection in the dorsal and ventral parts (Figure 2) and the cGMP level, assessed by radioimmunoassay in the dorsal and ventral horns (Figure 6) in the cervicothoracic segments. Only slight, non- significant differences were detected in both biochemical parameters forming NO/cGMP signaling pathway.

Our study demonstrates that NO represents a neuromodulator that may influence the NOS pools in remote parts of the spinal cord and supports the participation of NO in the processing of the information arising from the initial insult to affected neurons located far beyond the original lesion in both the anterograde and retrograde directions.

NO Formation in Acute and Chronic Spinal Cord Injuries

Although NO generated by constitutive NO synthases has a variety of physiological functions [Dawson & Dawson, 1994], under conditions of an inappropriate activation, NO may act as a mediator of NMDA receptor- induced toxicity [Dawson & Snyder, 1994; Strosznajder et al., 1994] providing the substrate for an abrupt formation of peroxynitrite. Such process can cause an irreversible damage resulting in a decrease or a complete loss of the neuronal viability. In order to investigate the mechanism of a delayed progressive or secondary neuronal damage after the spinal cord injury and the effects of experimental treatment, several models mimicking the acute and chronic mechanical compression of the spinal cord were developed in the rat, rabbit and dog [Nystrom et al., 1988; Sharma & Olsson, 1990; Maršala et al., 1995; Lukáčová et al., 2000; Takenobu et al., 2001; Vanický et al., 2001; Morino et al., 2003].

Hemisection at Thoracic Level Influences NO Pools Close to the Epicenter of Injury

With regard to the pathogenesis of the spinal cord injury numerous studies have been performed to explore the involvement of NO in areas located close to the epicenter of the impact site [Wu et al., 1994; Sharma et al., 1996; Callsen-Cencic et al., 1999; Chatzipanteli et al., 2002; Lukáčová et al., 2000; Lukáčová et al., 2001; Lukáčová et al., 2002b]. An upregulation of NOS expression was seen in interneurons located especially rostrally to the lesion [Wu et al., 1994; Sharma et al., 2000]. Spinal cord injury at Th7 segment caused a significant increase of cNOS activity in the lateral columns just below the site of constriction and a strong decrease in the ventral columns located above the primary injury site [Lukáčová et al., 2000]. Similarly, a time-dependent decrease in cNOS activity and an iNOS expression was noted in a traumatized Th10 segment and in segments located in close vicinity of the spinal cord injury [Chatzipanteli et al., 2002]. Our new data show a remarkable decrease of Ca^{2+}-dependent NOS activity in the injured site analyzed 1 and 7 days after spinal cord hemisection (Th9-Th10) in the rabbit (Figure 7). During the 7th day the enzyme activity returned approximately to a half of the control level. Inducible NOS enzyme activity was not detectable in sham-operated controls and it increased considerably at the injured site of the spinal cord in both postoperative times studied. It seems possible that a rapid formation and a release of NO in a close vicinity to the injury may accelerate neurodestructive events and lead to a secondary inactivation of constitutive NOS, possibly caused by NO binding to the hem iron of NOS [Rogers & Ignarro, 1992] and/or by the phosphorylation of NOS by proteinkinases [Dalkara & Moskowitz, 1997].

The changes in the constitutive and inducible NOS activities, NOS-IR, NOS protein and mRNA, observed after a focal trauma of the spinal cord within the injury and in the surrounding tissue point to a possibility that NO may represent a critical factor determining the extent of the functional impairment.

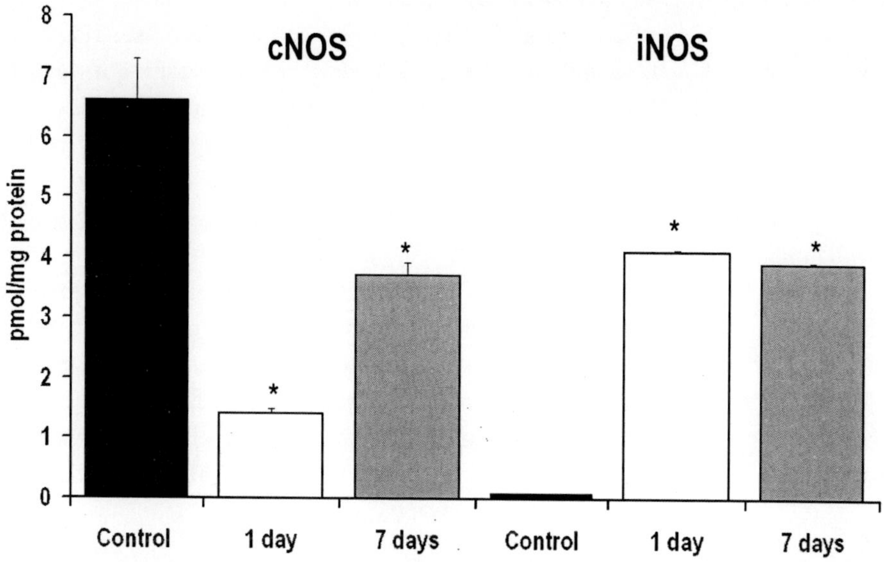

Figure 7. Time course of the constitutive (cNOS) and inducible (iNOS) nitric oxide synthase activity within the hemisection performed at Th9-Th10 segment of the spinal cord of the rabbit. The enzyme activity was detected by conversion of ^{14}[C]arginine to ^{14}[C]citrulline [Bredt et al., 1991] in the spinal cord segments of control and at 1 and 7 days of hemisection. For inducible NOS, $CaCl_2$ was ommitted from the incubation buffer. The results are expressed as pmol/mg protein. Data are given as means ± SEM (n=6); *$p<0.05$ with respect to control.

Midthoracic Constriction and NO/cGMP Signal Transduction System

Experimental evidence now suggests that the NO/cGMP signaling pathway may be implicated in the pathology of the CNS white matter [Garthwaite et al., 1999a,b; Pavel et al., 2001]. A surgically-induced Th7 constriction of 24 hour duration stimulated both the constitutive NOS activity and cGMP level by 120 % and 131 % in non-compartmentalized white matter of Th8-Th9 segments, located just caudally to the site of injury [Lukáčová et al., 2002b]. In the dorsal column, i.e. in the region of the spinal cord composed entirely of long ascending axons, NO/cGMP formation tended to increase in segments located just rostrally and caudally to the spinal cord constriction performed at Th7 level by the constriction of the dural sac produced by three loosely tied ligations with about 2 mm spacing (Figure 8). A significant increase of NO/cGMP formation in the lateral column below the site of the Th7 constriction speaks in favour of a decrease in the orthograde axoplasmic transport of ascending NOS immunoreactive axons, which clearly prevail in the lateral column in comparison with descending ones. The NOS-IR neurons of the intermediolateral cell column sending many NOS-IR dendrites into the lateral column [Maršala et al., 1998; Maršala et al., 1999] together with the NOS-IR axons of the lateral spinothalamic tract may, upon stimulation, also provide the substrate for NO/cGMP formation. A significant decrease in the rate of cGMP formation in the ventral column at Th5-Th6 segment level seems to correlate closely with a significant decrease of the constitutive NOS activity attributable to a slight constriction of the superficial portion of the ventral column at Th7 level. It also seems to correlate closely with a decreasing rate of the orthograde axoplasmic transport of the NOS affecting thick, highly NOS-IR axons [Lukáčová et al., 2000]. These axons form an ascending and, to a lesser extent, a descending propriospinal bundle located in the ventral column, connecting the sacral and lower lumbar segments with the cervical enlargement [Sherrington & Laslett, 1903; Barilari & Kuypers, 1969; Yezierski et al., 1980, Maršala et al., 2004]. The majority of NOS-immunoreactive and NADPHd-exhibiting axons passing through the ventral column of the white matter may originate from NOS-IR and NADPHd-exhibiting somata located in the medial and dorsomedial portion of the ventral horn, pericentral region and medial part of the intermediate zone of the sacral and lower lumbar segments [Maršala et al., 1998; Maršala et al., 1999; Maršala et al., 2001a]. Moreover, a considerably higher level of NOS activity assessed by a radioassay method in the ventral column of the lumbosacral segments in comparison to cervical ones supports the conclusions based on immunocytochemical and histochemical studies [Lukáčová et al., 1999; Lukáčová & Pavel, 2000].

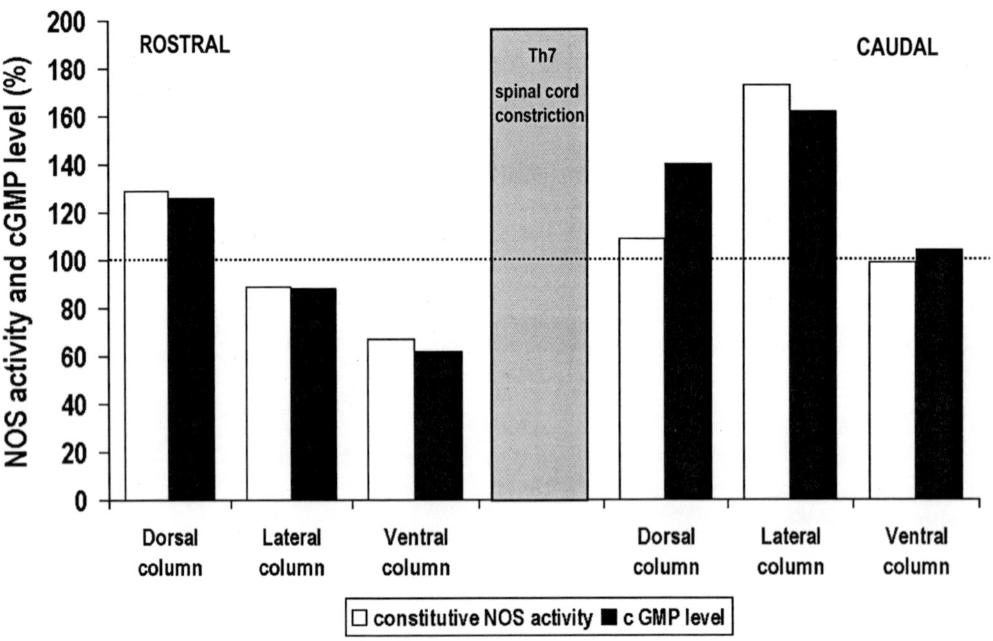

Figure 8. The percentage distribution of constitutive NOS activity and cGMP level in the dorsal, lateral and ventral columns of the white matter in the Th5-Th6 and Th8-Th9 segments of the rabbit´s spinal cord in control and in response to 24 h constriction of Th7 segment. The results of constitutive NOS activity (expressed as dpm/min/mg protein) and cGMP level (expressed as nmol cGMP/mg protein) in controls were taken as 100%. The values were statistically evaluated by ANOVA as well as the Tuckey-Kramer test; *$p<0.05$ with respect to competent controls; Data are given as means ($n=5$) ± SEM; NOS-nitric oxide synthase; cGMP-cyclic guanosine monophosphate.

Sources of NO/cGMP Formation in the White Matter after Midthoracic Constriction

There are some other sources that may account for the NO-dependent cGMP formation in the white matter located in close vicinity to the epicenter of the spinal cord injury. Very high levels of glutamate found at the site of the lesion in the spinal cord [Mc Adoo et al., 1997] and an excessive accumulation of intracellular Ca^{2+} [Lehotský et al., 1999] could result in stimulation of NOS thus creating suitable conditions for NO synthesis and its possible adverse effects on the white matter. An excessive production of NO and superoxide, appearing early after spinal cord injury near the compression site, may stimulate the peroxynitrite formation that in turn may break up, thus giving nitrogen dioxide and a compound having a hydroxyl-like reactivity leading to oxidation of protein, DNA and membrane-damaging lipid peroxidation [Lukáčová et al., 1994]. Moreover, the release of the free fatty acids, particularly arachidonic one in CNS injury [Horrocks et al., 1984; Halát et al., 1987; Nakano et al., 1990] is known to lead to an overexpression of NOS accompanied by an increase in cellular cGMP [Strosznajder & Chalimoniuk, 1996; Toborek et al., 2000]. Recently proposed data have suggested that in the primary cell cultures from the cerebellum of neuronal NOS knockout mice, norepinephrine is able to stimulate cGMP formation independently from sGC [Morton & Bredt, 1998]. Moreover, selective inhibitor of the soluble form of guanylyl cyclase points to a possibility that extracellular cGMP released under basal conditions in the brain might come from both NO-sensitive and NO-insensitive guanylyl cyclases [Vallebuona & Raiteri, 1993; Luo et al., 1994; Fedele et al., 1996]. The elevation of cGMP levels could be elicited by the atrial natriuretic factor, possibly localized in astroglial cells (De Vente et al., 1990). These non-neuronal elements form a dense perinodal processes approaching the node of Ranvier [Ransom et al., 1993] and, as such are located at sites known for a high rate of transmembrane ionic traffic between the axolema and the extracellular space [Stys et al., 1991; Stys & Steffensen, 1996].

Mechanisms of NO Mediated Neurotoxicity

Spinal cord neurons are particularly susceptible to toxic effects of excitatory amino acids. At the site of injury, the excitatory amino acid glutamate is markedly released from glutamatergic neurons to extracellular space and stimulates *N*-methyl-*D*-aspartate (NMDA) receptors located on nitrergic neurons, thus enhancing the release of NO [Prast & Philippu, 2001]. Both NO and glutamate when overstimulated can initiate a neurotoxic cascade.

In principle, the primary injury evokes an inflammatory response marked by generation of tumor necrosis factor-α, the accumulation and adhesion of neutrophils and the activation of endothelial cells and macrophages at the site of injury [Taoka et al., 1997]. These events, in turn, may contribute to neuronal death or injury by inducing the excessive formation of NO and oxygen free radicals [Merrill et al., 1993]. Another very important mediator of secondary CNS injury is oxidative-nitrosative stress [Hall & Braughler, 1993], induced by antioxidative depletion and/or an excess production of oxygen free radicals and NO at the site of injury. NO is able to promote oxidative damage by reacting with superoxide anion to form a strong oxidant, peroxynitrite ($ONOO^-$), and by perturbing ion metabolism [Beckman et al., 1990; Reif & Simmons, 1990]. The rate of peroxynitrite formation is three times faster than the dismutation of the superoxide anion to hydrogen peroxide, the reaction catalyzed by superoxide dismutase (SOD) [Dawson & Dawson, 1996]. Thus, at appropriate concentrations NO can effectively compete with SOD for superoxide anion. In addition, NOS might generate peroxynitrite directly by producing both NO and superoxide [Iadecola, 1997]. NO can oxidize lipids, proteins and nucleic acid [Dawson et al., 1994] causing spinal cord cytotoxicity either directly through NO or its derived species.

Taken together, spinal cord injury represents a serious clinical problem associated with a lifelong disability and far reaching consequences due to secondary complications. They depend on the stage of evolution of the damage and arise from the initial mechanical insult affecting the cells located far beyond the original lesion.

Colocalization of NOS in the Spinal Cord with other Neurotransmitters

Inasmuch spinal cord neurons possess a variety of receptors that are surrounded by numerous fibers from local, peripheral and supraspinal origins that release neurotransmitters, it is reasonable to suggest that spinal cord NOS is colocalized with several neuropeptides and neurotransmitters. Valtschanoff et al., [1992] have demonstrated the coexistence of NOS with GABA in the nerve terminals in the substantia gelatinosa of the dorsal gray matter of the spinal cord. Strong NOS/NADPH diaphorase stained neurons were detected in the intermediolateral cell column in the thoracic and upper lumbar segments [Vincent et al., 1989; Blottner & Baumgarten, 1992; Maršala et al., 1998; Maršala et al., 1999] and any of them were colocalized with choline acetyltransferase [Blottner & Baumgarten, 1992]. Indeed, it appears that most sympathetic and sacral parasympathetic preganglion neurons that are NOS/NADPHd positive [Anderson, 1992; Maršala et al., 1999] posses acetylcholine, the classical transmitter of all preganglionic neurons, cholecystokinin (CCK), calcitonin-gene related peptide (CGRP), leu-enkephalin and somatostatin.

In addition, some dorsal root ganglion neurons containing substance P and CGRP at lower thoracic levels, contain NADPH-diaphorase activity [Aimi et al., 1991] and NOS-IR (unpublished data). In a model of ligated and transected sciatic nerve, the expression of preprotachykinin gene was differently regulated in medium and large cells of the lower lumbar dorsal root ganglia. A newly synthetized substance P and NOS were demonstrated in the large dorsal root ganglion neurons after a sciatic nerve axotomy and nerve ligation [Noguchi et al., 1995; Lukáčová et al., 1993; Čížková et al., 2002]. Sciatic nerve transection exhibits CGRP, p75 or growth-associated protein-43 (GABA-43) in the rat spinal motoneurons without any NOS expression [Piehl et al., 1998]. After root avulsion, the expression of galanin message-associated peptide (GMAP) and NOS in spinal motoneurons is not associated with an upregulation of CGRP or a low-affinity nerve growth factor (p78) [Piehl et al., 1998].

The coexpression of NOS neurons with other neuropeptides or proteins occurring in the spinal cord after some spinal cord lesions seems to be very complex. However, much more research is needed to understand the intraspinal regulation and interrelation between NOS and other spinal cord injury-induced factors.

Conclusion

New results summarised here demonstrate that changes in NOS activity after spinal cord injury may have adverse effects on nitric oxide (NO) production in the white matter close to the site of constriction injury, thus broadening the scope of the secondary mechanisms that play a role in neuronal trauma and, in neurons located far away from the injury. An unexpectedly high number of NOS immunoreactive axons of varying thickness, confirmed also by a radioassay study, was identified in all white matter columns, i.e., the lateral, ventral and dorsal columns. Further experiments revealed that one portion of NOS immunoreactive axons located in the ventral and lateral columns of upper cervical segments (C1-C4) represents fact the bulbospinal respiratory descending pathway connecting monosynaptically the bulbar respiratory centers with the motoneurons of the phrenic nucleus located in the ventral horn of (C3-C5) segments. Second, a well recognized NOS immunoreactive axonal bundle took origin from NOS-IR somata located in the intermediate zone and medial part of the ventral horn of the lumbosacral segments, and, after trajectory in the ventral and lateral columns of the upper lumbar and thoracic segments terminates monosynaptically around the motoneurons of the ventral motor nucleus in C6-Th1 segments. Thus connecting functionally significant lumbar and cervical enlargement. Third, a component containing large NOS immunoreactive axons was found in the dorsal column of the lumbar and sacral segments and, with regard to fiber composition, orientation and intramedullar termination is thought to represent the monosynaptic Ia-motoneurons pathway forming the afferent limb of the stretch H-reflex and, as such, is responsible for proprioceptive afferentation originating peripherally from the muscle spindles and terminating around homonymous motoneurons.

Acknowledgements

The experimental work was supported by the VEGA Grants No. 2/2079/22, 2/5134/25, 2/3217/23 from the SAS and STAA Grant No. 51-013002.

References

Aimi, Y., Fujimura, M., Vincent, S. R. & Kimura, H. (1991). Localization of NADPH-diaphorase-containing neurons in sensory ganglia of the rat. *J. Comp. Neurol. 306,* 382-392.

Allaerts, W., De Vente, J., Markerink-Van Ittersum, M., Tuinhof, R. & Roubos, E. W. (1998). Topographical relationship between neuronal nitric oxide synthase immunoreactivity and cyclic 3',5'-guanosine monophosphate accumulation in the brain of the adult Xenopus laevis. *J. Chem. Neuroanat. 15,* 41-56.

Altman, J. & Bayer, S. A. (2001). Development of the Human Spinal Cord. *An Interpretation Based on Experimental Studies in Animals.* Oxford University Press.

Anderson, C. R. (1992). NADPH diaphorase-positive neurons in the rat spinal cord include a subpopulation of autonomic preganglionic neurons. *Neurosci. Lett. 139,* 280-284.

Archer, S. (1993). Measurement of nitric oxide in biological models. *FASEB J. 7,* 349-360.

Barilari, M. G. & Kuypers, J. M. (1969). Propriospinal fibers interconnecting the spinal enlargements in the cat. *Brain Res. 14,* 321-330.

Beckman, J. S., Beckman, T. W., Chen, J., Marshall, P. A. & Freeman, B. A. (1990). Apparent hydroxyl radical production by peroxynitrite: implications for endothelial injury from nitric oxide and superoxide. *Proc. Natl. Acad. Sci. U. S. A. 87,* 1620-1624.

Blot, S., Arnal, J. F., Xu, Y., Gray, F. & Michel, J. B. (1994). Spinal cord infarcts during long-term inhibition of nitric oxide synthase in rats. *Stroke 25,* 1666-1673.

Blottner, D. & Baumgarten, H. G. (1992). Nitric oxide synthetase (NOS)-containing sympathoadrenal cholinergic neurons of the rat IML-cell column: evidence from histochemistry, immunohistochemistry, and retrograde labeling. *J. Comp. Neurol. 316,* 45-55.

Bredt, D. S. & Snyder, S. H. (1992). Nitric oxide, a novel neuronal messenger. *Neuron 8,* 3-11.

Bredt, D. S., Glatt, C. E., Hwang, P. M., Fotuhi, M., Dawson, T. M. & Snyder, S. H. (1991). Nitric oxide synthase protein and mRNA are discretely localized in neuronal populations of the mammalian CNS together with NADPH diaphorase. *Neuron 7,* 615-624.

Brüning, G. (1992). Localization of NADPH diaphorase, a histochemical marker for nitric oxide synthase, in the mouse spinal cord. *Acta Histochem. 93,* 397-401.

Brustein, E. & Rossignol, S. (1998). Recovery of locomotion after ventral and ventrolateral spinal lesions in the cat. I. Deficits and adaptive mechanisms. *J. Neurophysiol. 80,* 1245-1267.

Callsen-Cencic, P., Hoheisel, U., Kaske, A., Mense, S. & Tenschert, S. (1999). The controversy about spinal neuronal nitric oxide synthase: under which conditions is it up- or downregulated? *Cell Tissue Res. 295,* 183-194.

Chatzipanteli, K., Garcia, R., Marcillo, A. E., Loor, K. E., Kraydieh, S. & Dietrich, W. D. (2002). Temporal and segmental distribution of constitutive and inducible nitric oxide synthases after traumatic spinal cord injury: effect of aminoguanidine treatment. *J. Neurotrauma 5,* 639-651.

Čížková, D., Lukáčová, N., Maršala, M. & Maršala, J. (2002). Neuropathic pain is associated with alterations of nitric oxide synthase immunoreactivity and catalytic activity in dorsal root ganglia and spinal dorsal horn. *Brain Res. Bull. 58,* 161-171.

Dalkara, T. & Moskowitz, M. A. (1997). Neurotoxic and neuroprotective roles of nitric oxide in cerebral ischaemia. *Int. Rev. Neurobiol. 40*, 319-336.

Dawson, T. M. & Dawson, V. L. (1994). gp120 neurotoxicity in primary cortical cultures. *Adv. Neuroimmunol. 4,* 167-173.

Dawson, T. M. & Snyder, S. H. (1994). Gases as biological messengers: nitric oxide and carbon monoxide in the brain. *J. Neurosci. 14,* 5147-5159.

Dawson, T. M., Bredt, D. S., Fotuhi, M., Hwang, P. M. & Snyder, S. H. (1991). Nitric oxide synthase and neuronal NADPH diaphorase are identical in brain and peripheral tissues. *Proc. Natl. Acad. Sci. U.S.A. 88,* 7797-7801.

Dawson, T. M., Dawson, V. L. & Snyder, S. H. (1994). Molecular mechanisms of nitric oxide actions in the brain. *Ann. N. Y. Acad. Sci.738,* 76-85.

Dawson, T. M., Dawson, V. L. & Snyder, S. H. (1992). A novel neuronal messenger molecule in brain: the free radical, nitric oxide. *Ann. Neurol. 32,* 297-311.

Dawson, V. L. & Dawson, T. M. (1996). Nitric oxide actions in neurochemistry. *Neurochem. Int. 29,* 97-110.

de Leon, R. D., Hodgson, J. A., Roy, R. R. & Edgerton, V. R. (1998a). Locomotor capacity attributable to step training versus spontaneous recovery after spinalization in adult cats. *J. Neurophysiol. 79,* 1329-1340.

de Leon, R. D., Hodgson, J. A., Roy, R. R. & Edgerton, V. R. (1998b). Full weight-bearing hindlimb standing following stand training in the adult spinal cat. *J. Neurophysiol. 80,* 83-91.

de Leon, R. D., Hodgson, J. A., Roy, R. R. & Edgerton, V. R. (1999). Retention of hindlimb stepping ability in adult spinal cats after the cessation of step training. *J. Neurophysiol. 81,* 85-94.

de Leon, R. D., Roy, R. R., Gomez-Pinilla, F., Motamedinia, P. & Edgerton, V. R. (2004). Interlimb coordination after a thoracic hemisection in rats. *Basic Appl. Myol. 14,* 215-222.

De Vente, J., Manshanden, C. G., Sikking, R. A., Ramaekers, F. C. & Steinbusch, H. W. (1990). A functional parameter to study heterogeneity of glial cells in rat brain slices: cyclic guanosine monophosphate production in atrial natriuretic factor (ANF)-responsive cells. *Glia 3,* 43-54.

Denninger, J. W. & Marletta, M. A. (1999). Guanylate cyclase and the NO/cGMP signaling pathway. *Biochim. Biophys. Acta 1411,* 334-350.

Dietz, V., 2002. Proprioception and locomotor disorders. *Nature/Neuroscience 3,* 781-790.

Dun, N. J., Dun, S. L., Förstermann, U. & Tseng, L. F. (1992). Nitric oxide synthase immunoreactivity in rat spinal cord. *Neurosci. Lett. 147,* 217-220.

Dun, N. J., Dun, S. L., Wu, S. Y., Förstermann, U., Schmidt, H. H. & Tseng, L. F. (1993). Nitric oxide synthase immunoreactivity in the rat, mouse, cat and squirrel monkey spinal cord. *Neuroscience 54,* 845-857.

Edgerton, V. R. & Roy, R. R. (2002). Paralysis recovery in humans and model systems. *Curr. Opin. Neurobiol. 12,* 658-667.

Egberongbe, Y. I., Gentleman, S. M., Falkai, P., Bogerts, B., Polak, J. M. & Roberts, G. W. (1994). The distribution of nitric oxide synthase immunoreactivity in the human brain. *Neuroscience 59,* 561-578.

Faraci, F. M. (1990). Role of nitric oxide in regulation of basilar artery tone in vivo. *Am. J. Physiol. 259,* H1216-H1221.

Fedele, E., Jin, Y., Varnier, G. & Raiteri, M. (1996). In vivo microdialysis study of a specific inhibitor of soluble guanylyl cyclase on the glutamate receptor/nitric oxide/cyclic GMP pathway. *Br. J. Pharmacol. 119,* 590-594.

Feldman, J. L. (1986). Neurophysiology of breathing in mammals. In *Handbook of Physiology Sect. I., The Nervous System,* Vol. IV, pp. 463-524. American Physiol. Soc.

Förstermann, U., Boissel, J. P. & Kleinert, H. (1998). Expressional control of the 'constitutive' isoforms of nitric oxide synthase (NOS I and NOS III). *FASEB J. 12,* 773-790.

Förstermann, U., Pollock, J. S., Schmidt, H. H., Heller, M. & Murad, F. (1991). Calmodulin-dependent endothelium-derived relaxing factor/nitric oxide synthase activity is present in the particulate and cytosolic fractions of bovine aortic endothelial cells. *Proc. Natl. Acad. Sci. U. S. A. 88,* 1788-1792.

Furchgott, R. F. & Zawadzki, J. V. (1980). The obligatory role of endothelial cells in the relaxation of arterial smooth muscle by acetylcholine. *Nature 288,* 373-376.

Garthwaite, G., Brown, G., Batchelor, A. M., Goodwin, D. A. & Garthwaite, J. (1999a). Mechanisms of ischaemic damage to central white matter axons: a quantitative histological analysis using rat optic nerve. *Neuroscience 94,* 1219-1230.

Garthwaite, G., Goodwin, D. A. & Garthwaite, J. (1999b). Nitric oxide stimulates cGMP formation in rat optic nerve axons, providing a specific marker of axon viability. *Eur. J. Neurosci. 11,* 4367-4372.

Garthwaite, J. & Boulton, C. L. (1995). Nitric oxide signaling in the central nervous system. *Annu. Rev. Physiol. 57,* 683-706.

Garthwaite, J., Charles, S. L. & Chess-Williams, R. (1988). Endothelium-derived relaxing factor release on activation of NMDA receptors suggests role as intercellular messenger in the brain. *Nature 336,* 385-388.

Geller, D. A., Nussler, A. K., Di Silvio, M., Lowenstein, C. J., Shapiro, R. A., Wang, S. C., Simmons, R. L. & Billiar, T. R. (1993). Cytokines, endotoxin, and glucocorticoids regulate the expression of inducible nitric oxide synthase in hepatocytes. *Proc. Natl. Acad. Sci. U. S. A. 90,* 522-526.

George, W. J., Polson, J. B., Ag, O. T. & Goldberg, N. D. (1970). Elevation of guanosine 3',5'-cyclic phosphate in rat heart after perfusion with acetylcholine. *Proc. Natl. Acad. Sci. U. S. A. 66,* 398-403.

Goshgarian, H. G. & Rafols, J. A. (1984). The ultrastructure and synaptic architecture of phrenic motor neurons in the spinal cord of the adult rat. *J. Neurocytol. 13,* 85-109.

Goshgarian, H. G. & Rafols, J. A. (1981). The phrenic nucleus of the albino rat: a correlative HRP and Golgi study. *J. Comp. Neurol. 201,* 441-456.

Halát, G., Lukáčová, N., Chavko, M. & Maršala, J. (1987). Effects of incomplete ischemia and subsequent recirculation on free palmitate, stearate, oleate and arachidonate levels in lumbar and cervical spinal cord of rabbit. *Gen. Physiol. Biophys. 6,* 387-399.

Hall, E. D. & Braughler, J. M. (1993). Free radicals in CNS injury. *Res. Publ. Assoc. Res. Nerv. Ment. Dis. 71,* 81-105.

Hecker, M., Mulsch, A., Bassenge, E., Förstermann, U. & Busse, R. (1994). Subcellular localization and characterization of nitric oxide synthase(s) in endothelial cells: physiological implications. *Biochem. J. 299,* 247-252.

Hecker, M., Walsh, D. T. & Vane, J. R. (1991). On the substrate specificity of nitric oxide synthase. *FEBS Lett. 294,* 221-224.

Hobbs, A. J. (1997). Soluble guanylate cyclase: the forgotten sibling. *Trends Pharmacol. Sci. 18,* 484-491.

Horrocks, L. A., Dorman, R. V. & Porcellati, G. (1984). Fatty acids and phospholipids in brain during ischemia. In: P. Bes, R. Braquet, R. Paoletti & B. K. Siesjö (Eds.), *Cerebral Ischemia Excerpta Medica* (Congress Series 654A, pp. 211-226). Amsterdam, Elsevier.

Iadecola, C. (1997). Bright and dark sides of nitric oxide in ischemic brain injury. *Trends Neurosci. 20,* 132-139.

Ignarro, L. J., Buga, G. M., Byrns, R. E., Wood, K. S. & Chaudhuri, G. (1988). Endothelium-derived relaxing factor and nitric oxide possess identical pharmacologic properties as relaxants of bovine arterial and venous smooth muscle. *J. Pharmacol. Exp. Ther. 246,* 218-226.

Iles, J. F. & Roberts, R. C. (1987). Inhibition of monosynaptic reflexes in the human lower limb. *J. Physiol. 385,* 69-87.

Lee, J. H., Price, R. H., Williams, F. G., Mayer, B. & Beitz, A. J. (1993). Nitric oxide synthase is found in some spinothalamic neurons and in neuronal processes that appose spinal neurons that express Fos induced by noxious stimulation. *Brain Res. 608,* 324-333.

Lehotský, J., Kaplán, P., Račay, P., Mézešová, V. & Raeymaekers, L. (1999). Distribution of plasma membrane Ca2+ pump (PMCA) isoforms in the gerbil brain: effect of ischemia-reperfusion injury. *Neurochem. Int. 35,* 221-227.

Lukáčová, N. & Pavel, J. (2000). Catalytic nitric oxide synthase activity in the white and gray matter regions of the spinal cord of rabbits. *Physiol. Res. 49,* 167-173.

Lukáčová, N., Chavko, M. & Halát G. (1993). Effect of stobadine on lipid peroxidation and phospholipids in rabbit spinal cord after ischaemia. *Neuropharmacology 32,* 235-241.

Lukáčová, N., Chavko, M. & Maršala, J. (1994). Ischemia/reperfusion-induced lipid peroxidation in the central nervous system. *Biologia 49,* 871-879.

Lukáčová, N., Čížková, D., Križanová, O., Pavel, J., Maršala, M. & Maršala, J. (2003). Peripheral axotomy affects nicotinamide adenine dinucleotide phosphate diaphorase and nitric oxide synthases in the spinal cord of the rabbit. *J. Neurosci. Res. 71,* 300-313.

Lukáčová, N., Čížková, D., Maršala, M., Jalč, P. & Maršala, J. (1999). Segmental and laminar distributions of nicotinamide adenine dinucleotide phosphate-diaphorase-expressing and neuronal nitric oxide synthase-immunoreactive neurons versus radioassay detection of catalytic nitric oxide synthase activity in the rabbit spinal cord. *Neuroscience 94,* 229-237.

Lukáčová, N., Čížková, D., Maršala, M., Lukáč, I. & Maršala, J. (2002a). The regional distribution of nitric oxide synthase activity in the spinal cord of the dog. *Brain Res. Bull. 58,* 173-178.

Lukáčová, N., Čížková, D., Maršala, M., Pavel, J., Jalč, P., Šulla, I., Kafka, J. & Maršala, J. (2000). Effect of midthoracic spinal cord constriction on catalytic nitric oxide synthase activity in the white matter columns of rabbit. *Neurochem. Res. 25,* 1139-1148.

Lukáčová, N., Pavel, J., Jalč, P., Čížková, D., Maršala, M., Lukáč, I., Chalimoniuk, M., Strosznajder, J. & Maršala, J. (2001). Effect of spinal cord compression on cyclic 3',5'-guanosine monophosphate in the white matter columns of rabbit. *Neurochem. Int. 39,* 275-282.

Lukáčová, N., Pavel, J., Jalč, P., Čížková, D., Maršala, M., Silvašiová, M., Lukáč I. & Maršala, J. (2002b). The effect of spinal cord constriction on NOS-cGMP signaling pathway in the white matter of rabbit. *Biologia 57*, 789-795.

Luo, D., Leung, E. & Vincent, S. R. (1994). Nitric oxide-dependent efflux of cGMP in rat cerebellar cortex: an in vivo microdialysis study. *J. Neurosci. 14,* 263-271.

Maršala, J., Lukáčová, N., Šulla, I., Wohlfahrt, P. & Maršala, M. (2005). The evidence for nitric oxide synthase immunopositivity in the monosynaptic Ia-motoneuron pathway. *Exp. Neurol., (in press).*

Maršala, J., Čížková, D., Kafka, J., Lukáčová, N., Lukáč, I. & Maršala, M. (2001a). Densitometric patterns of NADPH diaphorase staining in the spinal cord of dog. *Biologia 56,* 681-689.

Maršala, J., Čížková, D., Lukáč, I., Lukáčová, N. & Maršala, M. (2001b). Ascending nitrergic propriospinal connections in the spinal cord of the dog. *Soc. Neurosci. Abstr. 27,* Program No. 517.11.

Maršala, J., Kafka, J., Lukáčová, N., Čížková, D., Maršala, M. & Katsube, N. (2003a). Cauda equina syndrome and nitric oxide synthase immunoreactivity in the spinal cord of the dog. *Physiol Res. 52,* 481-496.

Maršala, J., Lukáčová, N., Čížková, D., Kafka, J., Katsube, N., Kuchárová, K. & Maršala, M. (2002). The case for the bulbospinal respiratory nitric oxide synthase-immunoreactive pathway in the dog. *Exp. Neurol. 177,* 115-132.

Maršala, J., Lukáčová, N., Čížková, D., Lukáč, I., Kuchárová, K. & Maršala, M. (2004). Premotor nitric oxide synthase immunoreactive pathway connecting lumbar segments with the ventral motor nucleus of the cervical enlargement in the dog. *J. Chem. Neuroanat. 27,* 43-54.

Maršala, J., Maršala, M., Lukáčová, N., Ishikawa, T. & Čížková, D. (2003b). Localization and distribution patterns of nicotinamide adenine dinucleotide phosphate diaphorase exhibiting axons in the white matter of the spinal cord of the rabbit. *Cell. Mol. Neurobiol. 23,* 57-92.

Maršala, J., Maršala, M., Vanický, I. & Taira, Y. (1999). Localization of NADPHd exhibiting neurons in the spinal cord of the rabbit. *J. Comp. Neurol. 406,* 263-284.

Maršala, J., Šulla, I., Jalč, P. & Orendáčová, J. (1995). Multiple protracted cauda equina constrictions cause deep derangement in the lumbosacral spinal cord circuitry in the dog. *Neurosci. Lett. 193,* 97-100.

Maršala, J., Vanický, I., Maršala, M., Jalč, P., Orendáčová, J. & Taira, Y. (1998). NADPH diaphorase in the spinal cord of dogs. *Neuroscience 85,* 847-862.

Mayer, B., John, M., Heinzel, B., Werner, E. R., Wachter, H., Schultz, G. & Bohme, E. (1991). Brain nitric oxide synthase is a biopterin- and flavin-containing multi-functional oxido-reductase. *FEBS Lett. 288,* 187-191.

Mc Adoo, J., Hughes, M. G., Xu, G. Y., Robak, G. & DeCastro, R. Jr. (1997). Microdialysis studies of the role of chemical agents in secondary damage upon spinal cord injury. *J. Neurotrauma 14,* 507-515.

Meller, S. T., Pechman, P. S., Gebhart, G. F. & Maves, T. J. (1992). Nitric oxide mediates the thermal hyperalgesia produced in a model of neuropathic pain in the rat. *Neuroscience 50,* 7-10.

Merril, J. E., Ignarro, L., Sherman, M. P., Melinek, J. & Lane, T. E. (1993). Microglial cell cytotoxicity of oligodendrocytes is mediated through nitric oxide. *J. Immunol. 151,* 2132-2141.

Miller, S. & van der Burg, J. (1973). The function of long propriospinal pathways in the co-ordination of quadrupedal stepping in the cat. In R. B. Stein, K. G. Pearson, R. S. Smith & J. B. Redford (Eds.), *Control of Posture and Locomotion* (Plenum, pp. 561-557).

Miller, S. & van der Meché, F. G. (1976). Coordinated stepping of all four limbs in the high spinal cat. *Brain Res. 109,* 395-398.

Miyazaki, M., Kayama, Y., Kihara, T., Kawasaki, K., Yamaguchi, E., Wada, Y. & Ikeda, M. (1996). Possible release of nitric oxide from cholinergic axons in the thalamus by stimulation of the rat laterodorsal tegmental nucleus as measured with voltammetry. *J. Chem. Neuroanat. 10,* 203-207.

Moncada, S., Palmer, R. M. & Higgs, E. A. (1991). Nitric oxide: physiology, pathophysiology, and pharmacology. *Pharmacol. Rev. 43,* 109-142.

Morino, T., Ogata, T., Horiuchi, H., Takeba, J., Okumura, H., Miyazaki, T. & Yamamoto, H. (2003). Delayed neuronal damage related to microglia proliferation after mild spinal cord compression injury. *Neurosci. Res. 46,* 309-318.

Morris, R., Southam, E., Gittins, S. R., de Vente, J. & Garthwaite, J. (1994). The NO-cGMP pathway in neonatal rat dorsal horn. *Eur. J. Neurosci. 6,* 876-879.

Morton, D. B. & Bredt, D. S. (1998). Norepinephrine increases cyclic GMP levels in cerebellar cells from neuronal nitric oxide synthase knockout mice. *J. Neurochem. 71,* 440-443.

Murad, F. (1994). The nitric oxide-cyclic GMP signal transduction system for intracellular and intercellular communication. *Recent Prog. Horm. Res. 49,* 239-248.

Murad, F., Mittal, C. K., Arnold, W. P., Katsuki, S. & Kimura, H. (1978). Guanylate cyclase: activation by azide, nitro compounds, nitric oxide, and hydroxyl radical and inhibition by hemoglobin and myoglobin. *Adv. Cyclic Nucleotide Res. 9,* 145-158.

Nakano, S., Kogure, K., Abe, K. & Yae, T. (1990). Ischemia-induced alterations in lipid metabolism of the gerbil cerebral cortex: I. Changes in free fatty acid liberation. *J. Neurochem. 54,* 1911-1916.

Noguchi, K., Kawai, Y., Fukuoka, T., Senba, E. & Miki, K. (1995). Substance P induced by peripheral nerve injury in primary afferent sensory neurons and its effect on dorsal column nucleus neurons. *J. Neurosci. 15,* 7633-7643.

Nystrom, B., Berglund, J. E. & Bergquist, E. (1988). Methodological analysis of an experimental spinal cord compression model in the rat. *Acta Neurol. Scand. 78,* 460-466.

Palmer, R. M., Ashton, D. S. & Moncada, S. (1988a). Vascular endothelial cells synthesize nitric oxide from L-arginine. *Nature 333,* 664-666.

Palmer, R. M., Rees, D. D., Ashton, D. S. & Moncada, S. (1988b). L-arginine is the physiological precursor for the formation of nitric oxide in endothelium-dependent relaxation. *Biochem. Biophys. Res. Commun. 153,* 1251-1256.

Pavel, J., Lukáčová, N. & Maršala, J. (2000). Regional changes of cyclic 3',5'-guanosine monophosphate in the spinal cord of rabbit following brief repeated ischemic insults. *Neurochem. Res. 25,* 1131-1137.

Pavel, J., Lukáčová, N. & Maršala, J. (2001). The effect of repeated sublethal ischemia on NO/cGMP signal transduction system in gray matter of the rabbit spinal cord. *Bratisl. Lek. Listy 102,* 273-277.

Piehl, F., Hammarberg, H., Tabar, G., Hökfelt, T. & Cullheim, S. (1998). Changes in the mRNA expression pattern, with special reference to calcitonin gene-related peptide, after axonal injuries in rat motoneurons depends on age and type of injury. *Exp. Brain Res. 119*, 191-204.

Pou, S., Pou, W. S., Bredt, D. S., Snyder, S. H. & Rosen, G. M. (1992). Generation of superoxide by purified brain nitric oxide synthase. *J. Biol. Chem. 267,* 24173-24176.

Prast, H. & Philippu, A. (2001). Nitric oxide as modulator of neuronal function. *Prog. Neurobiol. 64,* 51-68.

Ransom, B. R., Waxman, S. G. & Stys, P. K. (1993). Anoxic injury of central myelinated axons: ionic mechanisms and pharmacology. *Res. Publ. Assoc. Res. Nerv. Ment. Dis. 71,* 121-151.

Reif, D. W. & Simmons, R. D. (1990). Nitric oxide mediates iron release from ferritin. *Arch. Biochem. Biophys. 283,* 537-541.

Rogers, N. E. & Ignarro, L. J. (1992). Constitutive nitric oxide synthase from cerebellum is reversibly inhibited by nitric oxide formed from L-arginine. *Biochem. Biophys. Res. Commun. 189,* 242-249.

Rossignol, S. & Barbeau, H. (1993). Pharmacology of locomotion: an account of studies in spinal cats and spinal cord injured subjects. *J. Am. Paraplegia Soc.* 16, 190-196.

Saito, S., Kidd, G. J., Trapp, B. D., Dawson, T. M., Bredt, D. S., Wilson, D. A., Traystmann, R. J., Snyder, S. H. & Hanley, D. F. (1994). Rat spinal cord neurons contain nitric oxide synthase. *Neuroscience 59,* 447-456.

Scherer-Singler, U., Vincent, S. R., Kimura, H. & McGeer, E. G. (1983). Demonstration of a unique population of neurons with NADPH diaphorase histochemistry. *J. Neurosci. Methods 8,* 229-234.

Schmid, H. A. & Pehl, U. (1996). Regional specific effects of nitric oxide donors and cGMP on the electrical activity of neurons in the rat spinal cord. *J. Chem. Neuroanat. 10,* 197-201.

Schmidt, H. H., Gagne, G. D., Nakane, M., Pollock, J. S., Miller, M. F. & Murad, F. (1992). Mapping of neural nitric oxide synthase in the rat suggests frequent co-localization with NADPH diaphorase but not with soluble guanylyl cyclase, and novel paraneural functions for nitrinergic signal transduction. *J. Histochem. Cytochem. 40,* 1439-1456.

Sharma, H. S. & Olsson, Y. (1990). Edema formation and cellular alterations following spinal cord injury in the rat and their modification with p-chlorophenylalanine. *Acta Neuropathol. (Berl). 79,* 604-610.

Sharma, H. S., Westman, J., Gordh, T. & Nyberg, F. (2000). Spinal cord injury induced c-fos expression is reduced by p-CPA, a serotonin synthesis inhibitor. An experimental study using immunohistochemistry in the rat. *Acta Neurochir. Suppl. 76,* 297-301.

Sharma, H. S., Westman, J., Olsson, Y. & Alm, P. (1996). Involvement of nitric oxide in acute spinal cord injury: an immunocytochemical study using light and electron microscopy in the rat. *Neurosci. Res. 24,* 373-384.

Sherrington, C. S. & Laslett, E. E. (1903). Observations on some spinal reflexes and the inter connection of spinal segments. *J. Physiol. (London) 29,* 58-96.

Smolenski, A., Burkhardt, A. M., Eigenthaler, M., Butt, E., Gambaryan, S., Lohmann, S. M. & Walter, U. (1998). Functional analysis of cGMP-dependent protein kinases I and II as mediators of NO/cGMP effects. *Naunyn Schmiedebergs Arch. Pharmacol. 358,* 134-139.

Snyder, S. H. & Bredt, D. (1991). Nitric oxide as a neuronal messenger. *Trends Pharmacol. Sci. 12,* 125-128.

Steiner, A. L., Ferrendelli, J. A. & Kipnis, D. M. (1972). Radioimmunoassay for cyclic nucleotides. 3. Effect of ischemia, changes during development and regional distribution of adenosine 3',5'-monophosphate and guanosine 3',5'-monophosphate in mouse brain. *J. Biol. Chem. 247,* 1121-1124.

Strosznajder, J. & Chalimoniuk, M. (1996). Biphasic enhancement of nitric oxide synthase activity and cGMP level following brain ischemia in gerbils. *Acta Neurobiol. Exp. 56,* 71-81.

Strosznajder, J., Chalimoniuk, M., Samochocki, M. & Gadamski, R. (1994). Nitric oxide: a potent mediator of glutamatergic neurotoxicity in brain ischemia. *Ann. N. Y. Acad. Sci. 723,* 429-432.

Stys, P. K. & Steffensen, I. (1996). $Na^{(+)}$-Ca^{2+} exchange in anoxic/ischemic injury of CNS myelinated axons. *Ann. N. Y. Acad. Sci. 779,* 366-378.

Stys, P. K., Waxman, S. G. & Ransom, B. R. (1991). Na^{+}-Ca^{2+} exchanger mediates Ca^{2+} influx during anoxia in mammalian central nervous system white matter. *Ann. Neurol. 30,* 375-380.

Takenobu, Y., Katsube, N., Maršala, M. & Kondo, K. (2001). Model of neuropathic intermittent claudication in the rat: methodology and application. *J. Neurosci. Methods 104,* 191-198.

Taoka, Y., Okajima, K., Uchiba, M., Murakami, K., Kushimoto, S., Johno, M., Naruo, M., Okabe, H. & Takatsuki, K. (1997). Role of neutrophils in spinal cord injury in the rat. *Neuroscience 79,* 1177-1182.

Toborek, M., Garrido, R., Malecki, A., Kaiser, S., Mattson, M. P., Hennig, B. & Young, B. (2000). Nicotine attenuates arachidonic acid-induced overexpression of nitric oxide synthase in cultured spinal cord neurons. *Exp. Neurol. 161,* 609-620.

Vallebuona, F. & Raiteri, M. (1993). Monitoring of cyclic GMP during cerebellar microdialysis in freely-moving rats as an index of nitric oxide synthase activity. *Neuroscience 57,* 577-585.

Valtschanoff, J. G., Weinberg, R. J. & Rustioni, A. (1992). NADPH diaphorase in the spinal cord of rats. *J. Comp. Neurol. 321,* 209-222.

Vanický, I., Urdziková, L., Saganová, K., Čižková, D. & Gálik, J. (2001). A simple and reproducible model of spinal cord injury induced by epidural balloon inflation in the rat. *J. Neurotrauma 18,* 1399-1407.

Vincent, S. R. (1994). Nitric oxide: a radical neurotransmitter in the central nervous system. *Prog. Neurobiol. 42,* 129-160.

Vincent, S. R., Aimi, Y. & Kimura, H. (1989). The distribution of NADPH-diaphorase in the peripheral nervous system. *Neurosci. Abstr. 15,* 376.

Vizzard, M. A., Erdman, S. L., Erickson, V. L., Stewart, R. J., Roppolo, J. R. & De Groat, W. C. (1994). Localization of NADPH diaphorase in the lumbosacral spinal cord and dorsal root ganglia of the cat. *J. Comp. Neurol. 339,* 62-75.

Wang, X. & Robinson, P. J. (1997). Cyclic GMP-dependent protein kinase and cellular signaling in the nervous system. *J. Neurochem. 68,* 443-456.

Wood, P. L. & Garthwaite, J. (1994). Models for the diffusional spread of nitric oxide: implications for neuronal nitric oxide signalling and its pharmacological properties. *Neuropharmacology 33,* 1235-1244.

Wu, W., Liuzzi, F. J., Schinco, F. P., Depto, A. S., Li, Y., Mong, J. A., Dawson, T. M. & Snyder, S. H. (1994). Neuronal nitric oxide synthase is induced in spinal neurons by traumatic injury. *Neuroscience 61,* 719-726.

Xu, M., Ng, Y.-K. & Leong, S. K. (1998). Induction of microglial reaction and expression of nitric oxide synthase I in the nucleus dorsalis and red nucleus following lower thoracic spinal cord hemisection. *Brain Res. 808,* 23-30.

Xu, W., Liu, L., Smith, G. C. & Charles, G. (2000). Nitric oxide upregulates expression of DNA-PKcs to protect cells from DNA-damaging anti-tumour agents. *Nat. Cell. Biol. 2,* 339-345.

Yezierski, R. P., Culberson, J. L. & Brown, P. B. (1980). Cells of origin of propriospinal connections to cat lumbosacral gray as determined with horseradish peroxidase. *Exp. Neurol. 69,* 493-512.

Yezierski, R. P., Liu, S., Ruenes, G. L., Busto, R. & Dietrich, W. D. (1996). Neuronal damage following intraspinal injection of a nitric oxide synthase inhibitor in the rat. *J. Cereb. Blood Flow Metab. 16,* 996-1004.

Index

A

D

E

F

G

H

I

J

K

L

M

N

O

P

Q

R

S

T

U

V

W

Z